FROM LUCY TO LANGUAGE
REVISED, UPDATED, and EXPANDED

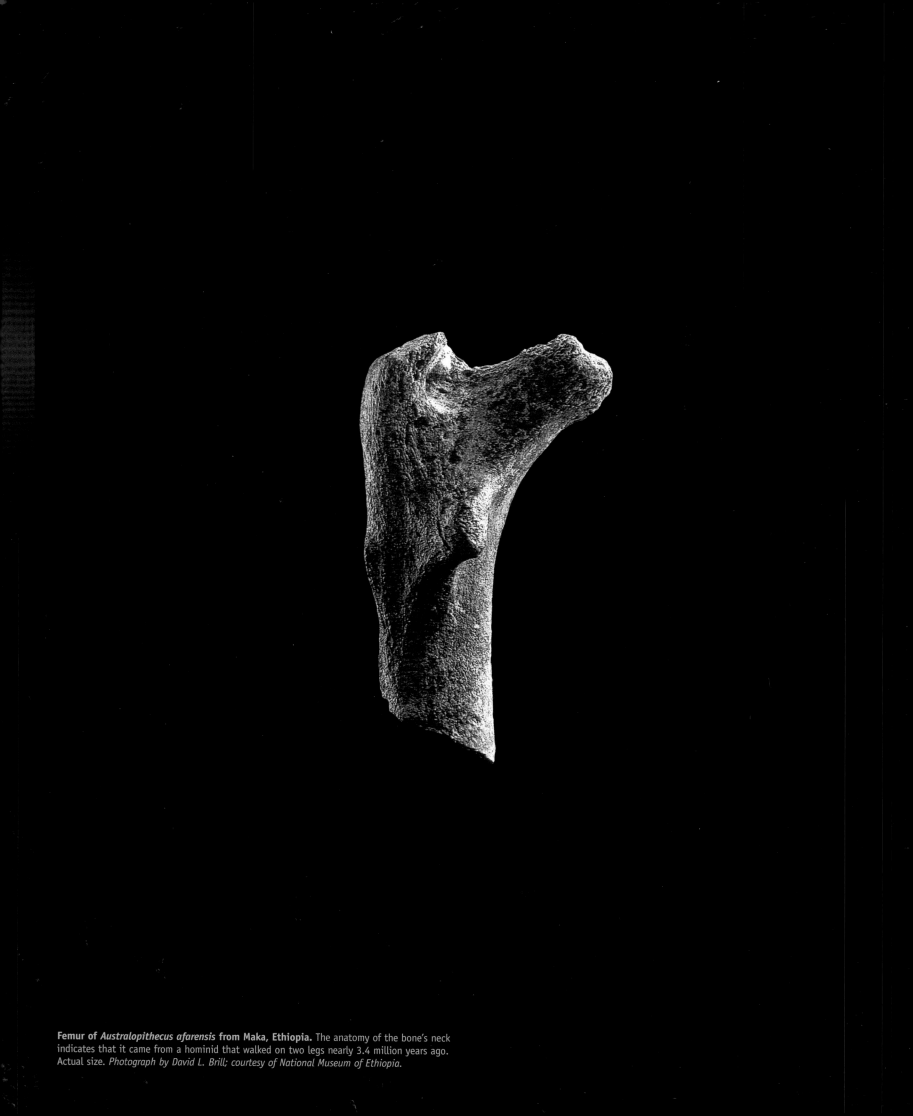

Femur of *Australopithecus afarensis* from Maka, Ethiopia. The anatomy of the bone's neck indicates that it came from a hominid that walked on two legs nearly 3.4 million years ago. Actual size. *Photograph by David L. Brill; courtesy of National Museum of Ethiopia.*

Australopithecus africanus, **Sts 5.** Dubbed *Plesianthropus transvaalensis* by Robert
Broom in 1938, this specimen picked up the nickname "Mrs. Ples." Actual size.
Photograph by David L. Brill; courtesy of Transvaal Museum.

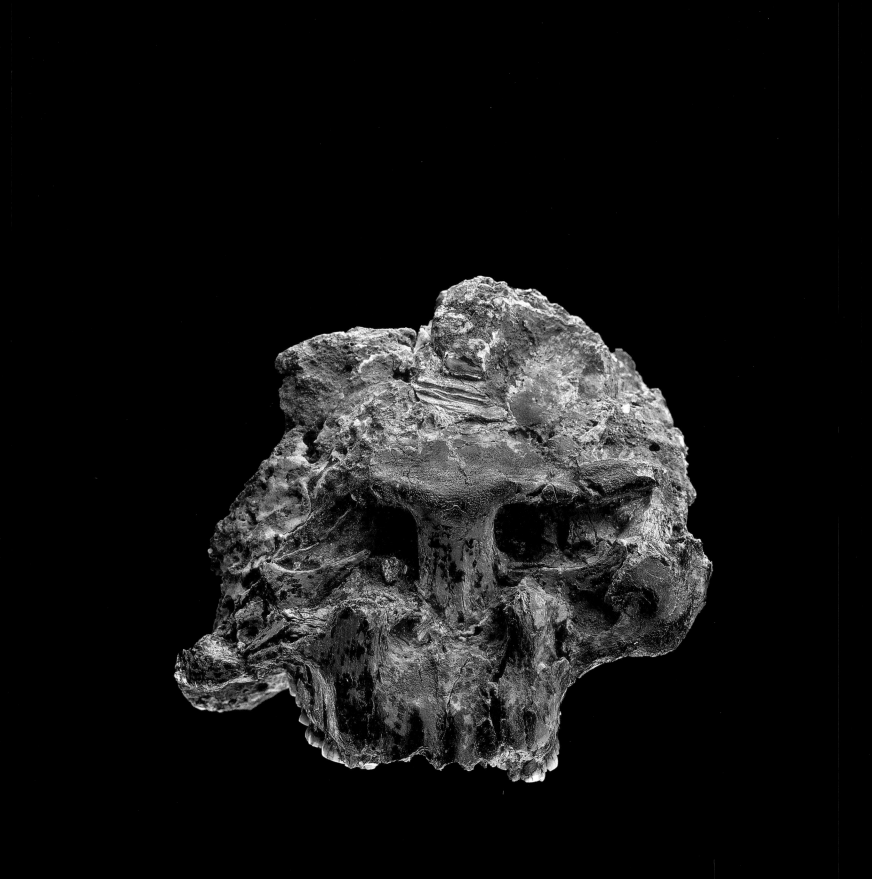

***Australopithecus robustus*, SK 79.** The tremendous pressure put on bones during the slow process of burial and fossilization is evident in this cranium from Swartkrans, South Africa. Actual size. *Photograph by David L. Brill; courtesy of Transvaal Museum.*

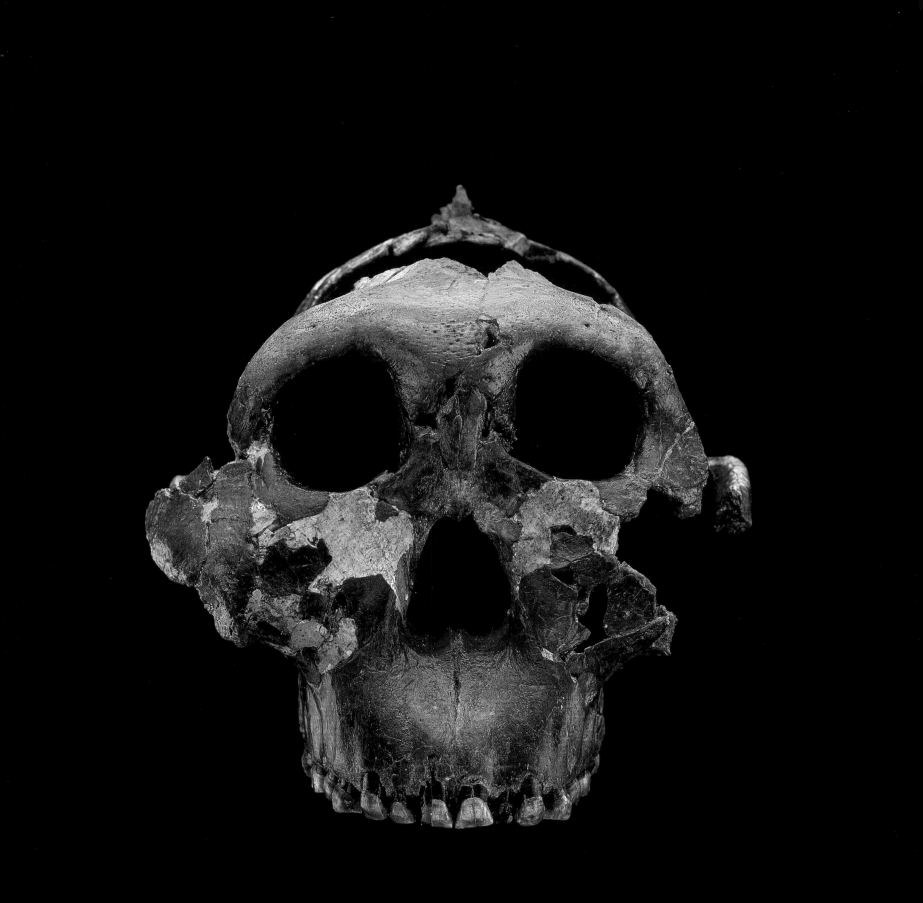

***Australopithecus boisei*, OH 5.** *A. boisei* was a truly impressive creature, dominated
by a massive skull with a broad, concave face—a face like no other in human evolution.
Actual size. *Photograph by David L. Brill; courtesy of National Museum of Tanzania.*

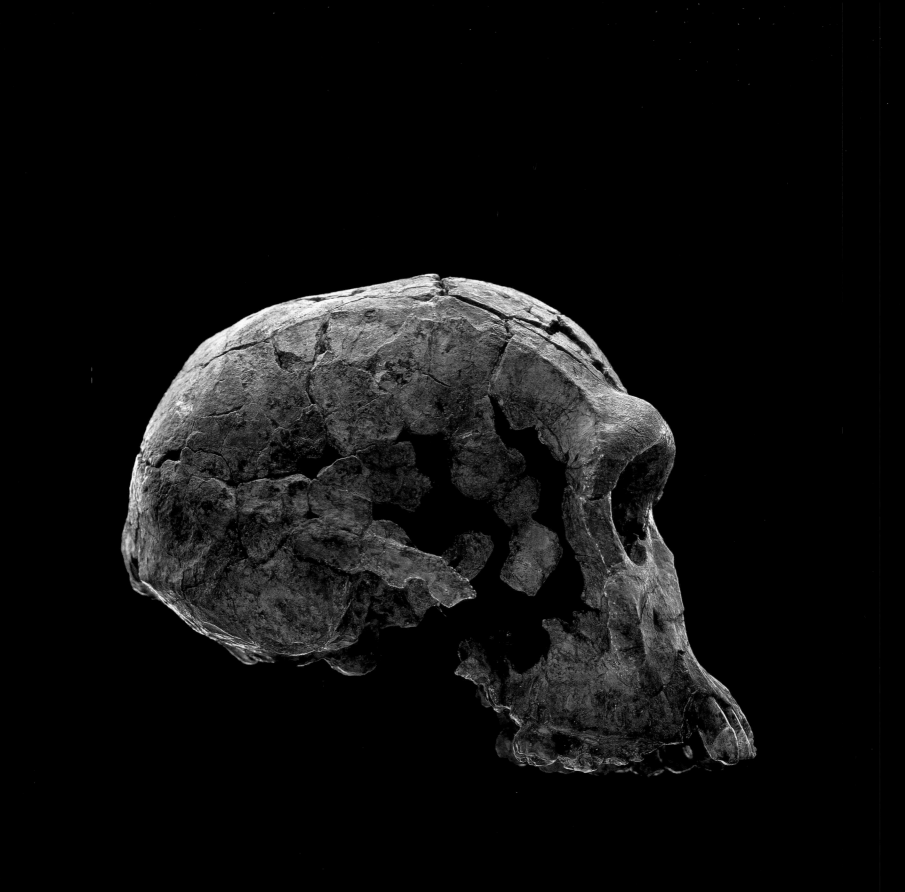

Homo habilis, **KNM-ER 1813.** Anatomy supports the idea that 1813 belongs in
habilis, and for some it is a candidate for the ancestor to later *Homo*. Actual size.
Photograph by David L. Brill; courtesy of National Museums of Kenya.

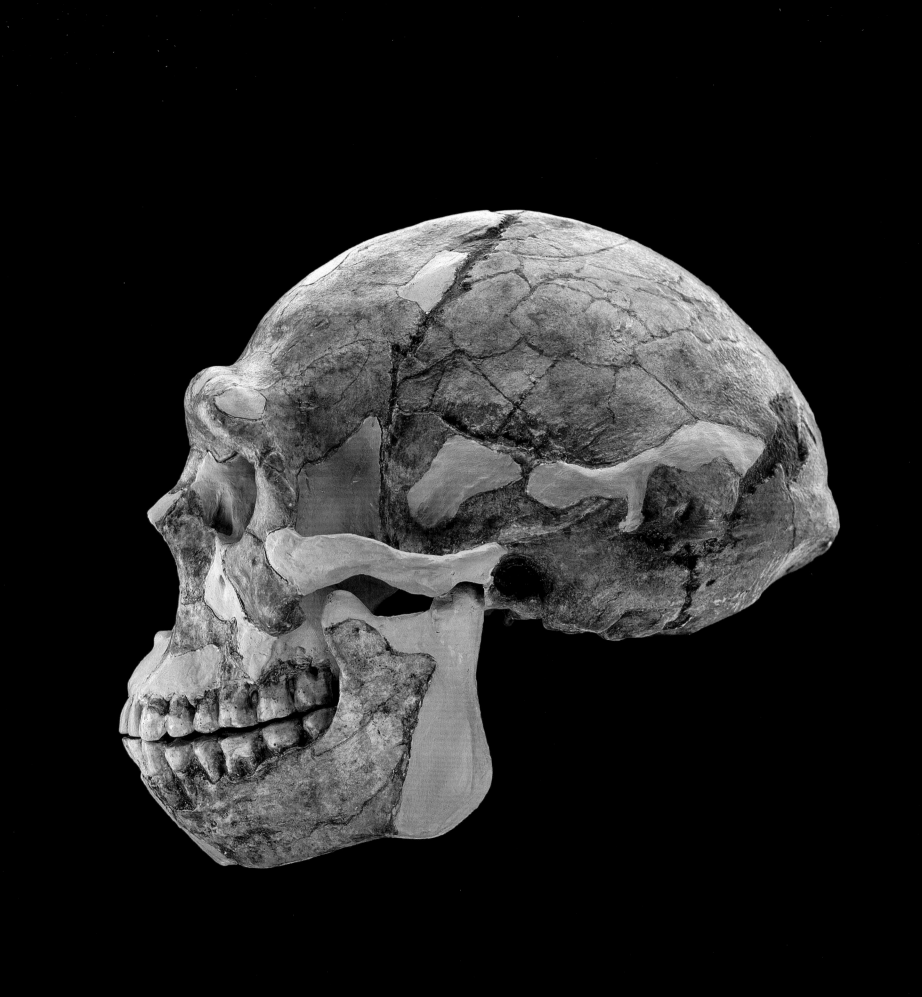

***Homo erectus,* reconstructed skull of Peking Man.** The cave of Zhoukoudian, or Dragon Bone Hill, about 40 kilometers south of Beijing, China, in which the original 400,000 to 500,000 year old fossils that were used to make this reconstruction were found, has also yielded early evidence of fire. Actual size. *Photograph by David L. Brill; courtesy of American Museum of Natural History.*

***Homo heidelbergensis,* Petralona 1.** With some Neandertal-like features and others that are much more primitive, this 300,000 to 400,000 year-old cranium underscores the difficulty of trying to pin a species designation to a poorly dated, singular fossil. Actual size. *Photograph by David L. Brill; courtesy of Paleontolgical Museum, University of Thesaloniki.*

Homo neanderthalensis, **La Ferrassie 1.** This 50,000 year-old cranium exhibits the "classic" Neandertal anatomy that evolved in glacier-covered western Europe midway between this species appearance and extinction. Actual size. *Photograph by David L. Brill; courtesy of Musée l'Homme.*

***Homo sapiens,* Skhul V.** Skhul V belongs to a population whose anatomy is overwhelmingly similar to that of modern humans but whose Mousterian technology is similar to the Neandertals with whom they periodically shared the Middle Eastern landscape for 40,000 years. Actual size. *Photograph by David L. Brill; courtesy of Peabody Museum of Archaeology and Ethnology, Harvard University.*

FROM LUCY TO LANGUAGE
REVISED, UPDATED, and EXPANDED

DONALD JOHANSON

& BLAKE EDGAR

Principal Photography David L. Brill

A Peter N. Nèvraumont Book

SIMON AND SCHUSTER
New York London Toronto Sydney

Horse head figurine from Isturitz, France. Bone carvings became a more common form of art in Europe during the Magdalenian period from 18,000 to 12,000 years ago, when some of the most spectacular decorated caves were also painted. Actual size. *Photograph by David L. Brill; courtesy of Musée des Antiquités Nationale.*

Other Books by Donald Johanson

Ecce Homo: Scritti in onore dell'uomo dell terzo millenio
(co-editor Giancarlo Ligabue)

The Skull of Australopithicus afarensis
(co-authors William H. Kimbel and Yoel Rak)

Ancestors: The Search for Our Human Origins
(co-authors Lenora Johanson and Blake Edgar)

Journey from the Dawn
(co-author Kevin O'Farrell)

Lucy's Child: The Discovery of a Human Ancestor
(co-author James Shreeve)

Lucy: The Beginnings of Humankind
(co-author Maitland Edey)

Blueprints: Solving the Mystery of Evolution
(co-author Maitland Edey)

SIMON & SCHUSTER
Rockefeller Center
1230 Avenue of the Americas
New York, New York 10020

A Peter N. Nèvraumont Book

Text Copyright © 2006 Donald C. Johanson & Blake Edgar
Photography © 2006 David L. Brill

For information about special discounts for bulk purchase, please contact Simon & Schuster
Special sales at 1-800-456-6798 or business@simonandschuster.com

10 9 8 7 6 5 4 3 2

Library of Congress Cataloging in Publication Data

Johanson, Donald C.
    From Lucy to language / Donald Johanson & Blake Edgar:
principle photography by David L. Brill.
            p.   cm.
            "A Peter Nèvraumont book."
            Includes bibliographical references and index.
                1. Human evolution. 2. Fossil man. 3. Australopithecines.
    I. Edgar, Blake. II. Title.
    GN281.J57 1996
    573.2--dc20   96-31576
                                        CIP

ISBN 978-0-7432-8064-8
ISBN 978-0-7432-8064-4

Printed and bound by Editoriale Bortolazzi-Stei, Verona, Italy

Designed by José Conde
Updated, Revised, and Expanded Edition Designed by John Barnett, 4 Eyes Design
Jacket Design by Cathleen Elliott

Created and Produced by
Nèvraumont Publishing Company
New York, New York

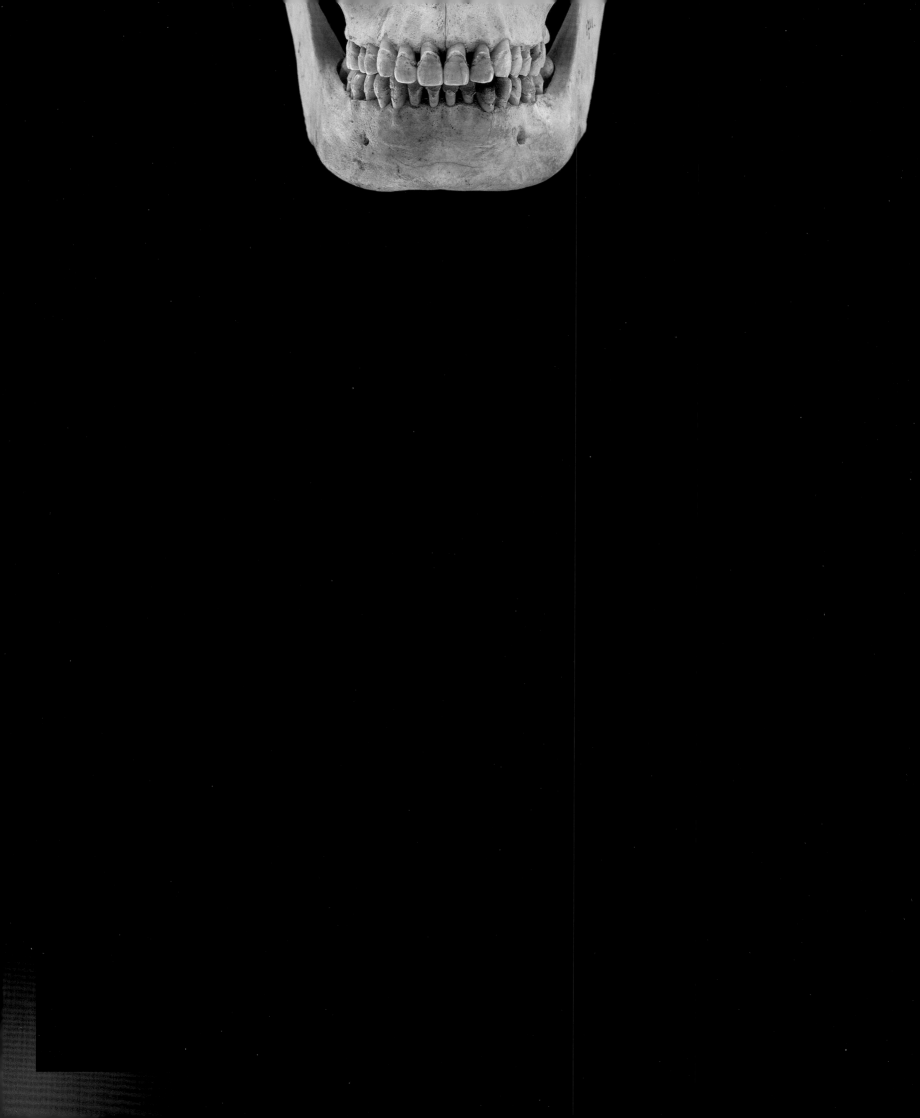

contents

PART 1

Central Issues
of Paleoanthropology

To Lenora and our son Tesfaye for support and understanding.

– D.C. Johanson

To Aidan, who couldn't wait. – B.E.

Part 1

CENTRAL ISSUES OF PALEOANTHROPOLOGY

1 · The Human Creature

A HUMAN is any member of the species *Homo sapiens* ("wise man"), the only living representative of the family Hominidae. The Hominidae, or hominids, are a group of upright-walking primates with relatively large brains (see page 37). So, all humans are hominids, although not all hominids could be called human. ❰ Next, all humans are primates. The mammalian order of Primates includes over 300 species of prosimians (such as lemurs, tarsiers, and lorises), monkeys, apes, and ourselves. Primates are unusual mammals, for they have evolved such distinctive traits as highly developed binocular vision (and a corresponding enclosed eye socket in the skull), mobile fingers and toes with flat nails instead of claws and with sensitive pads at the tips, a shortened snout with a reduced sense of smell, and large brains relative to body size. ❰ If primates are unusual for mammals, humans are even more unusual for primates. We are essentially elaborated African apes. We share more than 98 percent of our genetic material—the information that codes for our proteins, bones, brains, and bodies—with chimpanzees. Yet, despite such similarities, there are significant genetic and physical differences between humans and apes. ❰ Clearly, that distinctive portion of our DNA must involve some regulatory genes that code for our unusual features. Humans also possess 46 chromosomes to an ape's 48. Our version of the extra ape genes have been lumped together on chromosome 2; perhaps in that process some crucial mutations arose, but we would have to sequence and study all human and ape DNA to solve this paradox. Our close genetic affinity to apes has prompted some authorities, notably Jared Diamond in his book *The Third Chimpanzee*, to argue that it is fallacious to separate humans and apes into two separate zoological families: humans in the Hominidae and African apes into the Pongidae. ❰ We walk upright on two limbs, and to accommodate such a strange posture we have developed a specialized pelvis, hip and leg muscles, and an S-shaped vertebral column. We have tiny canine teeth and flat faces with a protruding nose. Males have a pendulous penis, while in females the physical signs of ovulation are concealed, something that happens in no other primate. ❰ Humans are highly social animals, a trait we inherited from our primate past, but we have taken it to new extremes through the development of complex written and spoken language which enables us to communicate nuances of feeling as well as information, and a material culture that includes symbolic art. We are also called a moral animal. Besides that strange habit of walking upright, perhaps it is our inventiveness and our introspective nature that truly distinguish humans among the primates. ❰ Our species, *Homo sapiens*, was first described in 1758 by the Swedish botanist Carl von Linné, whom we know better as Carolus Linnaeus. Most early descriptions concentrated on a very few traits, the most obvious being brain size. If we were to create a richer, more complete biological characterization of our species, many other traits would need to be included. Humans have a relatively long life span that begins with immaturity at birth and a prolonged infancy. Physical maturation is delayed during childhood, and then occurs quickly during the adolescent growth spurt. We are polytypic in morphology and skin and hair color but genetically very homogeneous. Our behavior is marked by habitual tool use, communication through spoken and written language, and the symbolic representation of objects. We are culturally adapted to survive in a broad range of physical environments, climates, and temperatures. We are omnivorous and share food extensively with others. Our body is relatively hairless except for the head and face, axilla, and pubic region. Skeletal features include a hand with an opposable thumb that endows us with a power grip and precise, fine hand movements; relatively straight and slender limb bones; a pelvis, lower limbs, and associated muscles specifically modified for bipedal locomotion; an enlarged hallux, or big toe, in line with the rest of our toes rather than opposed to them; and a foot with a weight-bearing arch to absorb the stresses of two-legged locomotion. ❰ These traits provide a hint of who we are biologically. Many other features of anatomy, behavior, and diversity in *Homo sapiens*, as well as in other ancestral hominids, provide the basis of content for this book. As we query the remains of our extinct relatives for clues to who we are and how we got that way, we will discover that we are much nearer to them than we think, even if separated by millions of years.

2 · The Quest for Origins

SINCE at least the Upper Paleolithic, some 40,000 years ago, every human society has devised a creation myth to explain how humans came to be. The need to explain our origins is one of the universals of being human. Creation myths are based on cultural beliefs that have, in one manner or another, been adopted as legitimate explanation by a particular society. To a large extent, creation myths glorify the specialness of humans. In the broadest view, such myths undertake to explain our differences from all other creatures—our humanness. ❰ In contrast to cultural myths about human origins, the science of paleoanthropology, which also tries to construct a narrative about how humans came to be, is rooted in the scientific method. This method, based on objective observation and evaluation, is governed by a set of rules that permit the testing of hypotheses, and the results of such tests may lead to the rejection or modification of the original construct. The success of paleoanthropology rests on integrating two different fields: Darwinian evolutionary theory and the study of Earth's geological history (see page 21). ❰ Much like a detective story, the quest for clues to our origins is exhilarating and filled with surprises. The goal, however, is not to figure out "who done it" but to understand why and how: why we differ from our closest relatives, the African apes, and how we became a bipedal, large-brained, culturally dependent animal. We are the last species in the zoological family Hominidae (hominids, in the vernacular), and to understand something of our place in nature (see page 111) we need to explore the lessons held by the past. ❰ As we learn more about our origins, it becomes apparent that although an ape ancestor became bipedal several million years ago, there was nothing in that development that ensured the eventual evolution of *Homo sapiens*. Bipedalism, a basic feature of hominids, did not make modern humans inevitable. Paleoanthropological discoveries make it clear that the human family tree is not

Donald Johanson, Hadar, Ethiopia.
Johanson holds an Oldowan stone artifact more than 2 million years in age.
Photograph by Enrico Ferorelli.

a single lineage in which one species succeeded another, leading relentlessly to the appearance of modern humans. Instead, the hominid fossil record suggests that our ancestry is better thought of as a bush, with the branches representing a number of bipedal species that evolved along different evolutionary lines. While all of those species were successful, sometimes for long periods, all went extinct. At the probable time of a common ancestor for humans and African apes, 6 to 8 million years ago, there was no guarantee that humans would evolve. Yet we did evolve, and because we turned out to be inquisitive creatures with the ability to reflect on our past, we have done so avidly. ❰ Paleoanthropology in part plays to that inquisitive, exploratory

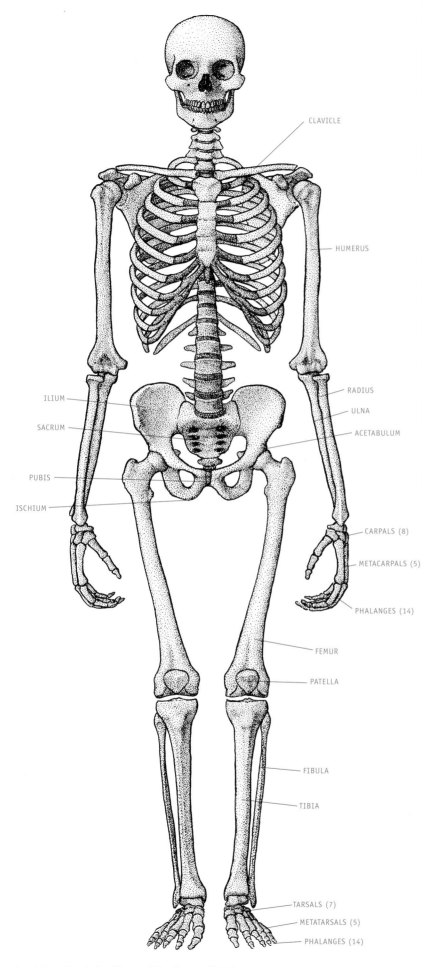

CLAVICLE

HUMERUS

ILIUM

SACRUM

PUBIS

ISCHIUM

RADIUS

ULNA

ACETABULUM

CARPALS (8)

METACARPALS (5)

PHALANGES (14)

FEMUR

PATELLA

FIBULA

TIBIA

TARSALS (7)

METATARSALS (5)

PHALANGES (14)

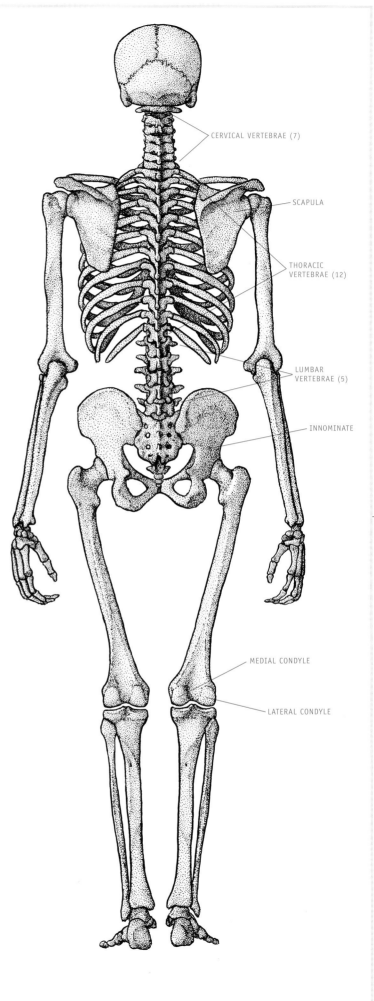

CERVICAL VERTEBRAE (7)

SCAPULA

THORACIC
VERTEBRAE (12)

LUMBAR
VERTEBRAE (5)

INNOMINATE

MEDIAL CONDYLE

LATERAL CONDYLE

Frontal and Posterior Views of the Human Skeleton.
The human skeleton generally contains a total of 206 bones, which are a type of living
tissue that grows and changes throughout life. *Illustrations by Diana Salles.*

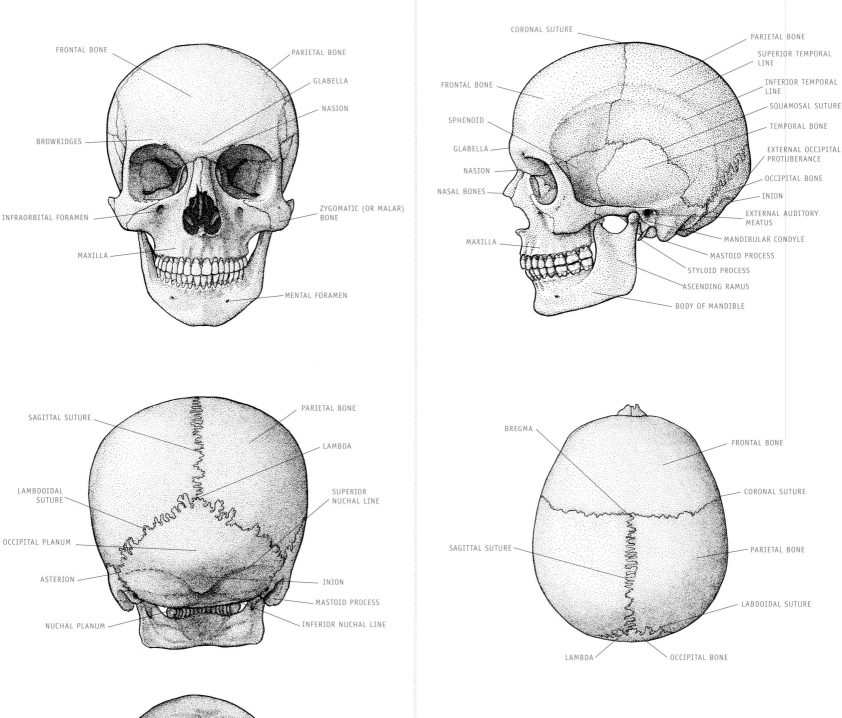

FRONTAL BONE
PARIETAL BONE
GLABELLA
NASION
BROWRIDGES
ZYGOMATIC (OR MALAR) BONE
INFRAORBITAL FORAMEN
MAXILLA
MENTAL FORAMEN

CORONAL SUTURE
PARIETAL BONE
SUPERIOR TEMPORAL LINE
FRONTAL BONE
INFERIOR TEMPORAL LINE
SQUAMOSAL SUTURE
SPHENOID
TEMPORAL BONE
GLABELLA
EXTERNAL OCCIPITAL PROTUBERANCE
NASION
OCCIPITAL BONE
NASAL BONES
INION
EXTERNAL AUDITORY MEATUS
MANDIBULAR CONDYLE
MAXILLA
MASTOID PROCESS
STYLOID PROCESS
ASCENDING RAMUS
BODY OF MANDIBLE

SAGITTAL SUTURE
PARIETAL BONE
LAMBDA
LAMBDOIDAL SUTURE
SUPERIOR NUCHAL LINE
OCCIPITAL PLANUM
ASTERION
INION
MASTOID PROCESS
NUCHAL PLANUM
INFERIOR NUCHAL LINE

BREGMA
FRONTAL BONE
CORONAL SUTURE
SAGITTAL SUTURE
PARIETAL BONE
LABDOIDAL SUTURE
LAMBDA
OCCIPITAL BONE

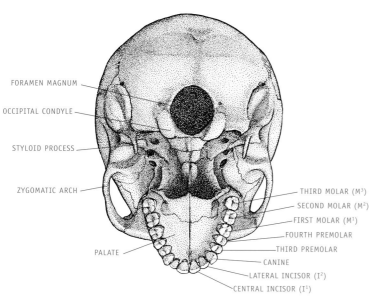

FORAMEN MAGNUM
OCCIPITAL CONDYLE
STYLOID PROCESS
ZYGOMATIC ARCH
THIRD MOLAR (M³)
SECOND MOLAR (M²)
FIRST MOLAR (M¹)
FOURTH PREMOLAR
THIRD PREMOLAR
CANINE
PALATE
LATERAL INCISOR (I²)
CENTRAL INCISOR (I¹)

Five Views of the Human Skull: Frontal, Lateral (left), Posterior, Superior, and Basal.
An adult human skull consists of a cranium and a mandible, or lower jaw. The cranium contains 27 bones, including six unpaired bones, eight paired bones, and a total of six inner ear bones. *Illustrations by Diana Salles.*

part of our makeup. Expeditions to remote terrains feature prominently in paleoanthropology, and whereas living for months in a tent, usually under desert conditions, is not to everyone's liking, the pursuit of our origins can be enjoyed in other ways. Each new find, it seems, receives front-page coverage in newspapers, and magazines featuring reconstructions of our ancestors on their cover become the best-selling issues of the year. Hominid fossils touch a responsive chord in people everywhere, who seem to have an inherent drive to know their beginnings. We want to know what the fossils have to say to us. ◖ There seems to be a magic in the fossilized bones that transcends time. Specimens like Lucy, a 3.2 million-year-old partial skeleton from Hadar, Ethiopia, have become touchstones for discussing human origins. Although older human ancestors have now been discovered, Lucy, with her affectionate name, has become a benchmark by which people judge new hominid discoveries. Even though distant relatives like Lucy lived very different lifestyles from us modern humans, the message they bring, after millions of years of suspended animation, is important to us all. ◖ Ultimately our fascination with the study of human origins nourishes our need for exploration and for understanding both our uniqueness and our close link to the natural world. Today more than ever, people are thinking about the future of the universe and the survival of humankind. For many, the lessons we can learn from our past give us a better perspective on ourselves, our place in nature, and how we view our future.

3 · Is Human Evolution Different?

FOR MOST of human evolution, cultural evolution played a fairly minor role and did not pick up speed until the Upper Paleolithic, some 40,000 years ago. Current archaeological evidence suggests that prior to 2.6 million years ago culture had a minor influence on the lives of the earliest hominids, who were constrained and directed by the same evolutionary pressures as the other organisms with which they shared their ecosystems. So, for most of the time during which hominids have existed, human evolution was no different from that of other organisms. ◖ Once our ancestors began to develop a dependence on culture for survival, however, a new layer was added to human evolution. According to late Sherwood Washburn, at the University of California, Berkeley, there is a definite relationship between biology and culture that he terms "biocultural feedback." Washburn suggests that the unique interplay of culture change and biological change could account for why humans have become so different. His basic premise is that as culture became more advantageous for the survival of our ancestors, natural selection favored the genes responsible for such behavior. Those genes that improved our capacity for culture would have had an adaptive advantage. The ultimate result of the interplay between genes and culture was a significant acceleration of human evolution, as manifested in, among other features, the growth of our brain and its mental capacity over the past 2 million years. ◖ Cultural and biological evolution contrast in a number of ways and are very different processes. Biological change, or evolution, is facilitated by the transmission of genetic information from one generation to a succeeding one by the configuration of DNA in genes. Cultural evolution is the passing on of information by behavioral means and involves the processes of teaching and learning. By these definitions, a bird might make a tool or a nest, but it does not learn how to do this; it is born with the genetic endowment which, at the appropriate time, "turns on," and a nest is constructed. Although humans are genetically equipped with basic biological imperatives, our sophisticated cultural behavior must be learned by teaching, and, most important, this learning is associated with a symbolic mode of communication, usually language. ◖ Information transmitted by DNA involves passing that information from one individual to another, which can only be done at a single point (conception) in the life span of that individual. Cultural evolution, on the other hand, is not passive but active and incorporates lifelong teaching and learning. Further, any one individual can teach one or many, and a single individual can learn from one or many. In cultural evolution, in contradistinction to biological evolution where the information is stored as a DNA sequence, information can be memorized, written, videotaped, audiotaped, and transmitted using sound, pictures, and words. ◖ Culturally transmitted information is behaviorally very flexible and not restricted. A bird can sing a mating song beautifully, but a tenor can sing many romantic arias in several different languages with noticeably distinct levels of passionate commitment. The plasticity of learned cultural information is a true hallmark of being human, as is evidenced by the myriad societies around the world. Cultural behavior is passed on by communication and therefore can spread much faster and to many more individuals than a genetic novelty that is transmitted only in the DNA. Biologically based behavior requires an enormous number of generations to spread, whereas cultural innovations, especially with the information revolution, spread exceedingly quickly. ◖ Human evolution is an intriguing interplay of biological evolution and cultural evolution. And in the view of sociobiologists Charles Lumsden of the University of Toronto and the noted Edward O. Wilson of Harvard University, who have dubbed this interaction "gene-culture coevolution," humans have been shaped through the synergetic interaction of genes and culture. In the final analysis, human evolution is different from the evolution of all other life on this planet. We are distinguished by our capacity for culture, which ultimately has biological roots.

4 · The Science of Paleoanthropology

PALEOANTHROPOLOGY calls on a broadly conceived and strategically implemented multidisciplinary approach to discover and interpret the evidence for human evolution. It is the responsibility of the paleoanthropologist to coordinate activities in the field and in the laboratory, carefully integrating knowledge contributed by specialists in geology, biology, and the social sciences. The goal is to understand, as thoroughly as possible, the process by which we became human. ◖ The major coordination of a field project is customarily under the leadership of an individual with a background in anthropology (the study of mankind). The paleoanthropologist (one who studies ancient humans) works closely with scientific colleagues to raise funding to support field projects, with a primary expectation being the recovery of the fossilized remains of our ancestors. The paleoanthropologist has the ultimate responsibility for overseeing the research from the planning stages, to actual fieldwork, to the recovery and in-depth laboratory study of particularly the hominid fossils, and finally to the publication of research results in the scientific literature. ◖ Once sites have been located (see page 23), an interdisciplinary team enters the field for weeks to months to undertake exploration and excavation focused on finding hominid remains. In the case of cave sites, the process of excavation is slow and tedious. In open-air sites such as those found in the Great Rift Valley in East Africa, exploration takes the form of foot survey, with teams of expedition members scouring the landscape in search of hominid fossils that have eroded from ancient geological strata. ◖ Hominids, however, are not all that a survey team might fruitfully uncover. In the past, hominids were a rather small and insignificant component of paleocommunities. Vertebrates such as elephants, antelopes, monkeys, pigs, hippos, and many others ranging in size down to small rodents were much more dominant elements on the landscape. The study of nonhominid fossil vertebrates is the purview of a cadre of

Fossil hunters at Hadar, Ethiopia. Finding bones of hominids and other vertebrates requires much legwork and a little luck. *Photograph by Enrico Ferorelli.*

paleontologists (scientists who study ancient animals), each concentrating on a particular group of animals. Some paleontologists work only in the laboratory, identifying and studying fossil specimens collected in the field by others; some make forays into the field to make their own collections. ❪ Paleontologists, by studying the composition of the fossil assemblages, help reconstruct the ancient ecology of the site. Further appreciation of the paleoenvironment is contributed by paleobotanists, who study fossil wood and fruits, and by palynologists, who identify past vegetation from fossilized pollen grains. ❪ In the field, the paleoanthropologist, while monitoring all aspects of the field campaign, often works with a team of highly experienced surveyors who have been trained to search for hominid fossils. The training, largely the result of extensive field experience, also benefits from instruction in osteology (the study of bones). Comprehensive and strategic foot survey in search of hominids demands keen eyesight and special dedication, for it is often conducted under very harsh and remote conditions. ❪ If artifacts, usually made of stone, are found, a field archaeologist undertakes painstaking and exacting excavation, carefully mapping and labeling each artifact or fossil uncovered. Such archaeological information aids in understanding how our ancestors may have used tools in hunting, scavenging, and food-processing activities. ❪ The paleoanthropologist, most often trained as a physical anthropologist with a specialized interest in some aspect of human anatomy such as the dentition, the postcranial bones, or skull morphology, works in the laboratory with other scientists who have their own area of anatomical expertise. During the anatomical study, following preparation and cleaning of the field specimen, each one is identified, measured, photographed, described, compared with other fossil specimens, and ultimately published. Experts in the area of biomechanics assist greatly in understanding the function of fossil hominid bones, often providing insights into the nature of the locomotor and masticatory systems and the possible use of hands for manipulation. ❪ Discoveries of hominid and nonhominid fossils and artifacts are of little use unless we know something about the context in which they were found and their geological age. Of overriding importance in the study of paleoanthropology are the geological sciences. One of the major undertakings of the geologist is to map the geological layers in which discoveries have been made. These stratigraphic geologists provide the depositional framework in which to place the discoveries, and often provide detailed descriptions of the sediments that assist in understanding the environment of deposition, such as a delta, river, lake, or back swamp. ❪ Geochronologists employ various methodolo-

gies for determining the exact age of the geological sequence. After field collecting appropriate samples for analysis, they spend long hours in the laboratory applying radiometric techniques to dating (see page 26). The strata of interest are usually, but not always, sediments of volcanic origin. ❪ At some sites where precise geochronological dating is impossible, some animal groups, such as pigs and elephants, are important indicators of a specific time period. The use of animals for dating a site is known as biostratigraphy, and, this may be the only way to estimate the age of the site. The precise dating of fossil hominid-bearing sequences is crucial when trying to place a find within the broader framework of human evolution. ❪ Mapping specialists assist in drafting accurate maps on which to record discoveries and employ aerial photographs and satellite imagery to assist in the search for potential new sites (see page 25). Highly mobile global positioning technology now permits the precise location of important sites and finds to within a few meters, or less, on Earth's surface. ❪ Field research at any hominid site is an ongoing process, whether it be in an open-air site in East Africa, a cave in the Middle East, a quarry in Asia, or a rock overhang in Europe. The paleoanthropologist must constantly refine the research strategy, thoughtfully integrate a diversity of disciplines, carefully phrase questions, apply new methodologies, and implement targeted survey and excavation to significantly augment our knowledge of human evolution.

HOW MUCH fossil evidence of our hominid ancestors has been recovered? To answer this question, one can consider either the number of individuals represented at different sites or the absolute number of fossil specimens (that is, individual bones or fragments) from those sites.

5 · The Early Human Fossil Record

An individual hominid can be represented by anything from a single tooth to a complete skeleton, but thinking in terms of the number of individuals gives a sense of the rarity of hominid fossils and the odds against their preservation. Hominids were certainly relatively rare animals on the landscape anyway, which only increases the challenge of finding their fossilized remains. ❪ The contents of any fossil sample will be biased by the specific conditions for preservation at a site. A host of variables, including the environment where an individual died, the potential for trampling by animals or transport of bones by water, carnivores, or other agents, the durability of the bones themselves, and damage or loss of the bones after exposure by erosion will influence what individuals will be successfully preserved as fossils, what fossils will be found, and their state of preservation. ❪ Compared to the extensive fossil of record some mammals—pigs, rodents, horses, and antelopes, for example—the fossil evidence of human evolution is relatively spotty. About half of the time span in the last three million years remains undocumented by any human fossils. From those periods that are documented, however, we have obtained a reasonably impressive sample that tells us much about the biology, ecology, and diversity of our ancestors. ❪ Until recently, from the earliest period of hominid evolution, more than four million years ago, only a handful of largely undiagnostic fossils had been found, including the mandible fragments from the Kenyan sites of Lothagam and Tabarin (see page 39). The situation improved starting with the discovery and naming of *Ardipithecus ramidus*, which lived about 4.4 million years ago. This species is known from 43 specimens of several individuals (see page 116), including a partial skeleton all from Aramis, Ethiopia. *Ard. ramidus* is also know from nine specimens at Gona, Ethiopia dated to between 4.3 and 4.5 million years ago. Seventeen specimens from the Middle Awash dating to between 5.2 and 5.8 million years ago have been assigned to *Ard. kadabba*. From the Tugen Hills, Kenya, 13 fossil specimens dated to 6 million years ago have been assigned to *Orrorin tugenensis* (see page 118). The oldest putative hominids are nine specimens, including a crushed skull, from 6-7 million-year-old deposits in Chad, attributed to *Sahelanthropus tchadensis* (see page 116). ◖ There are 78 specimens, predominately dental remains, of *Australopithecus anamnesis* (see page 129), a species from Kanapoi and Allia Bay in northern Kenya that lived just before 4 million years ago. The best represented early hominid is *Australopithecus afarensis*, which lived between 4 and 3 million years ago in eastern Africa. Its remains number close to 400 specimens, 9 percent of which derive from Hadar, Ethiopia. The second largest sample is from Laetoli, Tanzania where 26 specimens have been found. The best known specimens are the partial skeleton A.L. 288-1, or Lucy (see page 133), and more than 240 bones from 17 individuals from the single locality A.L. 333 (see page 135). Other fossils that most likely belong to this species include a 3.9-million-year-old partial frontal bone from Belohdelie, Ethiopia and some fragmentary remains from Omo Shungura and Mille, Ethiopia, and Koobi Fora, Kenya. ◖ In South Africa, the species *Australopithecus africanus* is represented by at least 120 individuals from Sterkfontein (see page 158) alone, plus the single individual from Taung (see page 154), and a few from Makapansgat. The most abundant South African hominid site is the cave of Swartkrans, where painstaking excavation of breccia has yielded about 200 *Australopithecus robustus* (see page 160) and six *Homo* fossils, which can attributed to a minimum of 85 individuals and six individuals, respectively. More than 50 specimens of *A. robustus* come from Drimolen. ◖ The earliest form of robust australopithecine in East Africa is *Australopithecus aethiopicus* ("southern ape of Ethiopia"). It appeared about 2.5 million years ago and is known from three specimens belonging to three individuals: the toothless partial mandible Omo 18, from Ethiopia, and from west Lake Turkana, Kenya, the partial mandible KNM-WT 16005 and the cranium KNM-WT 17000, better known as the Black Skull (see page 164). A maxilla lacking teeth has been recovered not far from Laetoli, Kenya and attributed to *A. aethiopicus*. Several specimens from the Pliocene of northern Kenya, dating to 3.5 million years ago have been assigned to *Kenyanthropus platyops* (see page 121). ◖ From between 2.4 and 1.2 million years ago, the period in which genus *Homo* evolved in Africa and began populating other continents, the hominid fossil record of East Africa consists of about 300 specimens of crania, mandibles, and teeth. A third of these belong to *Australopithe-*

Olduvai Gorge, Tanzania. Located in the Great Rift Valley, where eruptions, uplift, and erosion create ideal conditions for preserving fossils, this was one of the first sites in Africa to be explored for evidence of our ancestors. *Photograph by Emory Kristof: copyright National Geographic Society.*

cus boisei, such as the crania OH 5 and KNM-ER 406 (see pages 168 and 170), 85 belong to *Homo*, and the rest cannot be identified to a particular hominid. An additional 50 postcranial bones—those occurring "after the cranium," or below the skull—round out the hominid sample from this time period. ◖ African *Homo* fossils from this period can be assigned to the species habilis, *rudolfensis*, and *ergaster*. The sample of *H. habilis* fossils includes 34 specimens from Olduvai Gorge, including seven partial crania (see pages 182 and 184), four mandibles, assorted limb bones and teeth, and a fragmentary partial skeleton (see page 188). About half a dozen from Koobi Fora can be included in this species, the most complete being the cranium KNM-ER 1813 (see page 186). Several teeth, two mandibles, and a very fragmentary cranium from Omo may also belong to *H. habilis*. *Homo rudolfensis* is represented by several specimens from Koobi Fora, most notably the cranium KNM-ER 1470 (see page 189), which was the type specimen used to name this species in 1986, and the mandible of a separate individual, KNM-ER 1802. Another probable *H. rudolfensis* specimen is the 2.4 million-year-old mandible UR 501 from Malawi. *Homo ergaster* is better represented by several specimens from Koobi Fora and West Turkana, Kenya, including the type specimen mandible KNM-ER 992, the crania of KNM-ER 3773 (see page 194) and 3883, and the skeleton of KNM-WT 15000 (see page 196), as well as the partial cranium SK-847 (see page 198) from South Africa. ◖ Outside of Africa, the oldest hominid remains belong to *Homo erectus*. Several skulls, mandibles and postcranial bones assigned to *H. erectus* (sensu lato) derive from 1.7 million-year-old deposits at Dmanisi, Republic of Georgia (see page 192). Specimens from at least 48 *H. erectus* individuals—a third of the total worldwide sample of this species—have been found in Java, mainly at Trinil (see page 201), Sangiran (see page 205), Ngandong, and Modjokerto. These fossils include crania, skullcaps, and skull fragments from 30 individuals, jaws or jaw fragments from nine individuals, plus some limb bones and numerous teeth. Another third of the world's *erectus* fossils come from China, mainly from the single site of Zhoukoudian, where 45 hominid specimens from 15 individuals were recovered (see page 202). ◖ From the Middle Pleistocene, *Homo heidelbergensis* is represented by thousands of bones from at least 30 individuals from Atapuerca (see page 218), plus 20 others, usually partial or complete crania from elsewhere in Europe and Africa. Much more abundant are the Neandertals, the probable direct descendants of *H. heidelbergensis*. Remains of about 500 *H. neanderthalensis* individuals have been discovered (see pages 225-247), largely in western Europe, central Europe, and the Near East. An 18,000 year old partial skeleton with a complete skull comes from Flores, Indonesia and belongs to *H. floresiensis* (see page 248). ◖ From this brief inventory, it is clear that paleoanthropologists have accumulated a large sampling of fossil evidence from at least some hominid species, although disagreement as to how to classify and interpret this evidence persists. Only by gradually filling in the gaps in the fossil record can we hope to arrive at the more complete and cogent interpretation of our past.

CONTEMPORARY paleoanthropologists are greatly assisted in their search for fossil hominid sites by the use of sophisticated remote sensing data, such as satellite images, and computerized geographical information systems. In the

6 · Discovering Early Human Fossil Sites

past, the majority of sites were located by chance. Olduvai Gorge in Tanzania, for example, was found by a German entomologist exploring what was then German East Africa. While chasing a butterfly, so the story goes, Wilhelm Kattwinkel almost fell off a cliff into the gorge. During the course of Kattwinkel's natural history expedition he collected fossils at Olduvai and sent them back to

Berlin, where they aroused the interest of paleontologists. ❡ Many sites in eastern Africa were found as an adjunct to geological exploration of the Great Rift Valley, where the vagaries of volcanism, uplift, and erosion have created an ideal setting for the preservation and later the exposure of vertebrate fossils. An example is the site of Hadar, Ethiopia, where the partial skeleton of Lucy was found in 1974. In the late 1960s the French geologist Maurice Taieb, while mapping the geology of the little-known Afar depression, recognized abundant vertebrate fossils and stone tools eroding from ancient strata. He brought these to the attention of paleoanthropologists, and there began an important series of ongoing expeditions to the region. ❡ Sites in Europe have often been discovered by amateur archaeologists, who systematically search caves, rock overhangs, and fields where plowing brings up artifacts or fossils. In the absence of surface occurrences of artifacts or fossils, it is virtually impossible to predict where excavation might prove fruitful. ❡ Mining operations resulted in the recovery of the first Neandertal in Germany, in 1856, the first *Australopithecus* at Taung, South Africa, in 1924, and the Mauer mandible in gravel pits near Heidelberg, Germany, in 1907. The famous Cro-Magnon fossils in southwestern France were found during construction of a railroad. Speleologists discovered the remarkably productive cave of Atapuerca in Spain, as well as important caves with Paleolithic art such as those at Chauvet and Cosquer. One of the most spectacular finds by cavers is the 1993 discovery of the still unexcavated, virtually complete Neandertal skeleton at Altamura, Italy. ❡ Today, paleoanthropologists are aided in their search for new sites by space-based imagery. It was fortunate for the Paleoanthropological Inventory of Ethiopia that portions of the Ethiopian rift had been flown over in October 1984 by NASA's space shuttle, the *Challenger*. On board, a large-format camera recorded images on negatives roughly 23 by 46 centimeters that produced high-resolution stereoscopic photographs. The photographs were of great assistance in locating geological structures such as faults, as well as in differentiating certain rock types. They subsequently proved indispensable for research expeditions on the ground, navigating over terra incognita. ❡ Satellite-produced Landsat thematic mapper images further enhance the space-based imagery. Whereas photographs can register only visible and near infrared wavelengths, the satellite images are recorded in digital form and transmitted back to Earth, where computer enhancement can reveal hidden structures in the data. Multispectral scanners record a wide spectrum from blue to short-wavelength infrared. The brightness of one spot, or pixel, on Earth is recorded by a string of numbers. Pixel size has a spatial resolution of some 30 meters, but new images from the European remote sensing satellite SPOT has a pixel size of about 10 meters. ❡ Because different rock types have varying reflectance, Landsat images can be used to distinguish between volcanic rocks and sedimentary deposits. Shuttle photographs and Landsat images permitted the Paleoanthropological Inventory of Ethiopia team to predict the possibility of fossiliferous beds in a previously unexplored region of Ethiopian rift. The next step was to survey, on foot, those areas initially outlined using remote sensing. On the ground, rich fossil deposits older than 2 million years were encountered, and Acheulean tools were found in apparent association with vertebrate fossils. Satellite data have also proved useful in pinpointing fossil-bearing sediments at Fejej in southern Ethiopia, where in 1990 4 million-year-old *A. afarensis* specimens were collected. ❡ Now that remote sensing techniques have proved their usefulness in locating important paleoanthropological sites, they will probably be more widely employed. The finding of hominid specimens, however, will always rely on the untiring efforts of a dedicated ground survey team.

Skeleton of *Homo neanderthalensis* from Altamura, Italy. In 1993, spelunkers discovered this apparently complete skeleton with the skull partly obscured by a stalactite. *Courtesy of Bari State University.*

7 · Recovering the Remains of Early Humans

THE PROCESS of recovering fossil hominid remains is inherently destructive. As soon as a fossil is picked up from the ground or excavated from an ancient stratum, the specimen is no longer in its original context, and some information is lost. With awareness of this problem, modern recovery and excavation techniques are far more rigorous than the "treasure hunting" approach of earlier eras. The search for and excavation of hominid remains today incorporates a carefully articulated research design intended to preserve as much information as possible. ❡ Once a fossil-bearing deposits has been located, a strategic plan must be implemented to search for hominid remains. Fossils at open-air sites are rare and widely scattered. Accordingly, an intense and thorough foot survey with, at times, crawling on the outcrops is obligatory. The searchers must have good visual acuity and a thorough knowledge of vertebrate anatomy to sort out the hominid fragments from the background scatter of nonhominid fossil bones. A missed hominid fragment may wash away with the next rainstorm or disintegrate with additional weathering. ❡ Once a hominid fragment is recognized, a specific protocol is followed. Photographic and even video recording of the fossil may precede recovery of the specimen. A locality number is assigned and a general description of the locality is recorded in a field notebook. The exact location of the find is also identified by a pinprick on an aerial photograph and recording the precise coordinates using a global positioning system. ❡ If close examination of the specimen reveals that any breaks in it are fresh, other portions might be nearby. A larger area surrounding the initial find is delimited by a perimeter of string, and team members, on hands and knees, carefully search for additional fragments. If other bits are found, a string grid is established over the area most likely to yield more bone. Each square meter of the grid is intimately examined and all fossil fragments are collected, identified, numbered, and plotted on graph paper using a coordinate system. After this initial collection, the loose sediment is scraped up with a trowel or small shovel, placed in a bucket, carried away from the hominid locality, and sifted through a fine mesh screen. ❡ Stone matrix adhering to a fossil will often reveal the precise geological horizon from which the fossil came, because layers of sand, clay, or silt each have their own characteristic composition. Sometimes a decision will be made to excavate into a hillside in the hope of finding additional bones that have not yet weath-ered out of their matrix. Intially the sterile overburden is removed, using everything from a small backhoe, to a portable jackhammer and picks to loosen the compacted sediments, to small rock hammers and trowels as the fossil horizon is approached. Dental tools, sharpened nails, and other pointed digging tools allow slow, controlled excavation. When bones are encountered, dental picks and soft brushes are used to expose the limits of the fragment. ❡ After the specimen is exposed, it is not immediately removed. A small drawing is made of the specimen and its precise position is measured using metal tapes and plotted on graph paper. The compass orientation is noted and even the dip, or angulation, of the strata, which can be used to help establish the direction of water flow, for example. This information can direct searchers where to look for more fragments if the specimen had eroded out of its original setting and been transported by stream after a heavy rain. Because the horizons from which specimens are excavated are uneven, the depth of a specimen is recorded using a plumb bob or a surveyor's transit and stadia rod, or better yet a total station. For a datum point, a metal stake is driven into the ground. This serves as a reference point for horizontal and vertical measurements and may allow reconstruction of the spatial relationships of the find in a three-dimensional computer program. ❡ While the specimen is still in the ground, it may be treated with a preservative like Butvar, an acetone-soluble polyvinyl acetate, to prevent any further fragmentation. After the preservative is dry the specimen is lifted, labeled, put into a

Computer technology and fossil hunting. The Total Station, here at Mossel Bay, South Africa is connected to a handheld computer, which is used to plot all finds in three dimensions with great precision and accuracy. All the data are downloaded to a laptop for analysis and storage. *Photograph courtesy of Curtis Marean.*

plastic bag, and carefully transported back to the field laboratory. ◖ Every care is taken in the field to chronicle in notebooks as much information as possible. All stages of excavation must be written up immediately and in detail, including measurements, stratigraphic information, the names of the excavators, and the date and time of the discovery. Because excavation is destructive, the field notebooks become the new record for the original locality. The information contained in the notebooks is invaluable for writing up scientific articles and becomes a reference for future investigators. ◖ Some countries, anticipating future innovations in paleontological excavation and recording techniques, limit the extent of work at any given site. For example, in Israel only one third of any site may be excavated, with the remainder reserved for other expeditions. Preservation of portions of a site also plays a crucial role when the rethinking of a scientific problem can only be solved by renewed excavation at a specific site and a reformulation of the research strategy.

WITHOUT a way to estimate the antiquity of paleontological and archaeological discoveries, it is impossible to construct a chronology for the events in human evolution. Fortunately, several techniques are available, and usually at least one of them can be

8 · Dating Fossils and Artifacts

used at a particular site by geologists or geochronologists. ◖ Two broad categories of techniques exist: relative and absolute dating. Relative methods indicate whether a given fossil, artifact, or site is younger or older than another one but cannot pinpoint an age in years. For instance, by the nineteenth-century geological principle of superposition, rock or sediment layers closer to the top of a vertical sequence should be younger than the underlying layers if the layers have not been disturbed. A common technique of relative dating, used to compare the ages of distant fossil sites, is biostratigraphy, which relies on the fossil remains of wide-

spread animals that change distinctly during their evolutionary history. Rodents, elephants, antelopes, and pigs have proved particularly useful as biological markers of time in Africa or the Near East. Biostratigraphy has helped establish a chronology for the fossil-rich limestone caves in South Africa, such as Swartkrans and Sterkfontein, which lack volcanic or radiogenic rock material suitable for absolute dating methods. Biostratigraphy also provides a system of checks-and-balances for evaluating age as determined by the more technical absolute methods. For instance, it was analysis of pig teeth that revealed an error in the potassium-argon age estimate for the KBS Tuff at Koobi Fora in Kenya and forced a revision by more than a million years of the age for such hominid fossils as KNM-ER 1470 (see page 189). ◖ Absolute dating methods are a more powerful and precise tool for pinpointing evolutionary events. Many of the methods are radiometric: they rely on the constant rate of decay of certain radioactive isotopes—varieties of a chemical element—in rocks to act as a clock that reveals a specific age in years. These methods tend to be quite new, and even the earliest absolute techniques have had important refinements in the last decade or so. ◖ Perhaps the most famous absolute dating technique is the radiocarbon method, which was developed in the 1940s and first applied to date a piece of acacia wood collected from the oldest Egyptian pyramid, the step pyramid of Saqqara. Radiocarbon dating can be applied directly to bone or other organic matter. An age can be determined by measuring the amount of carbon-12 (the stable, predominant form of natural carbon) in an object and deducing how much carbon-14 (the rare, radioactive isotope) had been present when the object was living. Because carbon-14 decays into nitrogen at a known rate after death, it is possible to determine, after measuring the current amount of this isotope, an age for the object. The wood from Saqqara indicated that this first Egyptian pyramid was built 4,600 years ago. ◖ Radiocarbon isotope decay method only works for relatively young organic objects because the half-life of carbon-14 is 5,730 years (meaning that half of the carbon atoms in a sample will have disintegrated after the passage of this amount of time). Because a given sample of carbon-14 is reduced in size by half every 5,730 years, radiocarbon effectively runs out of time after 40,000 years. A newer approach, called accelerator mass spectrometry (AMS) radiocarbon dating, has the promise of extending the technique's viability further into the past, perhaps as far as 75,000 years ago. Because the AMS method can be applied to much smaller samples—a speck rather than a gram—than conventional radiocarbon decay methods, the new technique has been able to date directly for the first time such important archaeological discoveries as European Upper Paleolithic paintings in some of the decorated caves. ◖ For more ancient sites, geochronologists have turned to elements with much longer half-lives. Potassium is an abundant element in Earth's crust, and about 0.01 percent of all potassium is the radioactive isotope potassium-40, which has a half-life of 1.3 billion years. Potassium-40 decays naturally into argon-40. The more argon-40 a rock contains, the longer the rock clock has been ticking: the more argon that is present, the older the rock.

By measuring the amount of potassium present and the amount that has already decayed into argon, an age for the rock can be determined. ◖ Developed in the 1950s, the potassium-argon method was soon put to work dating the volcanic ash and tuffs that bracketed fossils or artifacts in Olduvai Gorge's layers of rock. Its application at Olduvai Gorge and later at many East African hominid sites transformed our concept of the span of human evolution. Since the 1980s the technique has itself been revolutionized through the use of lasers to melt individual rock crystals for dating, rather than relying on a larger sample that might be contaminated by older or younger crystals that would give a false date. Also, a single sample can be used to measure both the argon and potassium content, whereas the conventional potassium-argon technique required two separate samples. The new method, called single-crystal laser-fusion argon-argon dating, measures the ratio of argon-40 to argon-39, an artificial isotope created in a nuclear reactor from stable potassium-39 and which serves as a proxy to indicate the amount of potassium. This technique can also be carefully controlled to get a series of ages from the outside to the core of the rock sample. Recently, the argon-argon method was used to fine-tune the age of Lucy and the First

Dating volcanic ash from Hadar, Ethiopia. Geologist Jim Aronson collects a sample of from the BKT-1 tuff at Hadar for dating in a Berkeley laboratory. The single crystal laser-fusion argon-argon dating method relies on a laser to melt individual rock crystals containing a radioactive isotope that decays at a known rate and calibrates the rock's age. *Photographs by Enrico Ferorelli.*

Family fossils from Hadar to 3.2 million years. ◖ Several dating methods have been developed to cover the broad span of time that is either too old for the radiocarbon method or too young for potassium-argon to handle—that is, the period between about 300,000 and 40,000 years ago. Three of these methods—electron spin resonance (ESR), thermoluminescence (TL), and optically stimulated luminescence (OSL)—work by counting electrons trapped by flaws in the microscopic structure of crystals. ◖ Electrons accumulate over time as radiation released by radioactive elements in the rock and in surrounding sediment continues to punch holes into the crystal lattice. The electrons can be freed as light, in the case of TL and OSL, and the amount of escaping light can be carefully measured to estimate the time elapsed since a stone tool, for instance, was last heated by a campfire or exposed to the sun before burial. Rather than expel electrons from a rock, ESR excites them with a cascade of microwaves in a magnetic field. In response, the electrons flip their direction, or spin, through a process called resonance. The energy of resonance gives off a spectral signal—related to the number of trapped electrons—the magnitude of which can be measured and used to calculate an estimate of the sample's age. A sample can be tested repeatedly by ESR but only once with TL or OSL.   ◖ Neither TL nor OSL can be used to date bone directly, but they work well on burned flint tools, ceramics, or sediments. ESR works on a range of natural material including tooth enamel, coral, mollusk shells, and sand grains. Both TL and ESR have shed much light on the origins of modern humans, especially at cave sites in the Levant, where Neandertals and early modern humans overlapped in time for 50,000 years. OSL has been used at archaeological sites in Australia, where the results indicate that modern humans may have first arrived on that island continent as long as 60,000 years ago (see page 49). ◖ Still other dating methods have helped shed light on questions in human evolution. These methods include paleomagnetism, fission-track, uranium series, and amino acid racemization. So, paleoanthropologists have many tools at their disposal to estimate the age of their finds. No dating method comes without constraints or potential pitfalls, however, and great care must be taken in selecting appropriate samples in the field and preparing them in the laboratory. Ideally, as many methods as possible should be tried independently to provide a confident date for a fossil, artifact, or site.

LOCAL and even regional environmental and climatic changes apply pressures or open opportunities for flora and fauna alike. Species unable to adapt to the new conditions will perish. Others, which are preadapted or

9 · Climate and Human Evolution

can adapt more quickly to the new conditions, survive. Attempts have been made to link specific global-scale climatic events to environmental changes to several biological and cultural milestones in our evolutionary history. How close a fit do we find between climate change through time and the various speciations and extinctions traced by our fossil record? ◖ For example, the evidence for climate change in Africa during the span of hominid evolution comes from both land and sea: carbon isotopes in tooth enamel that document prevailing plant cover and soil, and dust from deep-sea drilling cores. Most plants, including trees, shrubs, and certain grasses, convert carbon into compounds with three-carbon chains in their chemical structure (and so are called C3 plants). Plants in dryer, more seasonal environments, especially temperate and tropical grasses (called C4 plants), build four-carbon chain compounds by an alternative process. Because C3 and C4 plants contain different proportions of stable carbon isotopes, the ratio of these isotopes can be studied to determine whether C3 or C4 plants—that is, whether woodland or grassland—predominated at times in the past. The same carbon isotope ratios show up in the tooth enamel of animals that have ingested and digested this vegetation. ❡ The enamel evidence suggests that between 6 and 8 million years ago in Africa, ocean surface temperatures dropped and a shift occurred from previously dominant C3 plants to dry-adapted C4 plants, the sort found in the savannas that cover nearly two thirds of Africa's land today. Although C4 plants began appearing in East Africa as long as 15 million years ago and occurred in patchy distribution mixed with woodland trees and shrubs, savanna grasses dominate the East African landscape as of 7 million years ago—about the time that the first hominid split off from its last common ancestor with African apes. ❡ Wind-blown dust from the African continent settles on the sea floor, where it gradually collects as an uninterrupted record of climate for the past several million years. Thicker dust layers form, for example, during particularly dry periods with strong winds. Seafloor dust accumulations indicate that after 2.8 million-years-ago, the ice sheets in the northern latitudes expanded vastly, causing a severe drying and cooling in Africa that favored arid-adapted plants and animals. The sediments from this time contain greater amounts of phytoliths, or bits of silica from grass. Similar cool, dry periods recurred 1.7 and 1 million years ago. ❡ According to the turnover pulse hypothesis proposed by paleontologist Elisabeth Vrba, the climate changes triggered new directions in evolution. Fossils of both large and small mammals—antelopes and rodents—show that these habitat-specific animals had a turnover of extinctions and speciations 2.5 million years ago in which the arid-adapted kinds survived in both southern and eastern Africa. Hominids, Vrba believes, responded in the same way to the swings in vegetation and, presumably, in rainfall and temperature. ❡ Sometime between 3 and 2.5 million years ago the hominid lineage split into two branches: one with the first member of genus *Homo* and a second with the first robust australopithecines. What is possibly the earliest fossil of *Homo*, a maxilla from Hadar, Ethiopia, was recently dated to 2.33 million years ago, about the time of the global cooling and drying. And the most ancient stone tools occur at this same time (see page 89). A provocative coincidence but the origins of *Homo* may go back even further in time, before the drop in temperature, or a species of *Australopithecus* may have actually made the first stone tools. ❡ A potential problem also arises with linking the emergence of robust australopithecines to this climate change. Antelopes switched from being woodland browsers to being grassland grazers at 2.5 million years ago, but the Black Skull, KNM-WT 17000 (see page 164), indicates that a primitive robust australopithecine had already evolved by 2.6 million years ago. ❡ Another cautionary tale comes from Konso-Gardula, Ethiopia, where the oldest known Acheulean artifacts, dated to 1.4 million years, have been found, along with a hominid mandible. This earliest evidence for a profound change in human technology falls in between the global cooling of 2.5 million years ago and the subsequent cooling that began a million years ago. By the time of the more recent cooling, which has been credited with spurring the migration of *Homo erectus* out of Africa, *H. erectus* had already been in Asia for almost 800,000 years. If the recently reported dates from Java (see page 205) are correct, *H. erectus* made huge leaps in its geographic range long before the alleged climatic catalyst. ❡ Establishing links between past climate change and evolutionary change depends on being able to pinpoint the first and last appearances of species in the fossil record, because these data theoretically should indicate, respectively, speciation and extinction events. But to determine these events reliably requires a thorough sampling of the fossil record across time, across different habitats, and across the range of a species. Then there is the challenge of accounting for the natural rarity of some species (and hominids were likely to have been relatively rare animals) as well as the problem of recognizing different species from fossil evidence. ❡ Although a broad-scale correlation exists between climate changes and changes in hominid species, there is no way to show that former triggered or drove the latter. We are gaining a fine-grained understanding of the

early climate, but the picture provided by scarce and fragmented human fossils remains too coarse-grained to allow definite connections to be drawn between the two lines of evidence. We should also remember that hominids seem to have been good ecological generalists, able to tolerate a wide range of environments, and as such they may have been less sensitive to climate change than habitat specialists.

10 · Teeth

BECAUSE teeth are capped by enamel, the most durable biological substance known, and because they possess a core made of a very hard mineralized tissue called dentine, they constitute the majority of fossil mammal specimens, hominid or otherwise. Teeth are extremely informative about the age, sex, diet, health, and taxonomic identity of extinct hominids, and we are fortunate to find them so plentiful in the fossil record. The informative power of teeth is the reason why paleoanthropologists are often well versed in dental anatomy and spend considerable time describing and analyzing remains of fossil dentitions. ❡ Hominids, like all Old World monkeys and apes, possess 32 adult teeth: 16 upper and 16 lower teeth. In each jaw there are two central and two lateral incisors, two canines, four premolars, and six molars. Each tooth type has its own function: incisors are for slicing, canines for grasping or piercing, and the cheek teeth, premolars, and molars for crushing and grinding. Our baby teeth, sometimes called the milk or deciduous dentition, number ten and consist of two central and two lateral incisors, two canines, and four molars in both the upper and lower jaws. Each hominid tooth has its own distinctive crown anatomy, permitting identification of its exact position in the dental arcade. It is relatively easy to tell right from left and upper from lower. Upper molars, for instance, have four cusps, while lower molars generally have five cusps. Detailed knowledge of dental anatomy also permits specific identification of the tooth's position-for example, first or second premolar. Each bump (cusp), groove, or crest has its own specific name, which permits detailed description of each tooth and facilitates easy comparisons between samples. The root number and structure also help identify teeth: upper molars have three roots and lower molars only two. ❡ Among mammals, teeth are so diagnostic of each species that often the taxonomic identity can be determined from a single tooth, usually a molar. For example, in 1935 the German paleontologist Ralph von Koenigswald purchased fossil mammalian teeth in a Chinese apothecary shop, where they were being sold as "dragon teeth" to be used for medicinal purposes. One tooth, an enormous lower third molar, caught his eye. It had a diagnostic groove pattern, common to all hominids and apes, called the *Dryopithecus* pattern. On the basis of this single specimen von Koenigswald erected a new genus, *Gigantopithecus* (giant ape), which remains the largest known primate ever to have lived. ❡ In virtually every diagnosis of a new species of fossil hominid, teeth play a vital role. This is not solely because teeth constitute the bulk of the fossil samples but because in each species there is a diagnostic set of anatomical features in the dentition. For example, each species of *Australopithecus* (see page 124-173) is distinguished by a unique constellation of dental traits. ❡ Insight into the diet of early hominids can be gleaned from studies of the shape of the cusps, the relative sizes of the teeth, the enamel thickness, and dental wear, including microwear. Thick dental enamel, such as that found in hominids, serves to extend the life of a tooth. This probably reflects an adaptation to a heavily masticated diet before the cooking of food became common. Robust australopithecines, which are thought to have subsisted on a diet of tough, low-quality food,

Teeth and jaws of *Australopithecus afarensis* from Hadar, Ethiopia. Dense and durable, teeth and jaws constitute the majority of the human fossil record and provide a range of vital information about an organism's identity and life history. Actual size. *Photograph by John Reader, Science Source/Photo Researchers.*

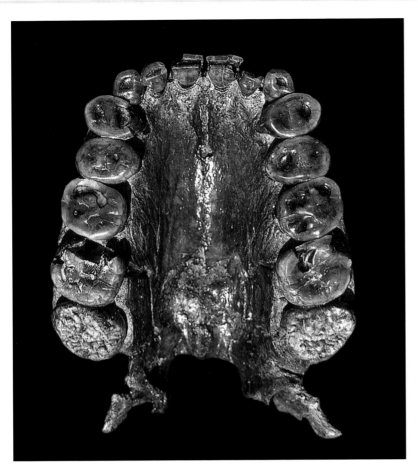

***Australopithecus boisei* maxilla from Olduvai Gorge, Tanzania.** The massive molars that characterize this species can be seen in this occlusal view of OH 5. Actual size. *Photograph by Donald Johanson, Institute of Human Origins; courtesy of the National Museum of Tanzania.*

had the thickest enamel of any primate. ◖ Within any species of hominid there is considerable overlap in the size of male and female teeth, but sometimes differences in canine size make it possible to assign a particular specimen to one sex or the other (see page 73). The approximate age at death of a hominid specimen can be determined by the state of dental eruption as well as by the degree of wear on the teeth. In Lucy's mandible, for example, the third molar, the wisdom tooth, is erupted and just beginning to show wear, leading to the conclusion that Lucy was fully adult when she died. ◖ Sometimes the health of an individual can be judged from the condition of the teeth. For instance, disease or poor nutrition can promote an interruption in dental development that is manifested as pits or grooves on the enamel surface—a condition know as hypoplasia. Because of the lack of refined sugar in the prehistoric diet, dental caries are rare but not unknown in early hominids. ◖ Microscopic study of the dental enamel adds substantially to what we can learn from the gross anatomy of the crowns. Dental enamel is composed of prisms of apatite crystals that are formed in an incremental manner. By looking at two different types of incremental markings, usually with a scanning electron microscope, it is possible to estimate the period of crown formation for individual teeth. Recent studies of hominid dental development suggest that australopithecines probably matured more quickly than modern humans, retaining a more apelike pattern of maturation (see page 78). ◖ Teeth were obviously important to the survival of our ancestors, being, to a large degree, the sole source of food preparation. For paleoanthropologists teeth are also crucially important, and odontology—the study of teeth—is sure to provide further insights into various aspects of the biology and behavior of our ancestors.

FOR MOST of the history of paleoanthropology, fossils and artifacts provided the only clues to the course of human evolution. Although these remain the most tangible and evocative sources of evidence, since the early 1960s significant contribu-

11 · Proteins, DNA, and Human Evolution

tions to the field have come from the analysis of proteins and genes in humans and other living primates. During the past four decades, data from the growing science of molecular evolution have in fact revolutionized our understanding of hominid evolution. ◖ Because all life shares an underlying biochemical foundation of DNA, studies of genetic molecules and the proteins they design permit comparison between very distantly related organisms in a quantifiable way. Molecules cannot tell you what a particular fossil is, but they can tell you what it isn't. This was the case with *Ramapithecus*, dated to 9 million years ago, the purported hominid whose taxonomic status had to be revised in the wake of molecular evidence. Before the molecular techniques were developed, estimates for the date when African apes and humans split from a common ancestor ranged from 4 to 30 million years ago, with most fossil experts choosing a date somewhere in the middle. But the biochemists came to a very different conclusion, namely, that African apes and humans had diverged much more recently, perhaps only 5 million years ago. ◖ That meant that *Ramapithecus* was far too old to qualify as a hominid, but paleontologists were loath to exclude this and other ancient fossils from the hominid family tree. Over time, and after much contentious debate, paleoanthropology has accepted this more condensed molecular time scale for human evolution. The cumulative evidence could not be ignored. Two techniques were primarily responsible for enhancing our understanding of the molecular relationships among primates: immunological reaction to blood proteins and DNA-DNA hybridization. (See page 41 for a discussion of how studies of mitochondrial DNA have contributed more recently to human origins research.) ◖ Proteins evolve over time and thus have ancestral forms, much as fossils reveal an ancestral or primitive anatomy. Because they are manufactured directly from a genetic blueprint, proteins have a structure that closely reflects the genetic makeup of an organism, so they can provide a good measure of genetic relatedness between different species. These two characteristics—that proteins evolve and have a structure that is genetically determined—form the basis of the immunological technique developed by biologist Morris Goodman in the late 1950s and early 1960s. ◖ The technique measures the magnitude of immunological cross-reaction as an indirect means of determining the degree of similarity between the proteins of two species. For example, a human blood protein, such as albumin, is injected into a rabbit, which will produce antibodies against any parts of the molecule that differ from its own albumin. The rabbit's anti-human albumin serum, rich with antibodies, can then be extracted and placed in test tubes. Albumins from a range of primates are added to the serum samples in the test tubes to measure the extent of immune reaction, visible as a precipitate that forms in the solution. The stronger the reaction, the more closely related are the human and the other species being tested. ◖ In 1962, Goodman announced that according to immunological data, humans, chimpanzees, and gorillas were equally related to one another. This countered the previously accepted belief that chimps and gorillas were each other's closest relative and that they had a common ancestor long after the first hominid split off from an ape evolutionary lineage. Few listened to Goodman at the time. ◖ But in that same year, biochemists Linus Pauling and Emile Zuckerkandl proposed that proteins evolved at a steady, determinable rate and thus had the constancy of a clock. The amount of genetic divergence must then be proportional to time, and so, knowing the amount of difference between molecules in two species and the rate of evolution in the molecule, one should be able to determine when the two species shared a common ancestor. Differences between the amino acid sequences in the same protein of two different creatures constitute the ticks in the molecular clock. The most accurate physical clock, however, is radioactive decay, so molecular divergence dates need to be calibrated based on radiometrically dated fossil or geological evidence. A common calibration point used in molecular studies of human evolution is the split between Old World monkeys and hominoids (humans, apes, and their ancestors), which, based on fossil evidence, is generally accepted as having occurred around 30 million years ago. ◖ In 1967, Goodman's technique was carried further by biochemists Vincent Sarich and Allan Wilson using the blood serum protein albumin. Their work confirmed Goodman's conclusion that humans, chimps, and gorillas were each other's closest living relatives. They found a stronger reaction between both human and chimp albumins and between human and gorilla albumins than between human and orangutan albumins. So the African apes were closer relatives to humans than the red Asian ape. Even more distantly related were the Asian lesser apes, the gibbon, and siamang. ◖ Then Sarich and Wilson added the dimension of time. Because they had previously demonstrated that albumin evolution in primates occurs at a steady rate, Sarich and Wilson suggested that the human and ape lineages had evolved at equal rates since their divergence and that albumin molecules could serve as a clock that had continued ticking along each lineage since that divergence. By establishing a relationship between immunological distance and time, they reached the conclusion that the human and African ape lineages diverged only 5 million years ago. ◖ The immunological technique has since been used with a host of proteins, including transferrin, hemoglobin, and cytochrome c. One drawback to this technique is that only a small amount of genetically coded information gets analyzed. Another molecular technique developed in the early 1960s, DNA-DNA hybridization, offered the promise of establishing evolutionary relationships using most of the genetic information contained within chromosomes. Like the immunological technique, this is an indirect method of ascertaining two species' similarity. ◖ In the DNA-DNA hybridization technique, DNA molecules, which consist of double strands wound into a helical shape, from two different species are boiled until the strands separate. Then the solution containing the separate strands is cooled, allowing the strands to anneal, often forming a hybrid of one strand from each species. A certain number of complementary bases from each strand will recognize each other and bond together. The more closely related the two species are, the more bases on their respective DNA strands will bond after the strands collide, and the more cohesive the resulting hybrid helix will be. The melting point—the temperature at which hybrid DNA separates again upon being reheated—is the quantitative measure of similarity: the lower the melting point, the less related are the two species under study. ◖ This technique was pioneered by Charles Sibley and Jon Ahlquist, who established the evolutionary relationships among birds before turning their attention to primates. Once they did, they found that DNA seems to be evolving at the same rate in all hominoids, so the average rate of evolution can be assumed to be steady along any two diverging lineages. Sibley and Ahlquist found a closer link between chimps and humans than either has with gorillas. This conclusion has been borne out more recently by other molecular studies (see page 32). As for specific dates, Sibley and Ahlquist put the chimp-human split at between 6.3 and 7.7 million years ago, while the common ancestor with gorillas lived 8 to 10 million years ago. ◖ Although these and other molecular techniques, and their revisionist conclusions, have become broadly accepted by anthropologists, elements of them remain controversial. The idea of a molecular clock, for instance, may hold for certain molecules, but there is no single, universal molecular clock, as evidence suggests that proteins and molecules

evolve at varying rates. If such obstacles can be overcome, the study of molecular evolution should continue to provide a vital counterpoint to the hard evidence obtained from fossils.

PALEOANTHROPOLOGY faces contentions from within and without. Often these arguments provide an opportunity to reexamine the evidence, shore up an argument, or educate the public in the issues that drive the field. Some debates originate in discoverers' egos, and their ramifications may be of interest only to the professional anthropologist. But the larger questions taken up in this book—where we come from, our relationship to the apes, how we should conduct ourselves in the future—touch hot nerves in American public life. The discussion of these issues has sent shock waves through our religious and educational institutions, and sometimes has to be resolved with the aid of our judicial system. ◖ The realization that humans, along with all other organisms on Earth, arrived here through a haphazard process that was guided by the natural selection of random variations ran counter to the belief that supernatural intervention guided human affairs. Darwin's exploration of biological rather than divine mechanisms as shaping human life, in exactly the same way those mechanisms shaped the lives of other creatures, struck a dissonant chord in many. Evolutionary theory was regarded as distasteful at best, sacrilege at worst. Most unsettling was the prospect that human existence was not inherently progressive and deliberate but the result of random forces—which would continue to operate through us, and on our progeny. ◖ Even today, with a vast storehouse of fossil evidence for human evolution, there are still considerable numbers of lay people and a few scientists who continue to embrace a creationist view of our origins. Yes, evolution is a theory, but a particularly robust and well-tested one. It seeks to understand and describe the biological and environmental mechanisms that drive evolution. Evolutionary theory passes the test of scientific hypothesis; creationism, currently repackaged as Intelligent Design, does not and is not science. As anthropologist Ashley Montagu once wrote, tongue in cheek, "Science has proof without certainty, creationism has certainty without proof." ◖ Controversy within the science of paleoanthropology is rampant and has always been, ever since the first remains of fossilized humans came to light in the mid-nineteenth century. The reasons for argument over a fragment of a jaw, a partial skull, a handful of teeth, or an uncertain geological date are not that difficult to understand. Often a scrappy fossil generates diverse interpretations of how it fits into the bigger picture of human evolution. Sometimes it is impossible to be certain about the actual significance of a fossil find, and it is better to postpone judgment until more complete specimens are recovered. ◖ Other reasons relate to the personalities in the discipline. The search for hominid fossils has attracted many strong-willed investigators who devote their lives to discovery and interpretation. It takes tremendous dedication to obtain funding and permits for exploration and to do fieldwork in remote areas without amenities. Even years of exploration do not guarantee discovery of significant fossils. The stakes are high. Hominid fossils are glamorous and bring scientific and popular success to those who find them, more so, say, than rodent fossils. The fossils themselves become revered and coveted. In some instances, unfortunately, access to fossils has been hindered because of a sense of proprietorship or differences of opinion between scientists. ◖ No scientist in any endeavor can say that he or she is entirely free of certain beliefs and preconceptions. We all carry with us opinions that play an important role in how we formulate our interpretations of the fossil evidence. The appearance of discordant evidence is sometimes met with an insistent reiteration of our original views. The discovery of

12 · Why is Paleoanthropology So Contentious?

a new fossil has the potential to alter our picture of human evolution, but it takes time for us to give up pet theories and assimilate the new information. In the meantime, scientific credibility and funding for more fieldwork hang in the balance. ❢ The science of paleoanthropology is still young. A number of investigators, however, are attempting to move the field onto a new plane by developing robust strategies for understanding what the fossils can tell us about our origins. The next step is to encourage cooperation, discussion, and the sharing of fossil discoveries with all colleagues. Paleoanthropologists who work together, freely engaging in discussion and debate, will reap the biggest rewards. ❢ Fossils are deeply fascinating. They speak from a remote past, and there is magic in them. The time has come to work collegially toward teasing out of each find information that will enhance our knowledge of our beginnings. Only in this way can we fully appreciate what messages were left behind by our ancestors eons ago. That is the real magic.

THOMAS HENRY HUXLEY first attempted to explain the relationship between humans and African apes from an evolutionary perspective in his 1871 book, *Man's Place in Nature*. Everything we have learned since

13 · Our Closest Living Relatives

then has only affirmed and elaborated on Huxley's intuition. Huxley conjectured that we did not descend from living apes, but rather that both humans and apes descended from some distant common ancestor. Unfortunately, when he made his arguments, Huxley had only a pair of Neandertal skulls to turn to for fossil evidence. We continue to dig deeper into the human fossil record, but the fossil record for living African apes remains almost nonexistent, probably because of the poor prospects for fossilization in the sorts of forest environment that apes have always favored. ❢ Nonetheless, African apes, we now know, are our closest living relatives, with whom we most likely shared a common ancestor within the last 6 to 8 million years. Without all the fossils to firmly establish this kinship, scientists have turned to behavioral observation (see page 83), comparative anatomy, and genetic studies. Despite the odd juxtaposition of significant anatomical differences and insignificant genetic differences between African apes and humans, both anatomy and genetics point toward the same result for our closest living nonhuman kin. ❢ Humans clearly depart from apes in several significant areas of anatomy. Many of our anatomical differences stem from our adaptation to bipedal locomotion. Because we walk on two legs, our vertebral column is more curved than an ape's, with the skull more centered at the top of the column. At the opposite end, our pelvis has become squatter and broader, with the ilia (the large bones that form the bowl shape of our pelvis) curving forward to support the upper body. The hip and knee joints, the bony heel, and the foot, on which our big toe no longer splays sideways but follows in line with the other toes, further reveal our anatomical departure from the apes. ❢ More differences come to light when we compare ape and human limb lengths and proportions. As might be expected for a biped, we humans carry much of our body weight in the lower limbs, which are longer than our upper limbs. The arms make up some 7 to 9 percent of our body weight, but the legs comprise 32 to 38 percent. In African apes the upper and lower limbs (or forelimbs and hind limbs) are much more similar to each other in this regard. In bonobos, for instance, the upper limbs account for 14 to 17 percent of the total body weight and the lower limbs account for 20 to 28 percent. In gorillas, the upper limbs account for 14 to 16 percent of body weight and the lower limbs for about 18 percent. The evenly balanced proportions of a gorilla's upper and lower limbs reflect this ape's adaptation to knuckle-walking on all fours, while chimps and bonobos will more readily take to two legs for short distances.

If we look only at relative body size and limb proportions, it appears that chimpanzees resemble humans more closely than gorillas do. ❢ An astonishingly close genetic similarity exists between African apes and humans. Chimpanzees and bonobos share 99.3 percent of their total nuclear DNA, so these are obviously extremely close relatives. But chimps and humans still share 98.4 percent of their nuclear DNA. Despite our substantial anatomical and cultural differences from them, we depart from chimpanzees in only 1.6 percent of our genetic makeup. Such a short genetic distance is typically found among sibling species of the same genus, yet chimps and humans have been arbitrarily separated into distinct biological families. Gorilla DNA differs from human DNA by only 2.3 percent. Given such close numerical kinship, it is likely that mutations in regulatory mechanisms that control how genes get expressed underlie our biological divide with the apes. These genetic differences also reflect the varying amounts of time since each species split off from a common ancestor. Chimpanzees (*Pan troglodytes*) in West and East Africa have existed in separate populations for about 1.5 million years, and bonobos (*Pan paniscus*) split off from other chimps around 2.5 million years ago. Gorillas split off from this ape line much earlier, at least 8 million years ago. Sometime after the gorilla lineage diverged, the common ancestor of all hominids split off from this line to begin the evolutionary journey traced in this book. ❢ Was this common ancestor more like a gorilla or a chimp? Fossil evidence from *Australopithecus afarensis*, *A. anamensis*, and what we know of the even earlier species *Ardipithecus ramidus* suggests that a chimp is the closest modern comparison to this earliest common ancestor. We can speculate that in appearance and behavior, this animal was a black-haired, knuckle-walking, large-brained, thin-tooth-enameled, forest-dwelling frugivore (fruit eater). The bulk of evidence from numerous nuclear and mitochondrial DNA studies now also supports a closer genetic kinship between *Homo* and *Pan* than between either *Pan* and *Gorilla* or *Homo* and *Gorilla*. The rates of genetic change appear quite similar among humans, chimps, and gorillas, and for some time it was thought that these species split off from a common ancestor over such a brief period that genetic studies might fail to resolve a closest related pair. But a recent and comprehensive review of the relevant sets of DNA sequence data concluded that 11 supported *Homo* and *Pan* as the closest related pair, while only two favored the *Pan-Gorilla* pair and one favored the *Homo-Gorilla* pair as closest relatives. In other words, among the apes, our closest kin are chimpanzees.

THE TERM "missing link" is widely used but greatly misunderstood. In 1871, Charles Darwin in his book *The Descent of Man* suggested that humans and the African apes were descended from a common ancestor.

14 · The Last Common Ancestor of Apes and Humans

Many have interpreted Darwin's idea of a common ancestor for humans and modern African apes as a species that stood halfway between human and ape, blending the characteristics of each. What Darwin meant was that modern humans and apes could be traced back in time, through a series of separate species that converge at a common ancestor. So, in a sense, the discovery of nearly every fossil hominid is the discovery of another link in the long evolutionary chain from the common ancestor to modern humans today. They were all, each and every one of them, missing links. ❢ Following Darwin, "we must not fall into the error of supposing that the early progenitor of the whole Simian Stock, including man, was identical with, or even closely resembled, any existing ape or monkey." We must never lose site of the fact that every one of our ancestral species was successful in its own right. It was not some imperfect version of modern humans or apes but a thriving organism with its own specific adaptations. ❢ It is often asked why, if we evolved from apes, are apes still here? This question misses two important points. First, humans did not evolve from any living ape but from one in the past that was more generalized in its anatomy. Second, and following on the first point, contemporary African apes, like humans, have evolved a set of adaptations, such as knuckle-walking, that may not have been not present in the common ancestor. Because the common ancestor was a generalized ape, more ancient hominids will be more apelike in their appearance. ❡ During the 1970s and 1980s it was widely held among paleoanthropologists that the common ancestor for hominids and apes must have lived deep in the Miocene, some 15 to 20 million years ago. This made us very comfortably distant from the African apes which pleased those committed to the view that human distinctiveness has deep geological roots. ❡ This view was the result of a 1932 discovery of a fragmentary jaw with teeth, in the Siwalik Hills of northern India. The discovery was made by G. Edward Lewis, a paleontology graduate student at Yale University, who named the specimen *Ramapithecus punjabicus*, after the Hindu deity Rama. At the time, Lewis assigned the 7 to 8 million-year-old maxilla to the Hominidae. Most scientists ignored this suggestion, but in the 1960s the idea that *Ramapithecus* was a hominid gained popularity when similarities between it and later hominids were pointed out. In addition to the thick dental anatomy, presumed short face, hominid dental eruption pattern, and small canine, it was the reconstructed parabolic shape of the tooth rows (apes have U-shaped tooth rows) that seemed to clinch its affinities with the hominids. ❡ Even older corroborating evidence came from Fort Ternan in western Kenya, where Louis Leakey found jaw fragments dated to 14 million years ago that appeared virtually identical to the Siwalik maxilla. Some paleoanthropologists were so enthusiastic about these putative early hominids that they speculated *Ramapithecus* was a bipedal tool-making creature that prepared food and whose offspring had an extended juvenile period necessary for prolonged social learning. ❡ Debate raged in the anthropological community until the early 1970s, when the hominid status of *Ramapithecus* came under heavy criticism. The most damaging blow was the realization that, because of a basic error in interpreting anatomy, the upper jaw of *Ramapithecus* had been reconstructed incorrectly. In order to reconstruct the correct shape of the dental arcade it was necessary to have a midline. The Siwalik maxilla lacked a midline, and as other fossils with a midline were recovered from Kenya, their dental arcade appeared more U-shaped—more apelike. ❡ Corroborating evidence for a more U-shaped dental arcade in *Ramapithecus* came when paleontologist Peter Andrews, at the Natural History Museum in London, and Alan Walker, anthropologist at Pennsylvania State University, published a reconstruction of the Kenyan *Ramapithecus* mandible. Other than the thick dental enamel, there was little to link *Ramapithecus* with later hominids. But a particular difficulty arose with using thick enamel as the sole feature suggestive of hominid status: in Pakistan, wherever *Ramapithecus* remains were found, a thick-enameled ape known as *Sivapithecus* ("Siva's ape") also turned up at the same site. Ultimately enough *Sivapithecus* remains were found to validate the notion that what had been called *Ramapithecus* was only a smaller version of *Sivapithecus*, an ape. ❡ The *coup de grâce* came in 1981, when Harvard University anthropologist David Pilbeam described an 8 million-year-old partial cranium and mandible of *Sivapithecus* that had been found in Pakistan. The resemblance of this specimen to a modern orangutan was so striking that some thought it should be placed in the genus of the orangutan, *Pongo*. Ironically, David Pilbeam, along with his mentor, Duke University paleontologist Elwyn Simons, were the two scientists who had proposed and staunchly defended the hominid status of the Siwalik *Ramapithecus* in the 1960s and 1970s. ❡ Further criticism of the alleged hominid status of *Ramapithecus* came from an entirely different quarter—the developing discipline of molecular anthropology. Very simply, it was thought that the more closely related creatures were, the more similarities they would share in their blood proteins. With the aid of such techniques as electrophoresis, which can separate individual protein molecules for identification, it was found that humans and African apes were astonishingly similar, and the Asian apes, just as the fossil evidence suggested, were significantly more dissimilar. In fact, a bold suggestion was made to place humans and the African apes in the same zoological family, the Hominidae (see page 18). ❡ To some researchers, such closeness in blood proteins implied that the common ancestor for modern humans and the African apes must be much more recent than the date assigned by the *Ramapithecus* evidence. In the late 1960s and early 1970s it was determined that the degree of difference in certain blood proteins between different primates was directly related to the length of time elapsed since they last shared a common ancestor. ❡ Calibrating the molecular clock—a method of determining the timing of evolutionary divergence based on relative rates of mutations (see page 30)—on the basis of the last common ancestor for Old World monkeys and apes, roughly 30 million years ago, led to the astonishing conclusion that humans and African apes shared a common ancestor no earlier than 8 million years ago! So emphatic were the researchers who arrived at these conclusions that one, Vince Sarich, anthropologist at the University of California, Berkeley, wrote, "one no longer has the option of considering a fossil older than about 8 million years as a hominid *no matter what it looks like*." ❡ Although the molecular clock does have some potential shortcomings, such as the assumption of a constant rate of genetic mutation, it appears that the DNA and molecular studies are in agreement that a common ancestor for African apes and humans is much more recent than the *Ramapithecus* episode led scientists to believe. Only the fossil record, however, will be able to tell us what the anatomy of that last common ancestor was, the kind of world it lived in, and what sorts of behavior characterized the species. With bipedalism being the defining characteristic of early hominid evolution, one of the most important controversies this fossil evidence can address is the locomotor mode of the common ancestor of humans and African apes. ❡ The numerous anatomical similarities (broad thorax, short lumbar region, placement of should blade on the back, etc.) of the human body Bauplan with the African apes reflect an arboreal mode of locomotion, which included a great deal of suspensory posture. But in a terrestrial habitat the African apes are quadrupedal knuckle-walkers and we are bipedal. Unlike other terrestrial primates that use the palm surface of their hands for support, the apes support their weight on the backside of their middle finger bones or phalanges. ❡ It has long been assumed that knuckle-walking was not present in the common ancestor to us and the African apes, but appeared after the evolutionary split in the common ancestor to the chimps and gorillas. Bipedalism arose independently in our direct ancestors. ❡ This explanation, based on anatomy, led to a family level distinction between us the Hominidae and the large bodied apes Pongidae (this included the orang). Now, however, with molecular and genetic evidence it has become clear that orangs are only distantly related to the African apes and us but most importantly human and chimpanzees are more closely related to each other than chimpanzees are to the gorilla. The implication is that the common ancestor for gorillas, chimps and hominids was a knuckle-walker, a trait retained in the apes and replaced by bipedalism in our ancestors. Controversial support for this view comes from the observation that one of Lucy's forearm bones (the radius) has a bony ridge which helps stabilize the wrist in knuckle-walkers. ❡ An alternative scenario posits a non-knuckle-walking ancestor to African apes and hominids, implying that knuckle-walking evolved independently in chimp and gorilla lineages. This view is less parsimonious since knuckle-walking, a highly unusual adaptation, would have to have arisen more than once. ❡ Accepting that we and chimps are closely related has implications for

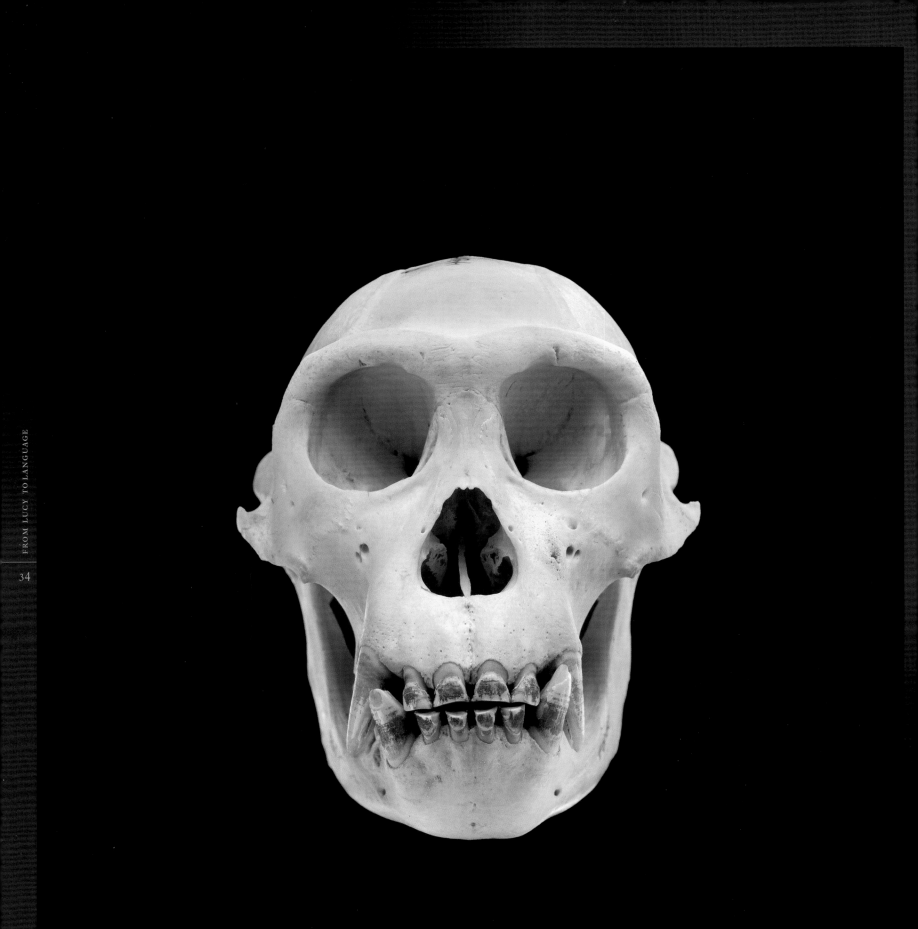

Skull of male chimpanzee, *Pan troglodytes*. Anatomical evidence and, more recently, much molecular evidence reveals that the common chimpanzee, is the living species most closely related to ourselves. Actual size. *Photographs by David L. Brill; courtesy of National Museum of Natural History.*

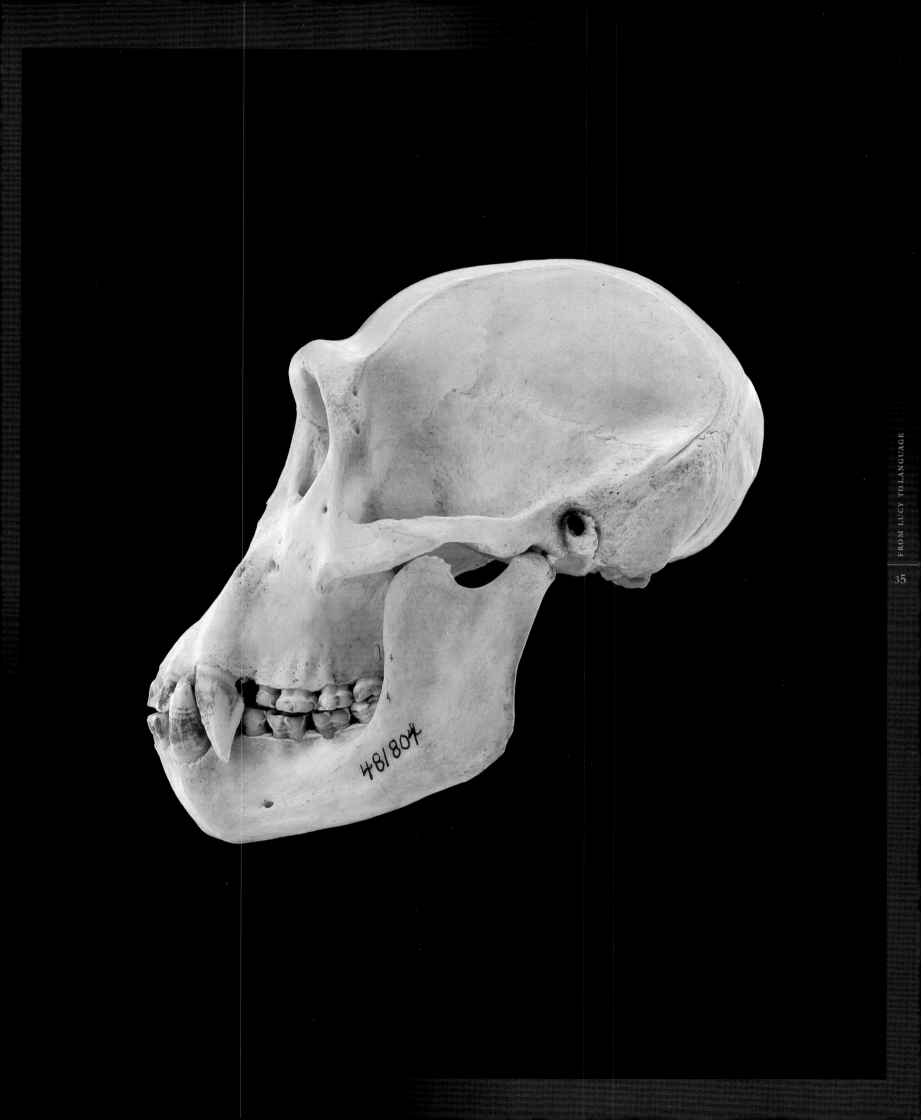

Skull of *Sivapithecus* from Potwar, Pakistan. Once considered to be in direct ancestry
to humans, fossils of *Ramapithecus* are now considered the same as those of *Sivapithecus*,
a likely ancestor of the orangutan. Actual size. *Photograph by David L. Brill; courtesy of
Peabody Museum of Archaeology and Ethnology, Harvard University.*

how we are classified. For an increasing majority of anthropologists the family Hominidae (hominid in the vernacular) includes bipeds, chimps and gorillas with bipeds placed in the Tribe Hominini (hominin in the vernacular). In this book we use hominid to refer to the human lineage after the split from the apes. ❦ The fall from grace of *Ramapithecus* as a hominid is an example of a hypothesis that was tested and ultimately discarded. Unfortunately, we know relatively little about hominid and ape evolution during the late Miocene making it thus far impossible to identify the last common ancestor. A few teeth, 6-12 million years old, from the Baringo Basin, Kenya suggest affinities with chimps and gorillas, and if true, the common ancestor may be much older than thought on the basis of molecular divergence estimates.

THE VARIETY of human family trees now cluttering the literature makes it virtually impossible to identify the most accurate tree because of the forest. A family tree is supposed to reflect genealogy; that is,

15 · Drawing the Human Family Tree

trees are constructed by tracing lines of descent back into the past through distinct ancestral species. Some scientists claim there is no infallible method for discriminating a bona fide ancestor from one that was only an extinct relative. Put the other way around, whereas we can be certain that we had ancestors, it is much more difficult to be certain that a particular fossil species actually left descendants, and, if it did, whether we are among them. ❦ We know that some fossil groups, like the robust australopithecines, which disappeared from the fossil record about 1 million years ago, were not on a direct line of descent to modern humans but were a separate group of dietarily specialized hominids. It is when we try to identify our direct ancestors that things become clouded. Although Lucy is widely embraced as one of our most ancient ancestors, we can never be positive that she herself left descendants. Her skeleton has no diagnostic anatomy to indicate that she even bore children, but *some* members of her species must have left descendants. ❦ Identification of the genuine ancestor becomes more complex when more than one possible candidate exists. This is especially true when there is temporal overlap of at least two kinds of co-existing hominids, making correct conclusions of ancestor-descendant relationships far from straightforward. This is the case for early *Homo*, which may contain three species: *H. habilis*, *H. ergaster*, and *H. rudolfensis*. In this instance a consensus as to the identity of the species ancestral to later hominids has proved problematic. A similar predicament exists for a much later time period, when Neandertals and anatomically modern humans overlapped in time. Depending on one's theoretical framework, Neandertals did or did not leave genes in modern human populations. ❦ The spate of hominid finds in the past fifty years, and with them evidence bearing out the Darwinian concepts of branching and diversification in evolutionary pathways, has made erecting our own family tree more difficult. The concept of a direct evolutionary relationship between fossil ancestors and modern humans was easy to defend when few fossils had been discovered. The unilinealists, for example, around the turn of the century saw a simple progression: apes to Java Man to Neandertals to modern humans. It was all very simple: we began with an apelike ancestor and, because of some general evolutionary trend—a particular inertia—light-skinned European hominids emerged. Such linearity, however, flew directly in the face of the Darwinian evolutionary processes. Why should human evolution be different from the evolution of any other organism? The notion that modern humans represent not only the ultimate goal of hominid evolution but the goal of evolution itself died hard. ❦ If the process of evolution is one of splitting and divergence, this has strong implications for the number of human species

that have existed (see pages 23 and 40). If, as some believe, there were many more species of human ancestors than we are able to recognize, then ancestor-descendant relationships become bewildering, making the building of evolutionary trees even more complex. Family trees become more similar to broad bushes with multiple branches of various lengths. ❦ In 1866 the German biologist Ernst Haeckel, a contemporary of Darwin, published the first evolutionary trees. With a fondness for coining scientific terms, such as ecology and ontology (the study of embryological development), he invented the word *phylogeny* for the study of ancestor-descendant relationships among organisms. So enthusiastic was Haeckel about the construction of family trees that he devised one for humans. On this tree he placed a missing link, calling it *Pithecanthropus alalus* ("speechless ape-man"). This was an inspiration to Eugène Dubois, who named his Java find *Pithecanthropus erectus* ("erect ape-man"). ❦ Following Haeckel, early attempts at classifying organisms and constructing family trees tended to consider only overall similarities and differences among various organisms. This method was suspect because it contained a great degree of subjectivity. More recently, in an attempt to bring more scientific rigor and objectivity to the process, a system known as cladistics was introduced. In cladistics, evolutionary relationships are examined by tabulating primitive versus derived character states for specific anatomical traits. Primitive character states are those that have not been altered through evolutionary change; derived character states are subsequent evolutionary novelties. Organisms are grouped into clades (branches) on the basis of shared derived characters inherited from a recent common ancestor, but not shared with more ancient ancestors. ❦ Anatomical characters are determined to be derived or primitive, and the cladogram that most economically explains the patterns of relationships of these features is considered the best. For example, among hominids, bipedalism is a primitive trait, meaning that it is a trait shared by all hominids and therefore of little value in differentiating one hominid species from another. This is why primitive traits are not of great value in assessing relationships between species, because all species possess these. In contrast, a character state such as significant brain expansion is a derived feature that can be used to separate *Homo* from *Australopithecus*. It is generally true, therefore, that the greater the number of derived characters shared by two species, the closer is their evolutionary relationship. ❦ The evolutionary relationships determined by cladistics are translated into a phylogenetic tree, which requires the addition of time and the supposed ancestor-descendant relationships of the different taxa (units in the classifications of organisms, in this instance, hominid species). A tree, therefore, is a visual representation of the pattern of hominid evolution through time. ❦ How a hominid family tree is drawn depends on the current state of knowledge of taxonomic diversity in the past. Although some species and time periods are moderately well represented by actual fossil finds, others are not. Because of the incomplete fossil sample, we have not constructed a hard-and-fast phylogeny but include dotted lines in an attempt to express alternative views. In some instances, because of the paucity of specimens, lack of detailed publication, or their long temporal separation from other taxa, dotted lines are absent or a question mark has been inserted, *e.g.*, *K. platyops*, *A. garhi*, *O. tugenensis*, *Ard. ramidus*, *Ard. kadabba*, and *S. tchadensis*. ❦ A comparison of the dental and jaw anatomy in *A. anamensis* and *A. afarensis* establishes a fairly plausible evolutionary connection. Although characters such as tooth shape and the buttressing of the anterior part of the jaw are different in these two species, the more primitive character state in *anamensis* makes it a good precursor to *afarensis*. If it turns out that *A. afarensis* overlaps in time with *A. anamensis*, the tree will get bushier in the 4 million-year time slice, indicating that perhaps the latter species diverged from the former and co-existed with it for a brief period of time.

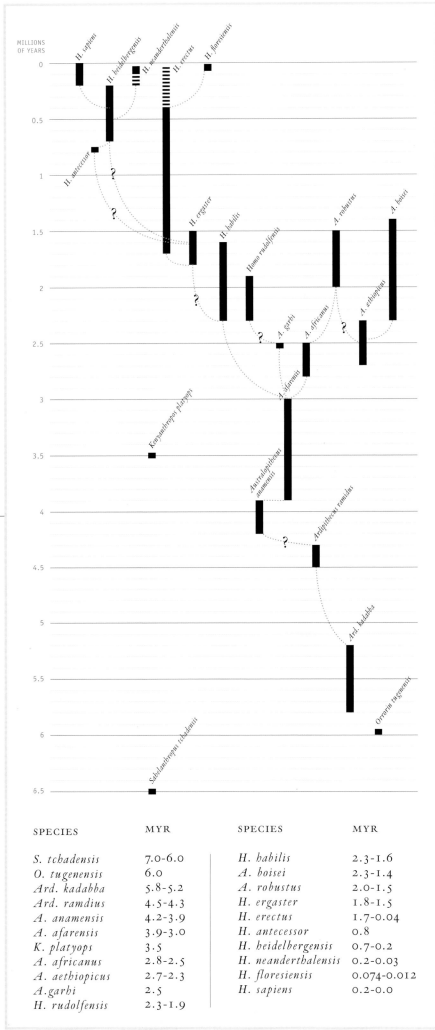

MILLIONS OF YEARS

H. sapiens
H. heidelbergensis
H. neanderthalensis
H. erectus
H. floresiensis
H. antecessor
H. ergaster
H. habilis
Homo rudolfensis
A. robustus
A. boisei
A. garhi
A. africanus
A. aethiopicus
A. afarensis
Kenyanthropus platyops
Australopithecus anamensis
Ardipithecus ramidus
Ard. kadabba
Orrorin tugenensis
Sahelanthropus tchadensis

SPECIES	MYR		SPECIES	MYR
S. tchadensis	7.0–6.0		*H. habilis*	2.3–1.6
O. tugenensis	6.0		*A. boisei*	2.3–1.4
Ard. kadabba	5.8–5.2		*A. robustus*	2.0–1.5
Ard. ramidus	4.5–4.3		*H. ergaster*	1.8–1.5
A. anamensis	4.2–3.9		*H. erectus*	1.7–0.04
A. afarensis	3.9–3.0		*H. antecessor*	0.8
K. platyops	3.5		*H. heidelbergensis*	0.7–0.2
A. africanus	2.8–2.5		*H. neanderthalensis*	0.2–0.03
A. aethiopicus	2.7–2.3		*H. floresiensis*	0.074–0.012
A.garhi	2.5		*H. sapiens*	0.2–0.0
H. rudolfensis	2.3–1.9			

❦ *A. africanus* is judged to be a sole ancestor to *A. robustus* and, because it displays a host of features derived in the robust direction, it is not viewed as an ancestor to *Homo*. Confirmation that *A. aethiopicus* shares a number of primitive features with *A. afarensis* suggests that a separate lineage leading to *A. boisei* evolved in East Africa, parallel to that seen in South Africa. A minority view holds that *africanus* went extinct without leaving descendants and that both *robustus* and *boisei* evolved from *aethiopicus*. ❦ The genus *Homo* presents several problems. First, there is a long gap in the fossil record between 2 and 3 million years ago where convincing intermediates between *A. afarensis* (the presumed ancestor to *Homo*) and earliest *Homo* are lacking. Second, some investigators have postulated as a second species at Sterkfontein a *Homo*-like contemporary of *A. africanus*, referred to *Homo* species novum. Third, *H. habilis*, as recognized by some, has made it increasingly difficult to nail down precise evolutionary relationships between taxa. *H. habilis* from Olduvai could be a dead end, leaving *H. rudolfensis* as an ancestor to *H. ergaster*. Some interpretations see *ergaster* as a common ancestor to *H. erectus* and *H. heidelbergensis*, while others see *ergaster* as ancestral to *erectus*, which in turn would be ancestral to *heidelbergensis*. There is increasing support for recognizing *H. neanderthalensis* as a separate species from *H. sapiens* and one that ultimately went extinct. *H. sapiens* would therefore have evolved from *H. heidelbergensis*, a common ancestor to Neandertals and to anatomically modern humans. ❦ This evolutionary tree stops with us—*Homo sapiens*. An intriguing question is, where will the tree lead? Will there be more speciations? Will we evolve into a different species— *Homo computerensis*? The answer is that the conditions that bring about speciation are currently absent in the modern world. Humans are no longer reproductively isolated, a necessary condition for sufficient differences to develop in populations to bring about speciation. Instead, we have become a highly mobile species aided in our movements by modern modes of transportation, and as a result of this mobility we have become homogenized as a species. Reestablishing the conditions for speciation would require isolating some part of the human population—through space colonization, say. If that island of humans in space remained out of contact with Earth long enough, then we might see the appearance of a new species of *Homo*.

16 · African Genesis

THE GREAT RIFT VALLEY in East Africa has yielded the most ancient fossilized remains of our ancestors and relatives. Although there is no reason to believe that early hominids were confined to eastern Africa, it is the unique geological history of the Rift Valley that is responsible for the concentration of finds. From what is known, however, it does appear that the earliest hominids were confined to Africa, since no remains of australopithecines have been found elsewhere in the world. ❦ Molecular studies confirm that African apes are more closely related to modern humans than either group is to the Asian orangutan. The African ape-human clade shared a relatively long common evolutionary history, subsequent to the time when orangutans followed their own evolutionary trajectory. There is a growing consensus, partly as a result of molecular studies, that the last common ancestor to the African ape-human clade will eventually be found somewhere between 6 and 8 million years ago, undoubtedly in Africa. ❦ Until recently, hominids older than 4 million years were limited to less than a half-dozen specimens. These specimens include a right mandible fragment with a first molar found at the site of Lothagam, in northern Kenya, and dated to approximately 5 to 6 million years ago. Although its anatomy is consistent with placement in the family Hominidae, the fragmentary nature of the Lothagam specimen makes assignment to

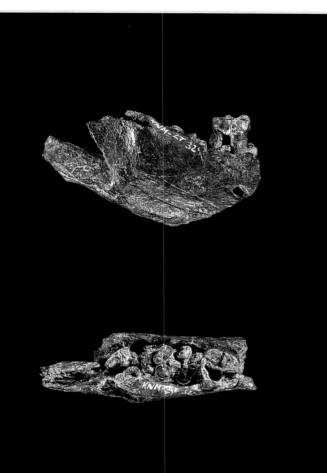

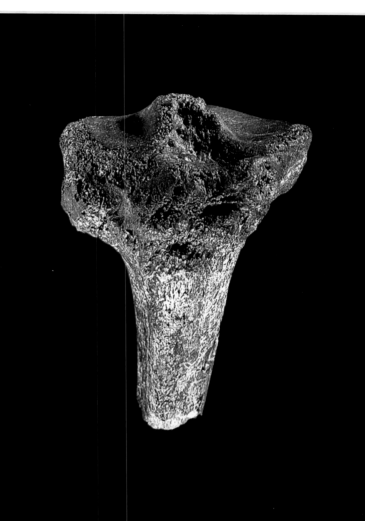

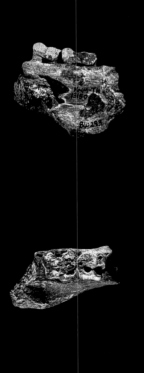

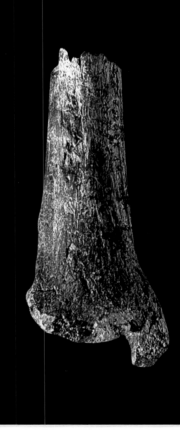

Mandible from Lothagam, Kenya. This fragmentary fossil exceeds 5 million years in age but is not complete enough to identify to species or even to classify as a hominid or an extinct ape. Actual size. *Photograph by Robert I.M. Campbell; courtesy of National Museums of Kenya.*
Mandible from Tabarin, Kenya, KNM-TI 13150. An age of between 4 and 5 million years makes this fossil broadly contemporaneous with Ethiopian fossils of *Ardipithecus ramidus*. This mandible has been identified as *Australopithecus praegens*. Actual size. *Photograph by Robert I.M. Campbell; courtesy of National Museums of Kenya.*

Tibia of *Australopithecus anamensis* from Kanapoi, Kenya. The earliest definitive evidence for bipedalism is this specimen, KNM-KP 29285. Both upper and lower ends were recovered, and the articular surface for the knee joint is clearly expanded from front to back as in a biped. Actual size. *Photograph by Robert I.M. Campbell; courtesy of National Museums of* Kenya.

genus or species impossible. A proximal humerus from Lake Baringo, Kenya, dated to approximately 4.2 million years ago, shares morphological similarities with *Australopithecus afarensis*. ❧ A mandible fragment from Tabarin, northern Kenya, with worn right first and second molars dates from between 4.15 and 5.15 million years ago. It has been designated a new species, *Australopithecus praegens*. The site of Fejej, in southern Ethiopia, dated to 4.2 million years yielded a handful of fossil hominid teeth that may be in *A. afarensis*. The somewhat isolated finds from Chad assigned to *Sahelanthropus tchadensis*, dated to 5 to 7 million years ago, as well as the 6 million-year-old *Orrorin tugenensis* from Kenya hint that the earliest occurrence of hominids, but their evolutionary relationship to later fossils is not currently well understood. ❧ With such a small collection of hominid fossils from deposits more than 4 million years old, the recent discoveries from Aramis, Ethiopia, and Kanapoi, Kenya, dated respectively to 4.4 and 4.2 million years, are a welcome addition to the very earliest part of our fossil ancestry. ❧ In 1994 a team of paleoanthropologists announced the discovery of fossil hominid specimens from Aramis, a site situated in the Middle Awash river valley in Ethiopia. Consisting of teeth, jaws, bits of skull, and postcranial bones, the Aramis material, radiometrically dated to 4.4 million years, possesses a unique constellation of anatomical features (see page 120), which stimulated the naming of a new species, *Australopithecus ramidus* ("ramid" is from the Afar people's word for root). After further consideration and the discovery of a nearly complete skeleton at Aramis, the entire collection was allocated to a new genus, *Ardipithecus ramidus* ("ardi" means ground or floor in the Afar language; thus "basal root ape"). A collection, mostly of teeth and jaws, from the Middle Awash dating to 5.2 to 5.8 million years ago may be ancestral to *Ard. ramidus* and have been assigned to *Ard. kadabba*. ❧ In August 1995 the discovery of 3.9 to 4.2 million-year-old hominid fossils from Lake Turkana in northern Kenya was announced. Consisting mostly of teeth and jaws, these finds came predominately from the site of Kanapoi. They show strong similarities with *A. afarensis*, but the Kanapoi hominids are characterized by a series of anatomical features (see page 129) that distinguish them as a new species, *Australopithecus anamensis* ("anam" is from the Turkana word for lake; thus, "southern ape of the lake"). A distal humerus found at Kanapoi in 1965 has now been attributed to this new species. Now that the 4 million-year barrier has been breached, even more ancient sites with early hominids are being discovered. ❧ The oldest finds from Chad, the Middle Awash, and Kenya are offering a tantalizing look at an early part of the hominid fossil record that was previously unavailable. These finds will no doubt generate debate, as happens with the announcement of any new hominid species, but finally we are able to delve deeper into the phylogenetic roots of our unique zoological family, the Hominidae.

17 · Early vs. Modern Humans

THE HUMAN family includes both living and extinct species of bipedal primates that have been evolutionarily distinct since their separation from a common ancestor with the African apes. The formal zoological category for this group of bipedal primates at the family level is the Hominidae, which in its more popular or vernacular form is abbreviated as hominid. ❧ The traditional classification, which places humans in the family Hominidae and the great apes (chimps, gorillas, orangutans) in the family Pongidae, has recently been challenged on the basis of biochemical and molecular studies. These studies have established the closeness of humans and African apes, to the exclusion of the orangutans. To more accurately reflect the evolutionary relationships between these apes and humans, some scien-

tists include them all in the Hominidae. To stress the close evolutionary connection between humans and African apes, they are put into the subfamily Homininae, while orangutans and their ancestors are placed into the subfamily Ponginae. Although these newer groupings more closely reflect the evolutionary history of apes and humans, the more traditional classification cited above is followed here, because it is widely used in the paleoanthropological literature. ❧ The family Hominidae is divided into early hominids and later hominids. In this book early hominids, with the exception of *Ardipithecus*, *Orrorin*, and *Sahelanthropus*, are placed in the genus *Australopithecus* ("southern ape") and collectively called australopithecines. Later hominids are all placed in the genus *Homo*. The Hominidae constitute a diverse zoological family, united by the ability to walk bipedally. Some of the greatest anatomical differences in hominid species become clear if we look at the differences between early and later hominids, that is, between australopithecines and *Homo*. ❧ The earliest hominids consist of three genera *Sahelanthropus tchadensis*, *Orrorin tugenensis*, and *Ardipithecus* with two species, *kadabba* and *ramidus*, followed by the genus, *Australopithecus*, which contains seven species: *anamensis*, *afarensis*, *africanus*, *garhi*, *aethiopicus*, *robustus*, and *boisei*. In this book the term australopithecines is employed as a collective term for all species of these early hominids. Some researchers object to this usage because it implies a subfamilial rank, Australopithecinae, and prefer the term australopith. ❧ Australopithecines, sometimes called early hominids (or even early humans), have thus far been found exclusively in African deposits ranging in age from more than 6 to nearly 1 million years old. Although their postcranial bones—those occurring "after the cranium," or below the skull—attest to committed terrestrial bipedalism, some species show features such as relatively long arms and curved finger and toe bones that are reminiscent of an arboreal ancestry. ❧ Australopithecines were vegetarian, and some species, the robust australopithecines (*A. robustus*, *aethiopicus*, and *boisei*), developed extreme masticatory adaptations such as very large postcanine teeth with thick, fast-growing enamel, thick mandibles, and massive chewing muscles such as the temporalis, sometimes associated with bony crests that anchor these muscles on the cranium. Faces range from being very prognathic (projecting) in earlier forms to flat and even dish-shaped in the later, robust forms. Australopithecines have postcanine megadontia, meaning that the molars and premolars are relatively large compared to the anterior teeth. ❧ The australopithecine skull has a small braincase, ranging from less than 400 cc to approximately 550 cc, and is dominated by a relatively large face. The cranial vault is relatively thin-walled and, when viewed from above, shows marked postorbital constriction. An anteriorly situated sagittal crest along the midline of the cranial vault, for attachment of the temporalis muscles, is sometimes present, particularly in robust males. Pronounced nuchal crests that anchor neck muscles occur in the occipital region, and foreheads range from flat and low to slightly more vertical. The mastoids are heavily pneumatized (full of air cells). ❧ Associated postcranial remains are not common, but australopithecines ranged in stature from 1 to 1.5 meters and in body weight from 27 to 45 kilograms (about the weight of a 10-year-old modern American child). Where sufficiently well known, species exhibit marked sexual dimorphism, with males substantially larger than females. ❧ *Homo*, the genus in which modern humans are placed, was named in 1758 by Linnaeus. It contains nine species: *rudolfensis*, *habilis*, *ergaster*, *erectus*, *antecessor*, *heidelbergensis*, *neanderthalensis*, *floresiensis*, and *sapiens*. The genus probably originated in Africa roughly 2.5 million years ago, but today it is global in distribution. *Homo* is distinguished from the australopithecines by a larger cranial capacity, ranging from roughly 530 cc in earlier species (417 cc if *H. floresiensis* is included) to upward of 2,000 cc in modern humans. Over time there has been an increase in stature (especially lengthening of the lower limbs) and body weight, and a reduction in sexual dimorphism. The entire facial skeleton with its associated masticatory apparatus, including the dentition, is reduced relative to that of the australopithecines. True browridges, absent in australopithecines, become strongly developed in some species. In general, muscle markings on the skull and on the post-cranial bones are reduced. The bones of the cranial vault are noticeably thicker than in australopithecines. ◖ Modern *Homo sapiens* is further distinguished by a number of anatomical features. Modern humans have fairly gracile skeletons with much reduced brow ridges and cranial buttressing. Our teeth and jaws are strongly reduced and our faces are vertical and flat, with a prominent nose. Our thinly built mandibles have a mental eminence (a chin). Our short skulls are distinguished by a high, vertical forehead and dominated by a very large brain. ◖ Perhaps the most distinguishing feature of our species is our dependence on culture for survival. Culture and symbolic language are central themes in distinguishing modern humans (beginning some 40,000 years ago). We communicate symbolically through language and have become dependent on culture for survival. Although some anatomical variation exists in *Homo sapiens*, making our species polytypic, these differences are insignificant and in no way confer separate racial status on any group of living humans.

IN 1987 a Berkeley biochemistry lab sent shock waves through the scientific community with the announcement that modern human genetic heritage had been traced back through time to our species' very origin. That origin occurred in Africa around 200,000 years ago. The study relied on a portion of our genetic reserves, called mitochondrial DNA (mtDNA) that is inherited only from one's mother, so the proposed founding mother of modern humanity naturally was dubbed "mitochondrial Eve," and this origins story became the Eve hypothesis. ◖ The scenario written in the mtDNA data had some staggering implications. Humans resulted from a relatively recent speciation event and rapidly spread out from Africa around the world. Moreover, all the more archaic populations of *Homo erectus* and *Homo heidelbergensis* already living outside Africa made no contribution to the modern human genome. Because the mtDNA failed to show any deep Asian roots for our species, for instance, Peking Man and Java Man were dead ends. ◖ The now-classic mtDNA study by the late Allan Wilson, Rebecca Cann, and Mark Stoneking examined gene diversity among 147 people from the United States, New Guinea, Australia, Asia, and Europe. The scientists homed in on DNA in the mitochondrion, an organelle outside the nucleus that generates energy for a cell. Mitochondrial DNA evolves much more rapidly than nuclear DNA, so it can be used to explore questions about recent evolutionary divergences by studying the differences between populations due to accumulated genetic mutations since some past divergence. In this case, the results showed that Africans had greater mtDNA diversity (reflecting a longer period of evolution) than other populations, and a human genealogical tree could be rooted in Africa with a founding mother between 140,000 and 290,000 years ago. ◖ Paleoanthropologists arguing for an ancient human lineage stemming directly from *erectus* populations on several continents couldn't abide by Eve. After the shock waves passed, the Eve hypothesis met with some stiff criticism: for using a less exact analytical method called restriction analysis instead of more detailed comparisons between complete genetic sequences; for using African Americans as a substitute for African populations; and for not including some sort of nonhuman genetic outgroup (a more distantly related species, such as a chimpanzee), to help root the human genetic family tree. In response, the study was refined and repeated, adding mtDNA from 189 people, including 121 Africans from six sub-Saharan regions. Chimpanzee mtDNA was examined to calibrate the molecular clock by comparing the differences between chimp

18 · Eve and Adam

and human mtDNAs accumulated since the two species diverged from each other. These results, published in 1991, also suggested that humans arose in Africa between 166,000 and 249,000 years ago. As before, the new Eve tree had two main branches, one leading to six African mtDNA types and the other to everyone else. Eve's support was shored. ◖ But not for long. The computer program used in both Eve studies, called PAUP, for Phylogenetic Analysis Using Parsimony, ranks different trees by the principle of parsimony: the tree requiring the fewest genetic mutations, or steps, to explain its pattern becomes the most plausible. It turns out, though, that the computer's results may depend on how data is entered, how many runs are made, and how many trees it analyzes. A study by geneticist Alan Templeton found 100 trees more parsimonious than the 1991 Eve tree, while another scientific team obtained 10,000 trees that were more parsimonious than the 1987 Eve tree and had a mix of geographic roots. In addition, the range of potential error in pinpointing a date for human origins meant that Eve could have lived anytime between 100,000 and a million years ago. Suddenly, the African origin for modern human mtDNA no longer appeared so clear cut. ◖ Although the Eve hypothesis remains controversial, the accumulated genetic evidence still suggests a recent origin, and probably an African one. A different segment of mtDNA, the cytochrome oxidase subunit II (CO II) gene being studied by Maryellen Ruvolo and others, indicates that, assuming chimps and humans shared a common ancestor 6 million years, modern humans apparently originated 222,000 years ago. And one study of the complete mitochondrial genome, albeit from only two humans, came up with an origin date of 176,000 years ago. And Africans have the greatest mtDNA and nuclear gene diversity of any population, which suggests that their genes have greater antiquity. ◖ If Eve has proves elusive, Adam has been even more so. A few teams of geneticists have undertaken the task of identifying appropriate portions of the Y chromosome, the male sex chromosome, for clues to when and where humanity's founding father(s) lived. Much of this chromosome never recombines with the X chromosome (the sex chromosome found in both males and females) and persists unchanged except for the occasional mutation, so like mtDNA the Y chromosome offers another potential way to trace human prehistory to its starting point. ◖ One Y chromosome study in the early 1990s concluded that Adam was a Pygmy living 200,000 years ago in the forest of what is now the Central African Republic. Three studies published in 1995 made further attempts to locate Adam. The first study using a small sequence of the ZFY gene, which may determine the development of sperm and testes, found an optimal date for Adam of 270,000 years ago, in close correspondence with the mtDNA dates for Eve. A pair of subsequent research papers offered slightly different stories. One pinned on to the ancestral human Y chromosome a date of 188,000 years ago and suggested a pattern of male migration from Africa to Europe and Asia; the other concluded that the modern human origin was much more recent, only 37,000 to 49,000 years ago. ◖ Nuclear DNA has also been employed to determine the timing of modern human origins. Some analyses suggest an African origin around the same time as indicated by the mtDNA, but increasingly the evidence points to an even more recent emergence. The analysis by Luigi Luca Cavalli-Sforza and colleagues of microsatellites in nuclear DNA from an ethnic sampling of 150 people concluded that Eve lived no longer ago than 133,000 years. A newer, more comprehensive nuclear DNA study by a large team of geneticists in six countries used two parts of the CD4 gene of chromosome 12 from a global sample of 1,600 individuals in 42 populations. Again, the most diversity occurred among sub-Saharan African populations, while a startling uniformity characterized all groups outside of Africa. After spreading throughout Africa, apparently, groups of these early modern humans embarked from the northeast corner of the continent into the Middle East about 102,000 years ago. From there

42

Fossils of *Homo sapiens* from Klasies River Mouth, South Africa. Four mandibles and several cranial fragments from these coastal caves date to between 80,000 and 100,000 years ago, supporting the idea that modern humans evolved recently in Africa. Actual sizes. *Courtesy of South Africa Museum.*

they entered Europe and continued spreading eastward into Asia and beyond. ◖ It is important to realize that mention of Adam or Eve does not mean that we are all descended from just one pair of individuals, as in the Biblical story of Eden. Mitochondrial Eve represents not the ancestral mother from whom all humans descend, but rather the specific carrier of an ancestral mtDNA molecule that gave rise to all modern human mtDNA. Many other men and women were Eve's contemporaries, and considering that mtDNA constitutes only 1/400,000th of our total genome, these contemporaries almost certainly contributed to other parts of our present genetic makeup. In order to have maintained our current level of genetic diversity, it is likely that populations of the first modern humans would have totaled at least 10,000. So our species did not derive from a single ancestral couple, an Adam and Eve, but did originate relatively rapidly from a founding population that lived in Africa sometime in the vicinity of 200,000 years ago.

IN the early part of the twentieth century, scientists like the English anatomist Sir Arthur Keith, who served as Louis Leakey's intellectual mentor, and the French paleontologist Marcellin Boule postulated that modern humans were of great antiquity.

19 · The Earliest Fossil Evidence of Anatomically Modern Humans

The pre-*sapiens* theory, articulated in large part by Boule and Keith, held that Neandertals could not be among the ancestors of modern humans not only because they were too brutish, but also because modern humans had already existed long before the Neandertals evolved. Boule strongly influenced his onetime student, French anthropologist Henri Vallois, who, after Boule's death in 1942, continued to champion the pre-*sapiens* theory well into the 1950s. ◖ The pre-*sapiens* theory influenced Boule's description of the Neandertal remains found at La Chapelle-aux-Saints. He stressed the skeleton's primitive anatomy, saying it was so apelike that no self-respecting European would want it in his ancestry. Keith's work on the Piltdown specimen, a skull of presumed great antiquity, reflected a similar theoretical framework. He gave the skull a larger cranial capacity than others had done, reflecting his view that brain expansion led the way in hominid evolution. He also turned a blind eye to any primitive features of the skull. It was from these two directions that Neandertals were squeezed onto a side branch. ◖ Strangely enough, Boule, Vallois, Keith, and others were correct in thinking that very ancient modern humans did exist, and most likely Neandertals were not our direct ancestors (see pages 56 and 246). They were, however, correct for the wrong reasons, because the specimens on which they based their conjectures were not as old as they thought. Piltdown Man was exposed as a hoax, a modern human cranium combined with an orang lower jaw and buried with ancient fossils. Also, skulls with modern anatomy, like the specimen from Galley Hill, England, were found to be burials of modern humans in older strata. Eventually, candidates such as the skeletons from Grimaldi, once thought by Boule to be ancient "Negroids," turned out to be geologically more recent than Neandertals. ◖ The real challenge to establishing the antiquity of anatomically modern humans was and remains dependable dating. The important events in the origin of modern humans occur in the time range between 200,000 and 40,000 years ago. During that period of time both classic Neandertals and anatomically modern humans made their appearance, but it is precisely that period that is not adequately addressed by traditional radiocarbon isotopic decay and potassium-argon dating methods, which are more suitable for fossils older or younger than the age range in question. The newer techniques of thermoluminescence, electron spin resonance, and uranium-thorium series dating (see page 26) do address the period of 200,000 to 40,000 years ago, and with their use

paleoanthropologists have been able to fill in some of the blanks in our early history. ◖ One persuasive case for early anatomically modern humans comes from Jebel Qafzeh, a cave located near Nazareth, Israel. Parts of more than 20 well-preserved skeletons with skulls, some fairly complete, are known from this site. With the application of both thermoluminescence and electron spin resonance dating techniques, the Qafzeh hominids have been dated to between 90,000 and 100,000 years ago. Although the Qafzeh hominids possess some primitive features, overall the skulls are astonishingly modern looking. They have high, short braincases with fairly vertical foreheads, only slight brow ridges, and an average cranial capacity of roughly 1,550 cc, comfortably within the range of modern humans. The mandibles have well-marked chins, further attesting to the modernity of the Qafzeh hominids. Additional evidence of *Homo sapiens* comes from the rockshelter at Skhul at Mount Carmel in Israel, a site roughly the same age as Qafzeh. Remains of at least ten individuals were recovered from Skhul, including several probable burials. ◖ In East Africa, in an area known as Omo Kibish in southern Ethiopia, just north of Lake Turkana, remains of a partial skeleton, Omo I (see page 252), were recovered in 1967. This skull associated with postcranial bones is very modern looking in having a rounded vault, an expanded parietal region, and a mandible with a prominent chin. A second individual, Omo II, appears more primitive, but is also considered to belong to *H. sapiens*. Recently the strata of Kibish have been redated using the argon-argon method, producing an astonishing age of 195,000 years for these fossils, making them the oldest evidence for *H. sapiens*. ◖ At the southern tip of Africa, the fossil specimens from Klasies River Mouth are associated with a Middle Stone Age archaeological assemblage. Age estimates based on uranium-thorium and electron spin resonance dating suggest that the hominids lived there between 120,000 and 75,000 years ago. Some specimens are more robustly built, but a partial mandible has a strongly developed chin and is relatively lightly built. A small portion of the lightly built frontal bone lacks a supraorbital torus and thus has a modern human morphology. ◖ Additional evidence for early *H. sapiens* in Ethiopia comes from the site of Herto that yielded partial crania of two adults and one child dated to 155,000 years ago. Because the adult specimens exhibit more robusticity than seen in living humans, these hominids are referred to a subspecies, *H. sapiens idaltu*. ◖ Outside of Africa, the evidence for anatomically modern humans is unconvincing until roughly 40,000 years ago. For example, in Europe the earliest appearance of modern humans, usually referred to as Cro-Magnon Man, is always associated with artifacts typical of the Upper Paleolithic (see page 274). In the Near East, remains of modern *Homo sapiens* are few and far between, but the earliest, from Ksar Akil, in Lebanon, is roughly 37,000 years old. Evidence for modern humans in eastern Asia is not very convincing owing to the scarceness of sites and fossils and the uncertainties of dating these finds. ◖ Based on what is currently known, *Homo sapiens* first appeared in Africa (biogeographically, the sites in present-day Israel are here considered part of Africa). One remarkable aspect of the earliest *sapiens* is that although they are anatomically modern, their associated stone tool technology, variously called the Mousterian or Middle Stone Age technology, suggests that they were not behaviorally modern. ◖ The implication from these finds is that *Homo sapiens* did not arise as a direct result of some major behavioral change, at least not one that can be detected in the fossil record. The postcranial bones of early *sapiens* were robustly built, suggesting that they were still dependent to a large degree on brawn and not on tools for many activities. ◖ This generates the obvious question of why a modern anatomy preceded a modern behavior. This is one of the key unanswered questions in paleoanthropology today. Is it possible that the brains of early *Homo sapiens* were simply not yet wired for sophisticated culture? The modern capacity for culture seems to have emerged around 50,000 years ago, and with it, behaviorally modern humans who were capable of populating the globe.

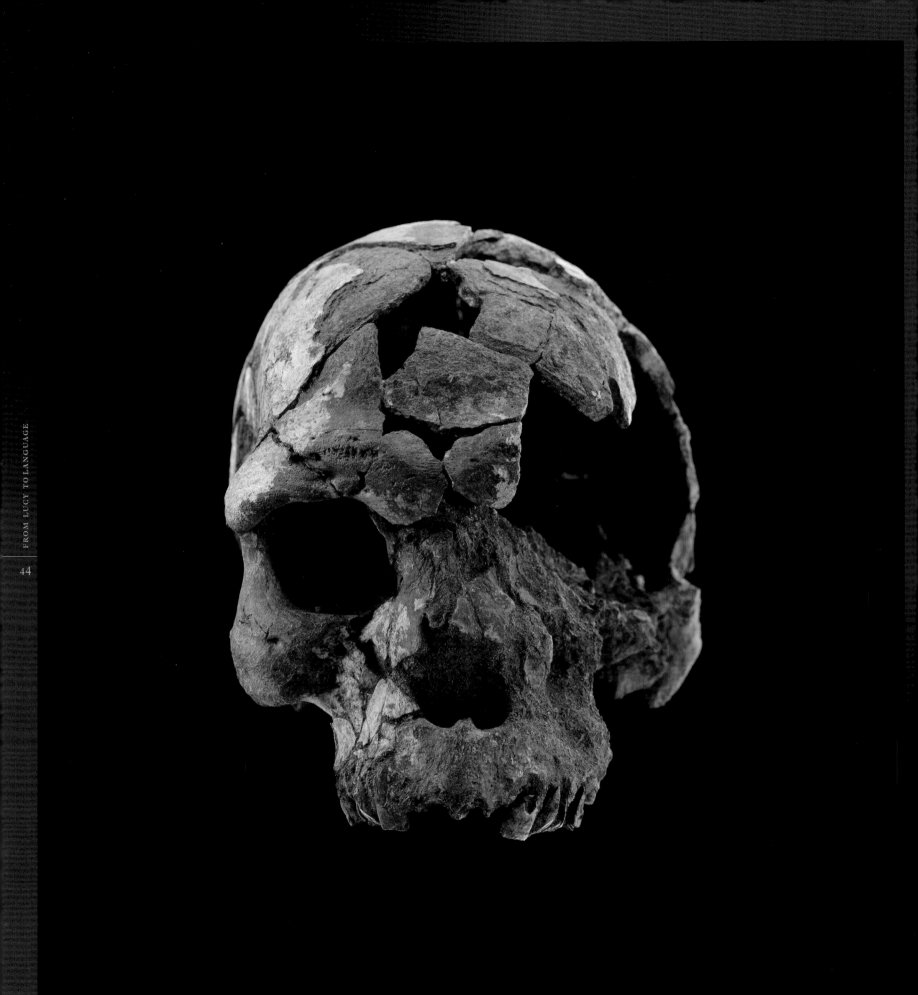

Cranium of *Homo sapiens* from Herto, Ethiopia. This specimen, BOU-VP-16/1, was recovered from the middle Awash region on the Bouri Peninsula. Dated to 155,000 years, this adult male is one of the most ancient occurrences of *Homo sapiens* thus far known. Actual size. *Photograph by David L. Brill; courtesy of National Museum of Ethiopia.*

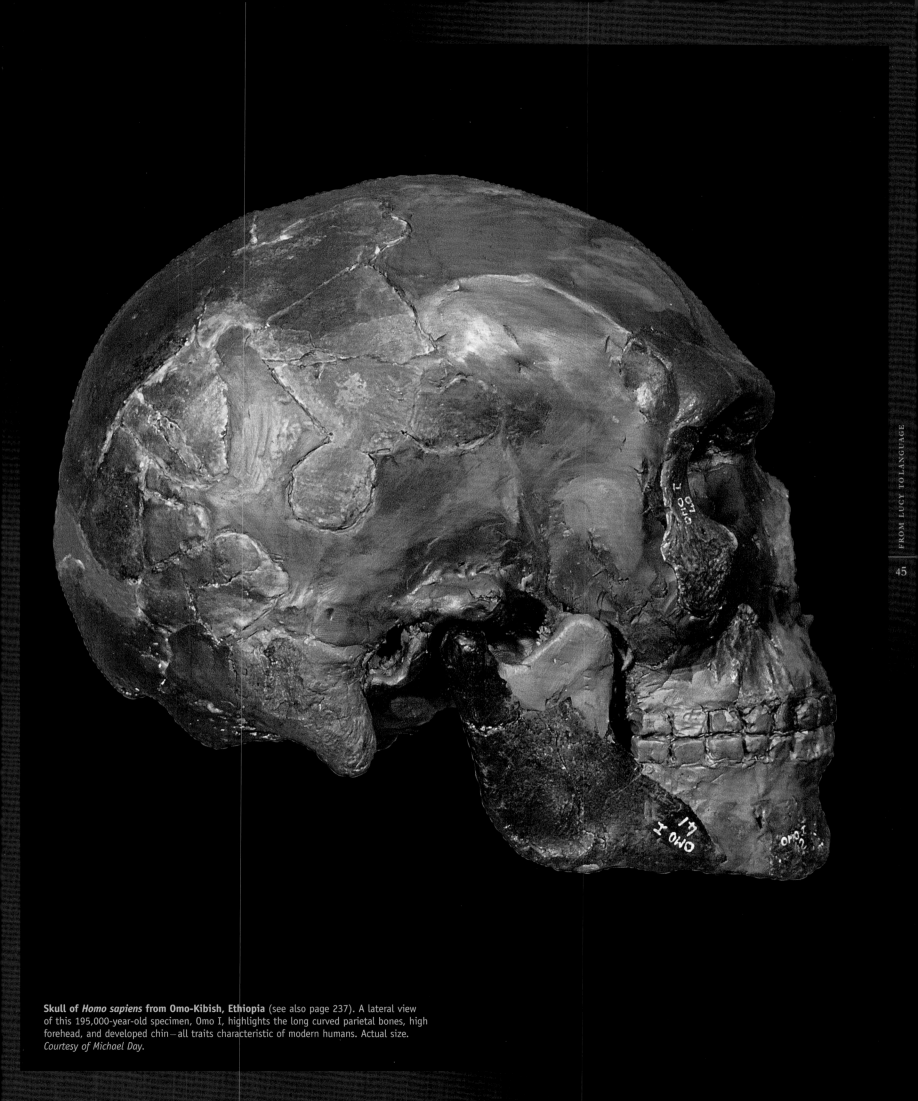

Skull of *Homo sapiens* from Omo-Kibish, Ethiopia (see also page 237). A lateral view
of this 195,000-year-old specimen, Omo I, highlights the long curved parietal bones, high
forehead, and developed chin—all traits characteristic of modern humans. Actual size.
Courtesy of Michael Day.

SEVERAL important discoveries in the 1990s have reopened the question of when our ancestors first journeyed from Africa to other parts of the globe. Traditionally, *Homo erectus* is credited as being the prehistoric pioneer, a species that left Africa about 1 million years ago and began to disperse throughout Eurasia. But recent evidence indicates that emigrant *erectus*, or perhaps its ancestor *Homo ergaster*, made a much earlier departure from our continental cradle. ❰ In 1994, reports of new dates for some old bones shocked anthropologists and made the cover of *Time* magazine. The bones in question come from Modjokerto and Sangiran, two sites on Java, the island where the first remains of *H. erectus* anywhere were found in 1891 (see page 205). The partial skull lacking a face or teeth from Modjokerto, discovered in 1936, belonged to a five- or six-year-old child *erectus*. Its geological age had been estimated at about a million years. With recent redating, however, it is thought that the child lived at least 1.8 million years ago. Other *erectus* fossils from Sangiran, including the face and back of two different skulls and a lower jaw, revealed an age of over 1.6 million years. ❰ The relatively new high-precision argon-argon dating method (see page 26), which uses a laser to determine the amount of argon gas inside single rock crystals, provided the age estimates for sediments at the two sites. Some have questioned whether the dated sediments come from the same spots as the fossils, with the suggestion that the fossils may be much younger. Nonetheless, the fact that sediment from two sites 240 kilometers apart had similar ages lends support to the earlier dates. These dates remain striking due to the absence of any other firm evidence for early humans in East Asia prior to a million years ago, and the individuals from Modjokerto and Sangiran would have certainly traveled through this part of Asia to reach Java. The earliest *erectus* specimens in east Asia include the cranium from Gongwangling (Lantian), China, which may be a million years old or closer to 800,000 years old, and some isolated teeth from Yuanmou and other Chinese sites that probably date to around 700,000 years ago. ❰ If *erectus/ergaster* left Africa nearly a million years earlier than anyone had suspected, then what enabled this hominid to move onward? It had been argued that a new tool technology, the Acheulean industry, attributed to *erectus* and typified by hefty hand axes and cleavers, gave this species an ecological edge in expanding its range. But the situation cannot be that simple, especially if *erectus* arrived in Asia before the first Acheulean tools appeared in Africa around 1.7 million years ago (which explains the long-noted absence of Acheulean tools in eastern Asia). ❰ Some of the best evidence for the first hominids to leave Africa comes from the medieval east Georgian town of Dmanisi. Remains of at least six individuals were recovered—four crania and three mandibles—which are dated to 1.8 million years ago by argon and paleomagnetism. The specimens show affinities to *H. habilis* and *H. ergaster*, but the unique amalgam of anatomical features prompted some to assign these hominids to

20 · Out of Africa

a distinct species, *H. georgicus* (see page 192). ❰ Early African emigrants may have skirted the eastern edge of the Mediterranean Sea before heading east. In Israel's Jordan Valley, the site of 'Ubeidiya has yielded stone tools from both an early chopper-core industry and crude Acheulean-like hand axes dated to between 1.2 and 1.4 million years ago by three different methods. The 'Ubeidiya tools resemble those of comparable age excavated from middle and upper Bed II at Olduvai Gorge, Tanzania. At the same Jordan Valley site were found part of a hominid skull and some teeth. Acheulean hand axes and cleavers dating to 600,000 years ago have also been found at Gesher Benot Ya'aqov, another site in Israel's Jordan Rift Valley, a northern extension of Africa's Great Rift Valley that may have offered similar lake and woodland ecology to hominid habitats in East Africa. ❰ One lingering mystery is did *Homo erectus* only head east into Asia, altogether bypassing Europe? No unequivocal fossils of *erectus* have emerged in Europe. Although some specimens, including the crania from Arago, France and Petralona, Greece (see pages 212 and 214, respectively) have been proposed as *erectus*, they belong to the more recent species *Homo heidelbergensis*. ❰ Many paleoanthropologists believed that no early humans entered Europe until 500,000 years ago, citing evidence such as a lower jaw from Mauer, Germany, the Heidelberg Man, the type specimen of *H. heidelbergensis*, which was discovered in 1907 (see page 210). In 1994, the announcement of a tibia, or shin bone, from Boxgrove, England, confirmed that by 500,000 years ago, early humans had reached Great Britain, when the island was a peninsula of Europe. ❰ The discovery of new fossils from Spain, published in 1995, secured a more ancient arrival of early humans in Europe. Three dozen bone fragments and a hundred stone tools were collected in the cave of Gran Dolina in the Atapuerca Mountains from sediments dated by the paleomagnetism method to 780,000 years ago. The fossils—several teeth, jaw and skull fragments, and hand and foot bones—come from four individuals, includ-

Fossils of *Homo antecessor* from Gran Dolina, Spain. The earliest Europeans may be represented by these 780,000-year-old remains of at least four individuals, including an adolescent and child. Actual sizes. *Photograph by Javier Trueba, Madrid Scientific Films; courtesy of Juan-Luis Arsuaga.*

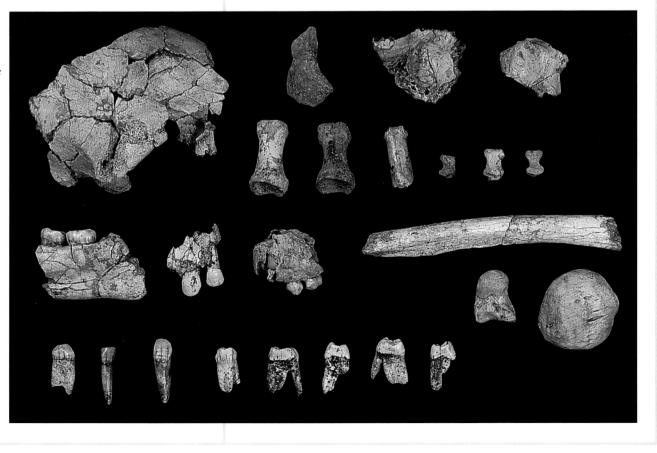

ing a child and an adolescent. Unlike either *erectus* or the later Neandertals in form, the Gran Dolina bones belong to a more primitive species call *H. antecessor*. Whether some cave harbors the remains of an even earlier European remains to be seen.

21 · The First Americans

IN 1932 near Clovis, New Mexico, while exploring the shore of a dried-up lake, archaeologists found some unusual stone tools between the ribs of extinct mammoths. Nothing like the elegant, finely flaked spear points had been found before. The base of the point was fluted: a large rectangular flake had been knocked off each side, permitting the stone to be hafted onto a wooden shaft that could then be hurled at an animal. The tool came to be called a Clovis point. The Clovis people were big game hunters who may have been lured to North America by the migrations of huge herds of herbivores, such now extinct animals as mammoths, mastodons, musk oxen, camels, horses, and ground sloths. Excavation at the site uncovered the remains of four mammoths associated with more Clovis points, all beneath layers containing smaller Folsom points, which until the Clovis discovery had been the oldest known technology in the Americas. ◖ Other sites with Clovis points made from chert and chalcedony turned up subsequently in each of the lower 48 states, southern Canada, and as far south as Tierra del Forego. Radiocarbon dates have consistently placed these artifacts within the time period of 11,000 to 11,500 years ago. No widely accepted archaeological sites of greater antiquity have been found, which suggests that modern human Paleo-Indians first entered the Americas at this time, crossing the Bering Sea land bridge from Asia around the end of the last Ice Age. ◖ This date, often called the Clovis barrier, has been rammed many times by evidence purporting to be a little or a lot older, but it has yet to crumble. Either the evidence for the antiquity of human bones and artifacts or the evidence that tools were made by human hands tends to evaporate. Perhaps nothing in American archaeology receives more critical scrutiny than the latest pre-Clovis claim, but a few sites have weathered the criticism better than the rest. ◖ One such site is Meadowcroft Rockshelter, in the Ohio River basin of Pennsylvania. Sediments excavated at Meadowcroft contained numerous artifacts, animal bones, and plant remains interlaced with charcoal from fire pits that provided organic material for a series of more than 50 radiocarbon dates. The most ancient deposits at the site date to between 31,000 and 21,000 years and contain no artifacts. But evidence of humans appears soon afterward. The oldest artifact, apparently a small piece of a bark basket or mat, has been dated to 19,600 years ago, plus or minus 2,400 years. A cut and charred chunk of deer antler was dated to around 16,175 years ago. ◖ Finished stone tools and flakes come from the same occupation level in the site, including retouched knives, blades, and other slicing implements. A single short, leaf shaped projectile point may date to around 13,000 years ago. None of the tools closely resemble Clovis technology, but the unusual blades have broad similarities with blade tools from the European Upper Paleolithic. ◖ Another enduring pre-Clovis contender comes from the Chilean site of Monte Verde in South America. Unlike Meadowcroft, Monte Verde is an open-air, creek-side marsh site, where a peat bog preserved stones and bones as well as wooden tools, logs, and branches arranged to form foundations of huts, and even a piece of mastodon flesh. These peat deposits have been dated to around 13,000 years ago. The remains of a separate structure had a foundation of sand and gravel mixed with animal fat where the occupants apparently stored food and used tools of stone, bone, and wood to prepare

Stone artifacts from Monte Verde, Chile. Two basalt projectile points and a possible drill made of black slate were excavated from an archeological layer dated by the radiocarbon method to 12,500 years ago. *Actual sizes. Courtesy of Tom D. Dillehay, University of Kentucky.*

Cave paintings from Monte Alegre, Brazil. Among the oldest known paintings in the Americas, these were made some 11,000 years ago by a hunter-gatherer culture contemporaneous with Clovis people, the earliest generally accepted culture in North America. *Photograph by Linda S. Weichelt; courtesy of Field Museum of Natural History.*

(left) **Projectile point from Meadowcroft Rockshelter, Pennsylvania.** The leaf-shaped Miller lanceolate comes from the top of the earliest artifact-bearing level, indicating human hunters lived at this site at least 13,000 years ago. Actual size. *Photograph by J.M. Adovasio; courtesy of Mercyhurst Archaeological Institute.*

(right) **Quartz crystal spear point from Monte Alegre, Brazil.** Excavated artifacts and charred remains of tropical fruits and small animals from this rockshelter demonstrate human occupation of Amazonia between 10,000 and 11,000 years ago. Actual size. *Photograph by Romulo Fialdini; courtesy of Museo Goeldi and Banco Safra.*

meat and hides. A few dozen people may have lived at Monte Verde at one time, surviving on a broad diet of gathered plants and hunted animals. ◀ Although archaeologists may be willing to accept an age of 13,000 years for Monte Verde, deeper deposits at the site have sparked greater controversy. These deposits contain more ambiguous artifacts associated with charred wood that was dated by the radiocarbon method to 33,000 years ago. Even the excavators remain somewhat skeptical of this earlier occupation date, which would precede the burials of Cro-Magnon in France. ◀ Pre-Clovis critics claim that the charcoal in some of the Meadowcroft Rockshelter sediments has been contaminated, skewing the radiocarbon results toward an earlier age. Monte Verde's critics question not the accuracy of the dates but the authenticity of the artifacts and their association with the ancient sediments (suggesting that younger tools may have filtered down into an older sediment layer). ◀ Other sites with possible evidence of an even earlier human arrival in the Americas have received much attention recently. Pendejo Cave in New Mexico contains two dozen living floors that may date as far back as 39,000 years ago. The evidence includes purported pebble and flake tools made of stone transported to the cave from a distant source, charcoal, a human hair, and apparent fingerprints impressed into clay some 30,000 years ago. The site does contain an amazing array of mammals that lived in the area between 30,000 and 40,000 years ago, but most archaeologists are withholding judgment about whether humans were among them. ◀ Further south, excavators of the Pedra Furada Rockshelter in Brazil contend that humans occupied the site beginning nearly 50,000 years ago—when Neandertals still prevailed in Europe—until 14,000 years ago, and again between 10,000 and 6,000 years ago. The youngest layers show clear signs of human use: finely made stone knives, stone hearths, and abundant rock art. Paintings on the walls of Pedra Furada—one of about 200 decorated rockshelters at the base of the same sandstone cliff—are thousands of years old. ◀ Beyond 14,000 years ago, however, the organic remains at this site become scarce, hearths become ambiguous, and art disappears. Several hundred crude quartzite cobble "tools" constitute the main evidence of human presence. A trio of American archaeologists—including the lead excavators of Meadowcroft and Monte Verde—recently examined the earlier evidence at Pedra Furada firsthand and concluded that the cobbles had broken naturally after falling from the cliff above the site. ◀ For now, Meadowcroft and Monte Verde remain the best bets for a pre-Clovis peopling of the Americas. If the

antiquity of either holds up, then Paleo-Indians must have entered North America long before 12,000 years ago in order to reach these remote locations. ◀ Further clues to a somewhat earlier arrival in the Americas come from the Caverna da Pedra Pintada, a cave site in the Brazilian Amazon. A battery of radiocarbon dates from burnt wood and seeds plus thermoluminescence and optically stimulated luminescence dates of burnt quartz tools and quartz grains, respectively, place human occupation at this site between 11,200 and 10,000 years ago. And these people had a culture quite distinct from that of the Clovis hunters: the early Amazonians painted cave walls, crafted triangular stemmed spear points, and subsisted on a broad diet of tropical fruits, fish, small mammals, and other game. Although these cave dwellers were approximate Clovis contemporaries, perhaps the first Americans had a similarly broad foraging strategy that only later shifted to the big-game hunting associated with Clovis culture. The scarce and scattered nature of solid pre-Clovis contenders should still give pause, however. If humans were abundant in the Americas before the almost ubiquitous makers of Clovis culture, why did they leave so little trace of their existence?

AFRICA was the sole setting of human evolution for the first 4 or 5 million years of hominid existence. It is only within the past 2 million years that any hominid exodus occurred (see page 46). The first species of *Homo* to spread into new continents

22 · Peopling the Globe

made an appearance in Asia around 1.8 million years ago and in western Europe around 1 million years ago. But the travels of these early explorers pale by comparison to the global colonization achieved by our species, which now occupies almost every terrestrial habitat on the planet in numbers approaching 6 billion. Indeed, the ability to colonize new habitats, prompted by such cultural developments as clothing, shelter, and boats, is a key human trait. We got to everywhere, eventually, because we could get there, and the trek took as few as 5,000 generations. ◀ Genetic differences in the nuclear DNA of human populations on various continents can be used to estimate when one population split off from others, or essentially when each continent first became populated. The scattered migration of individuals tends not to leave any genetic traces and in fact will reduce total genetic diversity by blending traits that would otherwise distinguish larger, more isolated populations. Mass migrations of people, however, do tend to increase diversity by stimulating genetic drift—in which novel mutations will more likely arise and be inherited by subsequent generations—and adaptation to new environments. It is the genetic traces of these mass migrations that inform us about when we came to occupy the globe. According to this gene geography, Africans became separated from other groups around 100,000 years ago. Groups in southeast Asia and Australia split off around 60,000 years ago. Asia and Europe became occupied by 40,000 years ago, and northeast Asia and the Americas harbored humans only within the last 30,000 years. In the vast Pacific, those parts closest to Australia and Sahul were colonized first, while Polynesia has probably been populated only for about 3,500 years. ◀ A separate nuclear gene study of two segments from chromosome 12 suggests a recent modern human migration out of northeast Africa about 100,000 years ago. The gene patterns then followed a distinctive eastward cline, as though these first human emigrants then entered the Middle East and Europe and continued east through Asia, across the Pacific, and into the Americas. ◀ From the Eurocentric perspective of paleoanthropology, Australia was long seen as a final frontier that was inhabited fairly recently. New dating evidence from the Malakunanja II rockshelter in Arnhem Land, near the northern tip of Australia, suggests that humans first arrived there 60,000 years ago. The site of Nauwalabila I in the Northern

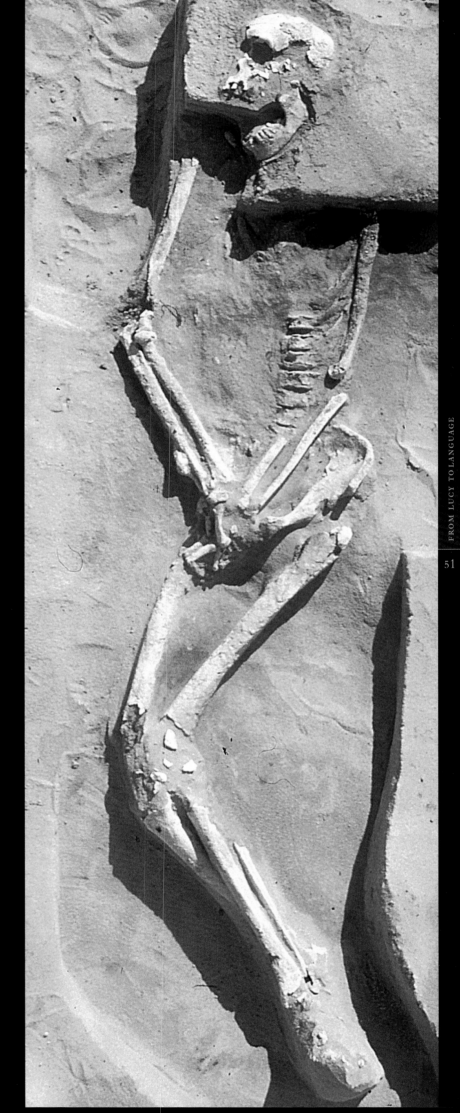

Rock painting from Kakadu National Park, Australia. The earliest Aboriginal art dates
to at least 40,000 years ago and was part of a tradition of rock painting that continued
until nearly the present day, including such recent works as these from northern Australia.
Courtesy of Donald Johanson, Institute of Human Origins.
***Homo sapiens* burial at Lake Mungo, Australia.** The Mungo III male skeleton dates to
nearly 30,000 years ago, but other evidence from northern Australia indicates that people
first arrived on the continent 20,000 years earlier. The body was sprinkled with red ochre,
a pracitice known from burials of similar age in Europe. *Courtesy of Alan Thorne, Australian
National University.*

Territory's Deaf Adder Gorge contains hematite used for pigment at a level dated to 53,000 years. Archaeological sites on the southern side of the island continent show that humans had spread across Australia as long as 40,000 years ago. ❦ Australia reveals in microcosm what became a global phenomenon after 50,000 years ago: a burst of colonization that put people all over the Old World, initially, and then across Australasia and the Western Hemisphere. Humans suddenly learned how to cross expanses of water and ice and how to find food in high mountains and parched deserts, even frozen tundra. What sparked this intrepidness? Intriguingly, this increase in colonization coincides roughly with the first archaeological signs of behaviorally modern humans; anatomically modern humans had existed for at least the previous 150,000 years, but 50,000 years ago there appear the first signs of art, of diversified tools for specific functions, and other clues to enhanced culture. Increasingly intricate and far-flung networks of trade and co-operative social alliances between groups probably facilitated human range expansion. ❦ With the advent of agriculture and animal domestication just 10,000 years ago, people could export not only themselves but the skills for cultivating staple crops. Human societies became more sedentary and structured, and a far greater population size could be maintained than under any hunter-gatherer lifestyle. Human population densities today exceed those at the end of the Paleolithic by as much as 10,000 times. But it is only recently that our numbers have reached such staggering sums. The planet gained its billionth human around 1800. It took only 130 years more for the population to double to 2 billion and little more than three decades to add a third billion. In the last 30 years alone, the population has nearly doubled again. ❦ Today we worry, with good cause, about our cumulative effect on the global environment. Many threats today stem from our sheer numbers and our demand for space and resources, but our impact on wildlife seems to have begun with the colonization of new continents. In the past 40,000 years, 85 genera of large mammals have gone extinct, including such charismatic creatures as woolly mammoths and rhinos, club-tailed glyptodonts, giant ground sloths, and saber-toothed cats. Some strong correlations exist between the time these animals disappeared and when humans entered their habitats. North America may have experienced a Pleistocene overkill of large mammals not long after Clovis-point-wielding PaleoIndian hunters arrived from Asia. Birds also experienced severe extinctions, especially on islands, where many flightless species lived. The Pacific islands, for instance, may have lost 2,000 bird species within the short span of human occupation. Whenever we reached new places, extinctions of previously secure animals apparently followed. Knowing this history, can we afford to find out just how many humans ultimately can occupy Earth?

23 · Defining Human Species

SINCE the time of Linnaeus, in the mid-eighteenth century, species, the lowest category in the Linnaean hierarchy of organisms, have been known as the basic unit of biological classification. The prevailing view during Linnaeus's time was that species were fixed, unchanging types that had been created through divine intervention. Darwinian thinking of the following century, by contrast, emphasized "descent with modification." Species were not static but dynamic—they changed, and they came and went. ❦ In the late twentieth century further definitions of species have appeared. The biological species concept, a product of the Modern Evolutionary Synthesis in the 1940s, states that a species is a group of organisms that is reproductively isolated from other such groups. The recognition species concept, which enlarges on this statement, holds that species are the most inclusive group that shares a fertilization system. This definition implies that members of a species have a specific mate-recognition system so that species breed with their own kind—lions do not mistake cheetahs, for example, as potential mates. These views of species have added strength to the idea that species are real, distinct, and readily identifiable entities in nature. ❦ Certain mating behaviors can be observed in living species, and genetic analysis can further strengthen the case for the validity of a particular species, but we do not have these advantages in studying fossils. How can we ever know if *A. boisei* and *A. robustus* were reproductively isolated from one another, or whether their different mate-recognition systems would have prevented them from mating? It seems that the biological species concept and the recognition species concept simply are not very useful for recognizing paleospecies. Even in the case of living organisms it is not always easy to ascertain what characteristics in the mate-recognition system keep individuals of the same species together. Moreover, with few exceptions, the morphological features that reinforce reproductive isolation are not found in the hard parts, which are all that we have available in the fossil record. ❦ Another aid to recognizing a living species is that its representatives live at the same time (*synchronic*) and in the same geographic location (*sympatric*). Hominid fossils are patchily distributed over millions years of time and spread across the globe. Very seldom do we have instances such as at Olduvai Gorge, where two distinct hominids, *Australopithecus boisei* and *Homo habilis*, not only overlapped in time but apparently occupied the same landscape. ❦ How, then, do we recognize paleospecies? How do we take a group of newly discovered fossils, such as those recently found at Kanapoi, and determine whether they belong to an already recognized species, such as *A. afarensis*, or to a new species? As a general guide, we look at modern species of primates that we know belong to a reproductively isolated group. For example, a series of features in the skeletons of bonobos and common chimpanzees, which do not interbreed (they are reproductively isolated from each other), are taken as a measure of species differences in the genus *Pan*. If we find, as was true for the Kanapoi hominids, that the fossil assemblage can be distinguished from other species of early hominids by a series of analogous morphological features, we can conclude that it is a distinct species—in this case, *A. anamensis*. ❦ This procedure is not always clear-cut, which is why paleoanthropologists continue to disagree over the number of hominid species that have ever existed. Evaluating the significance of morphological differences is not always easy. Only first-hand experience with the skeletal anatomy of numerous mammalian groups, especially primates, provides the necessary perspective from which to distinguish between variation within a species, or *intraspecific* variation, and variation between species, or *interspecific* variation. For instance, understanding the pattern of variation in a primate species that results from sexual dimorphism will prevent us from placing the males of a fossil assemblage into one species and the females into another species. ❦ Early workers in paleoanthropology tended to place every new fossil specimen found into a new species. This approach, known as "splitting," led to such a bewildering array of names that it was impossible to understand and discuss hominid evolution in a meaningful manner. In part, the splitter mentality was due to discoverer ego, but it also reflected the view that differences seen between fossils mirrored actual species distinctions. With the development of the Modern Evolutionary Synthesis and an increasing appreciation of population variability, the pendulum began to swing in the other direction. At that time, rather than emphasizing the individual variation seen in fossils, the idea that morphology could vary within a species began to influence the way in which fossil samples were partitioned. Perhaps as an overreaction to the previous splitting of fossil specimens into several species, the number of taxa was severely reduced, by those now known as lumpers. ❦ In the view of some researchers, the lumpers may have overreacted and, in so doing, underestimated the number of hominid species that existed and thereby oversimplified the complexity of hominid evolution. In fact, we will probably always underestimate the number of species in the hominid fossil record. One reason is that fossil samples tend to be fragmentary, and it may not be possible to garner sufficient evidence to justify establishing a new species. But, and even more important, there seems to be little correlation between species and morphological differences. Some groups of primates—for example, lemurs—consist of numerous species that are distinct on the basis of behavior and skin and hair coloration. These aspects do not fossilize, and if all we had were the skeletons, we would severely underestimate the number of lemur species. Two fossil femurs that look identical might not, in fact, belong to the same species—a sobering thought for those dealing with paleospecies. ◀ The problem of recognizing paleospecies is linked to a more fundamental problem in taxonomy—determining how species should be segregated. If an evolving lineage is continuous, as it must be, at least until it terminates in a final extinction, then dividing one species from another would appear to be arbitrary. If species A evolves into species B, and so on, how is this evolving lineage subdivided into species? Such thinking has given rise to the extreme view that *Homo sapiens* should be projected into the past to include *Homo* fossils dated to 2 million years ago. Rather than giving different species names when morphological changes are seen, it was suggested that different stage designations be given. Such a rendition contributes little to our understanding of change and diversity in human evolution. ◀ When the nature of speciation became better understood through an increasingly complete fossil record for a number of animal species, it became clearer that species appear very suddenly in the geological record. Throughout their life span as a species, they seem to be in a state of stasis, exhibiting little or no change, and then they disappear rather abruptly. It was easy to ascribe the suddenness of species' appearances and departures to the incompleteness of the paleontological record. ◀ The key to resolving this quandary is that an evolutionary lineage should not be likened to a continuous ladder of evolution but to a bush that reflects divergence and branching. The manner in which speciation, the process whereby new species are born, operates is that some splinter group of a species becomes geographically isolated and develops into a new species. The geographically isolated population, unable to exchange genes with the parent population, begins to diverge as a result of genetic drift and mutations. Presumably it is in these smaller outlying populations, with smaller gene pools, that speciation occurs. ◀ The Darwinian view of phyletic gradualism, which posits slow, minute change over extremely long periods of time, has been severely challenged by the more recent concept of punctuated equilibrium. The basic tenet of punctuated equilibrium is that new species arise by splitting off from ancestral species. This splitting is presumed to occur rapidly and sporadically. Therefore, the absence of fossils showing intermediate, smooth transitions between species is explained by this model as a result of sudden splitting, and the incompleteness of the fossil record need not be invoked to account for missing intermediate forms. A corollary to the punctuated equilibrium model is that descendant species closely resemble the ancestral species only for a short period of time. ◀ One final point requires consideration. Punctuated equilibrium does not ignore natural selection or adaptation. In this model, natural selection and adaptation operate differentially on various local populations of a species. Insofar as environmental and climatic fluctuations occur over relatively short periods, there are probably many opportunities for speciation—another reason to expect a greater number of species than is currently recognized for the Hominidae. ◀ Turning to the hominid fossil record, it is interesting to note that from its first appearance, roughly 4 million years ago, to its disappearance 1 million years later, *A. afarensis* (in its skeletal remains) appears relatively unchanged. Although we can make this statement about the hard parts,

which fossilize, nothing can be said about possible changes in behavior or the soft tissue. It is likely that when other taxa are as well known as *afarensis*, this pattern of long periods of stasis will prove to be typical of hominid evolution. ◀ Hominid evolution is similar to that of other organisms. Splitting and divergence would seem to be the rule, and in this case, as more hominids are found, the tree will undoubtedly become bushier. From a theoretical point of view, because morphological change and speciation do not necessarily go hand in hand, it is more than likely that some species will go undetected. ◀ Paleoanthropologists often wonder how many hominid species have ever existed. The answer is that we do not know, and it is highly likely that we will never know.

THERE is a theoretical concept in ecology called the competitive exclusion principle. According to this principle, only one species of a particular kind of animal can inhabit a specific ecological niche (which can be thought of as that animal's place and profession) at any particular time, because the existence of any contemporaneous species filling an identical niche in the environment would cause intense competition for food and other resources. Eventually all but one of the competing species would be forced out. This principle, once it was applied to human evolution, generated what became known as the single species hypothesis. ◀ As intelligent, large-bodied bipedal primates with presumably similar ecology and a dependence on tools and culture, it was thought, hominids could only have existed one species at a time. Only one hominid species lives today, of course, but was this always the case? ◀ Although fossil discoveries from Olduvai Gorge, Tanzania in 1959 and the early 1960s suggested that two distinct kinds of hominids might have occupied the environment at the same time, confirmation of this conjecture had to wait another decade, when in 1975, a nearly complete cranium of *Homo ergaster*, KNM-ER 3733 (see page 194), was found in sediments at Koobi Fora, in northern Kenya. Six years earlier, another complete cranium known as KNM-ER 406 (see page 170), a classic example of *Australopithecus boisei*, had been unearthed from the same sediment layer. That meant the two specimens, representing two distinct species, had lived, broadly speaking, at the same time. The simple story of only one hominid species living at any one time throughout human evolution was clearly wrong. ◀ If two distinct kinds of hominids shared the same landscape in East Africa nearly 2 million years ago (and apparently did so in South Africa as well), scientists needed to consider how those species would have divided the resources in the environment to permit their co-existence. In the case of these two species, *H. ergaster* may have been a wide-ranging omnivore with a large carnivorous component to its diet. The heavy-duty tools of *ergaster* suggest that this hominid was a butcher par excellence of hunted or scavenged game meat. Because its jaws and teeth were specially designed for crushing and chewing, *A. boisei*, is often characterized as a herbivore, focusing on a diet of hard foods, perhaps including nuts and seeds. ◀ But have more than one hominid species always co-existed? For the earliest period of human evolution, the existing evidence indicates that perhaps only a single species lived at any one time. More than 4 million years ago, we have evidence of *Ardipithecus ramidus* and *Australopithecus anamensis*, the former preceding the latter in time. Yet between 4 and 3 million years ago, in East Africa, there may have been at least two kinds: *Australopithecus afarensis* and *Kenyanthropus platyops*. ◀ Presumably between 2.5 and 3.0 million years ago a split in the family tree occurred, resulting in the robust australopithecines and early *Homo*. Many anthropologists recognize three species of robusts—*A. robustus*, *A. boisei*, and *A. aethiopicus*—and three species of *Homo*—*H. habilis*, *H. rudolfensis*, and *H. ergaster*. ◀ The most compelling

24 · Co-Existing Human Species

54

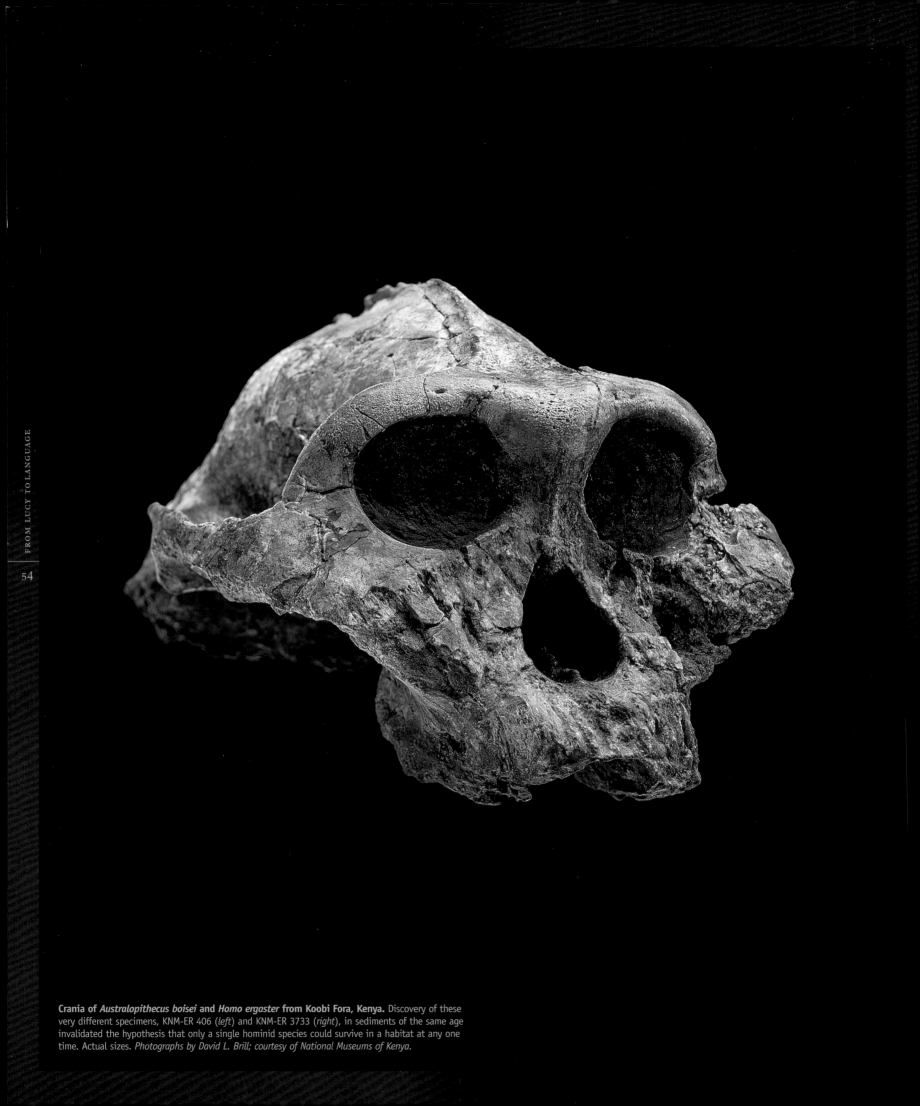

Crania of *Australopithecus boisei* and *Homo ergaster* from Koobi Fora, Kenya. Discovery of these very different specimens, KNM-ER 406 (*left*) and KNM-ER 3733 (*right*), in sediments of the same age invalidated the hypothesis that only a single hominid species could survive in a habitat at any one time. Actual sizes. *Photographs by David L. Brill; courtesy of National Museums of Kenya.*

case of hominid co-existence, however, involves ourselves as one of the characters. We now have ample evidence from the Near East, particularly from several caves on Mount Carmel in Israel, that Neandertals, or *Homo neanderthalensis*, and modern *Homo sapiens* overlapped in time and space for about 50,000 years. Did these similar species live together and interact, or did they lead essentially separate lives? Scenarios range from the evolution of Neandertals directly into modern humans, to co-existence and hybridization between these distinct populations, to the replacement of Neandertals by their more modern counterparts through violent confrontation or more subtle ecological competition. The first idea seems untenable given the length of time that moderns and Neandertals overlapped in the Near East. Hybridization is unlikely, if the two groups were biologically distinct, as suggested by their disparate anatomies. And although modern humans ultimately outlasted Neandertals, there is no hard evidence of violent or rapid replacement anywhere (although modern humans do appear rather suddenly in the archaeological record of western Europe). ❧ Additional support for the co-existence of human species derives from two locations, southeast Asia and Europe. The *H. erectus* remains from Ngandong, Java are now dated to approximately 40,000 years, making them contemporaneous with *H. sapiens* and *H. neanderthalensis*. Also from southeast Asia are the remains of diminutive hominids that lived as recently as 18,000, but may go back to 90,000 years, which were found on the Island of Flores, designated *Homo floresiensis*. In Europe two penecontemporaneous species have been recognized. One, *Homo antecessor* from Spain, 780,000 years, and second, *Homo cepranensis* from Italy dating to approximately 700,000 years. ❧ The single species idea is, however, not dead and has been resurrected, in a somewhat modified form, by Maciej Henneberg and Francis Thackeray. They concluded that variability in body size or cranial capacity of any hominid species did not exceed that seen in modern humans. This was prior, however, to the recovery of the Flores specimens, contemporaries of *H. erectus*, *H. neanderthalensis*, and *H. sapiens*, which have a stature of only three and half feet and cranial capacities of 417 cc! ❧ It is true that despite the underlying genetic homogeneity, modern humans have noticeable variability. Comparing fossil populations to the more variable modern ones, however, obscures the distinctive features that mark separate species. Looking at the *H. floresiensis* and a contemporary like *H. sapiens*, it is difficult to imagine how such different creatures could ever be accommodated in a single species.

AROUND the world, humans appear very different from each other. To some extent, we are. There are differences in skin color, body size, limb proportions, hair texture, and other physical features. Certainly, there are differences in dress and countless

25 · Human Diversity Today

cultural customs. But despite our visible variety of sizes, shapes, and features, humans are remarkably similar beneath the surface of our skins—"tediously uniform," in the words of one geneticist. In fact, two random people may differ by only one to three letters per thousand in the alphabetic code of their DNA. Put another way, the genetic sequences of two unrelated individuals may be as much as 99.9 percent identical. ❧ Human populations do differ in the presence or frequency of certain genetic alleles and their associated, tangible traits. Of the human genetic variation that exists, however, some 85 percent occurs between individuals of the same nationality. Most of our genetic variation occurs within the same population group rather than between two given groups. If all but one population were wiped out, much of our species' genetic diversity would still survive. ❧ Because our species is so homogeneous, it is not possible to delineate any solid genetic boundaries that correspond

with our culturally constructed categories of race. Minor genetic differences underlie about three quarters of the color variation seen in our skin, hair, and eyes, with the rest of the variation due to such factors as sunlight exposure and nutrition. ❧ But why are we so depauperate in genetic variation compared with our closest living relatives, the African apes? We split off from the apes around 6 to 8 million years ago, yet they seem to have accumulated much more genetic variation than us since that time. One possible answer is that our present limited genetic diversity stems from a relatively recent origin for modern humans. Perhaps we went through a genetic bottleneck early in our evolutionary history—a drastic reduction in total population that winnowed much of our previously accumulated diversity. Given the rapid rate at which humans migrated around the globe and our current population size, such a scenario seems hard to fathom. We continue to probe the genetic legacy contained in present populations to find the answers to our evolutionary history.

RACE is an elusive and controversial concept. On one hand there are racialists who assert that race is the defining characteristic for determining the value of individual humans. According to this concept, whole civilizations rise and fall depending on the racial purity of

26 · What is Race?

their citizenry. On the other hand, esteemed organizations such as the American Association of Physical Anthropologists and UNESCO assert that race as such does not exist, that it is a null category. ❧ The problem in trying to sort through these disparate points of view is that anyone who walks the streets of any large city encounters peoples who do not look like her or him—peoples who exhibit a diverse range of skin colors, hair textures, and facial physiognomies. If this isn't an example of race, what is it we are seeing? In fact, at first glance these different types seem to fit into distinct racial categories. When we look closer, however, it becomes difficult to impose hard and fast boundaries on such variety. Who qualifies, for instance, as "black": all African Americans, African Brazilians, Caribbean peoples, and Africans? And what about indigenous Australian peoples? They share many physical characteristics with peoples of sub-Saharan African ancestry, and yet are widely separated from them geographically and genetically. ❧ Historically humans have tended to choose mates from people living in similar social groups in the same area, and certain traits have been preserved in certain human populations. So, populations do possess different physical features that result from a combination of genetic inheritance and geographic location. Consider, for example, skin color, the most commonly cited signature of racial differences. We now know that fewer than ten genes control the range of variation in human skin color, from the darkest to the lightest tones. These genes occur in every human to produce melanin, skin pigments that can be black, brown, yellow, or red. All humans have the genetic capacity to become black or white—or any shade in between. It is the action or inaction of the enzyme tyrosinase that actually determines how light or dark our skin will be, and that depends on the environment where we live. Aboriginal Australians, with a history on the island continent that goes back perhaps 60,000 years, for instance, have evolved very dark skin, but the recent European immigrants to Australia have not had enough time to adapt to its harsh sun and thus suffer from epidemic rates of skin cancer. ❧ On the near horizon, we can perceive a genetics of eye and nose shape and of hair texture that will allow us to understand how differences in those features actually come about in the interaction of genes and environment. Though many genes have multiple effects, a phenomenon known in genetics as "pleiotropy" (meaning "many ways"), none of the known genetic differences, involving skin color or facial shape, have been connected in any significant way with neurobiology. If outward physical differences were to correlate with cognitive ability or personality types, these genetic differences should ultimately connect to neurobiological properties. Though they may yet do so, the odds are increasingly lengthening against that possibility. ◖ Populations themselves are the real units of human diversity. Originally there was one modern human population that emerged in Africa approximately 200,000 years ago. This population spread and diversified throughout Africa, and then approximately 100,000 years ago groups of humans began to leave the African continent. Due to geography, climate, and the sparsity of numbers, these groups tended to become isolated from one another, and evolved the basic physical differences we see today. While there is some question whether modern humans and Neandertals could mate and produce viable offspring, there is absolutely no evidence to suggest that the differences among populations of early modern human prevented success-ful interbreeding between them. In fact, since the end of the last Ice Age some 12,000 years ago, the global process of urbanization has resulted in reducing the differences between human populations. So while physical dif-ferences continue to exist among various human populations, those human traits that appear to have biological value do not show up more frequently in any particular population. Studies of the genetic diversity within and between traditional racial groupings have documented that over 90 per cent of human genetic diversity is contained within single populations with less than 10 per cent due to "racial" differences. Such data strongly reinforce the notion that from an evolutionary point of view, humans have always been a single race. The meaning we attribute to the differences we perceive among humans is political, not biological.

WE cannot directly measure the weight of our ancestors, but estimates of body weight can be obtained by comparing the fossilized remains of early humans with modern apes and living humans of similar dimensions. To begin these comparisons, we must first decide what parts of a fossil hominid would be

27 · The Size of Early Humans

diagnostic for our purposes. Until recently such attempts were hampered by the availability of only fragmentary skeletal material but a plenitude of teeth. Initially it was thought that there must be a close relationship between tooth size and body size. For example, reconstruction of body weight, based on its massive molars, for an australopithecine found in South Africa yielded an estimate of 72 kilograms. When researchers switched to the ankle bone as the reference point, however, a more realistic estimate of 34 kilograms was obtained. Thus, although fossil teeth constitute the vast majority of early hominid remains, using them to reconstruct body weight can produce unsatisfactory results. ◖ Common sense suggests why estimates of body weight based on measurements of teeth would be incor-rect. Body weight is not reflected in the size or mass of teeth. Furthermore, we have learned that australopithecines in particular were large-toothed relative to body size, a condition known as megadontia (see pages 40 and 131). Disproportionately large teeth in small bodies lead to estimates of body weight that are too high. ◖ Rather than measuring teeth, it is more appropriate to examine the weight-bearing postcranial skeleton (everything below the skull) for clues to body weight. Our body mass is transmitted to the ground through the pelvis, lower limbs, and feet—the skeletal segments that must be able to support our weight. Here researchers encountered another problem: not only have postcranial remains been quite fragmen-tary until recently, but the postcranial anatomy is very similar in different species of australopithecines and even in early *Homo*, making it difficult to distinguish species using only postcranial elements. Both shortcomings have been overcome in part by the recovery of a number of more complete skeletons such as Lucy. ◖ With postcranial skeletons at hand, and with the postcranial skeleton selected as the element to be compared in an estimate of body size, we can now turn to modern taxa. For comparison we have

modern humans and the living apes. The living apes must now be put to one side, because they evolved a significantly different method of locomo-tion that led to different limb ratios than in humans. Because modern apes are quadrupedal and transfer some weight through their forelimbs, their hind limbs account for proportionately less of their body weight than in humans. So, estimates of the body weight of our ancestors are best based on comparisons with contemporary humans, for both are bipeds. ◖ Next, the postcranial bones of extant humans of known weight are measured. Statistical methods establish a close correlation between measurements of postcranial bones and body weight. For example, the diameters of the femoral head and top and bottom ends of a tibia closely correlate with body weight. The greater the body weight, the larger are these diameters in modern humans. ◖ The associated skeletal elements of the Lucy speci-men offered an unprecedented opportunity for evaluating the accuracy of predicting body size using a variety of skeletal measurements. When mea-surements taken on a human sample, including some small individuals such as pygmies, were used for comparison, Lucy's weight was reconstructed as 27.3 kilograms. Confidence in this estimate is high, since measurements of seven of Lucy's individual bones from shoulder to ankle and compari-sons with the human sample gave estimates ranging from 24.2 to 30.2 kilograms. As might be expected for a biped, the best predictors of body weight were the lower limb joints. ◖ A cursory glance at the postcranial bones assigned to *Australopithecus afarensis* reveals that there is consider-able range in size. In fact, there appear to be two size groups: one large and one, like Lucy, small. Assuming the smaller individuals are female and the larger ones male, reconstruction of body weight in this species pointed to an intriguing aspect of the biology of *A. afarensis*: a large size discrepancy between males and females. Males, at about 45 kilograms, are roughly 40 percent heavier than females, which average 29 kilograms. The relatively small size of females and the rather pronounced sexual dimorphism have important implications for behavior (see page 73). Stature estimates for *A. afarensis*, based on comparisons with modern humans, also confirm sexual dimorphism, with males averaging about 1.5 meters tall and females about 1 meter tall. ◖ A number of postcranial specimens are now known from both eastern and southern Africa for the time range between 4 and 1 mil-lion years ago. Because, however, species of *Australopithecus* and *Homo* were contemporaries for nearly 1.5 million years and because their postcranial anatomy is nearly identical, it is not an easy task to assign isolated finds to a particular species. Therefore, postcranial elements can be positively iden-tified only when diagnostic dental material is associated with the find. Fos-sil postcrainial bones, however, are only rarely associated with the kind of dental material need for species identification, and so, for instance, *A. boisei* estimates of weight are based on only one specimen of each sex. Estimates for *A. robustus* and *A. africanus* are derived from several individuals, thanks to the larger samples from southern Africa. Estimates of body weight and stature are given in the chart below.

	BODY WT. (kg)		STATURE (m)	
	male	*female*	*male*	*female*
A. afarensis	45	29	1.5	1.1
A. africanus	41	30	1.4	1.1
A. robustus	40	32	1.3	1.1
A. boisei	49	34	1.4	1.2

There is certainly room for improvement in the estimates of body weight and stature for early humans, and perhaps greater accuracy will be forth-coming with the recovery of additional individuals. A number of obser-vations, however, can be made based on what is already known. Male australopithecines are generally larger than females, an observation that supports sexual dimorphism in all australopithecine species. Although

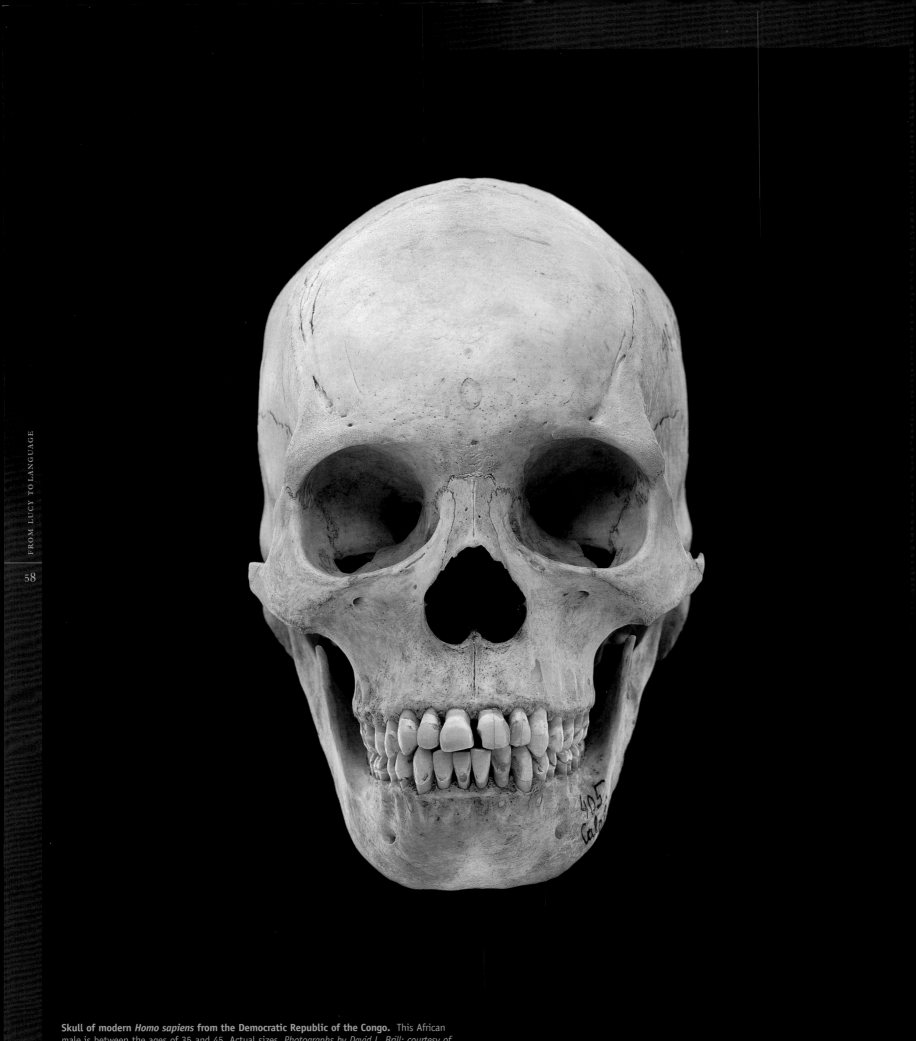

Skull of modern *Homo sapiens* from the Democratic Republic of the Congo. This African male is between the ages of 35 and 45. Actual sizes. *Photographs by David L. Brill; courtesy of American Museum of Natural History.*

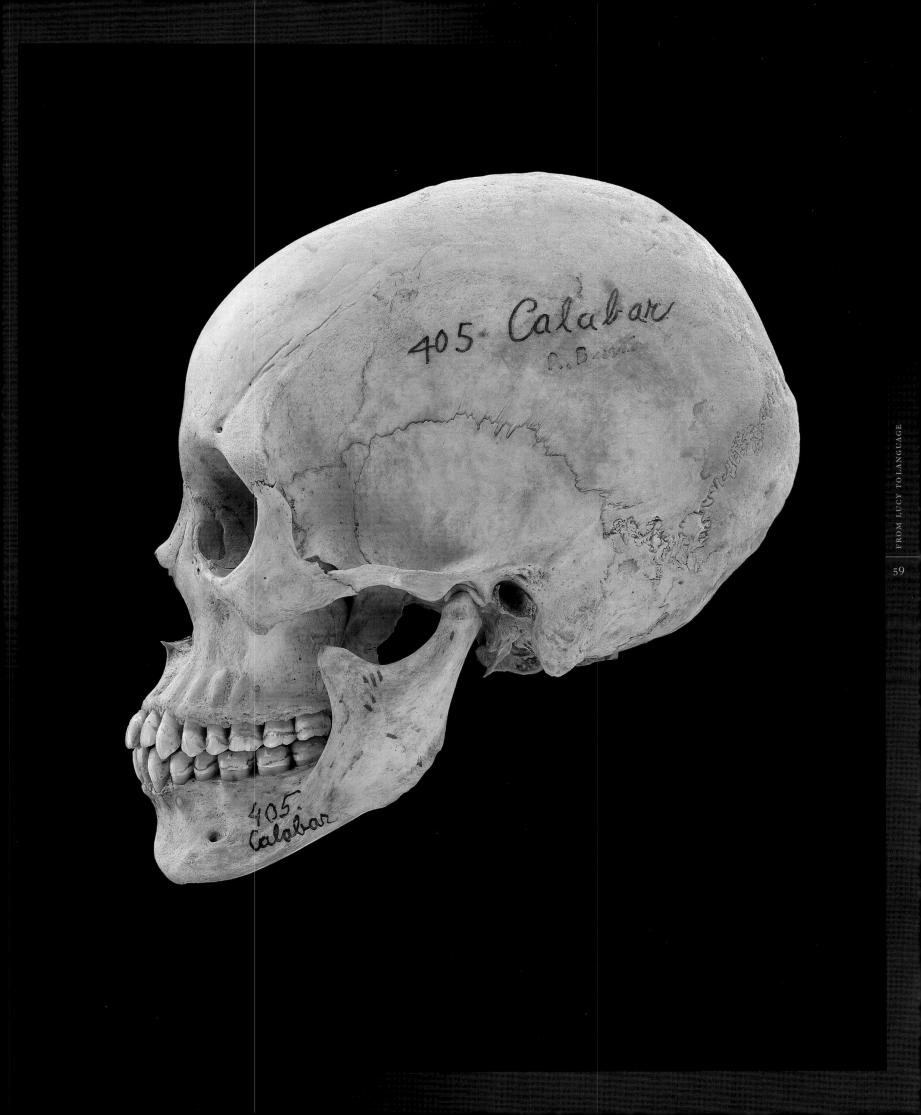

405. Calabar
R. Burton

405.
Calabar

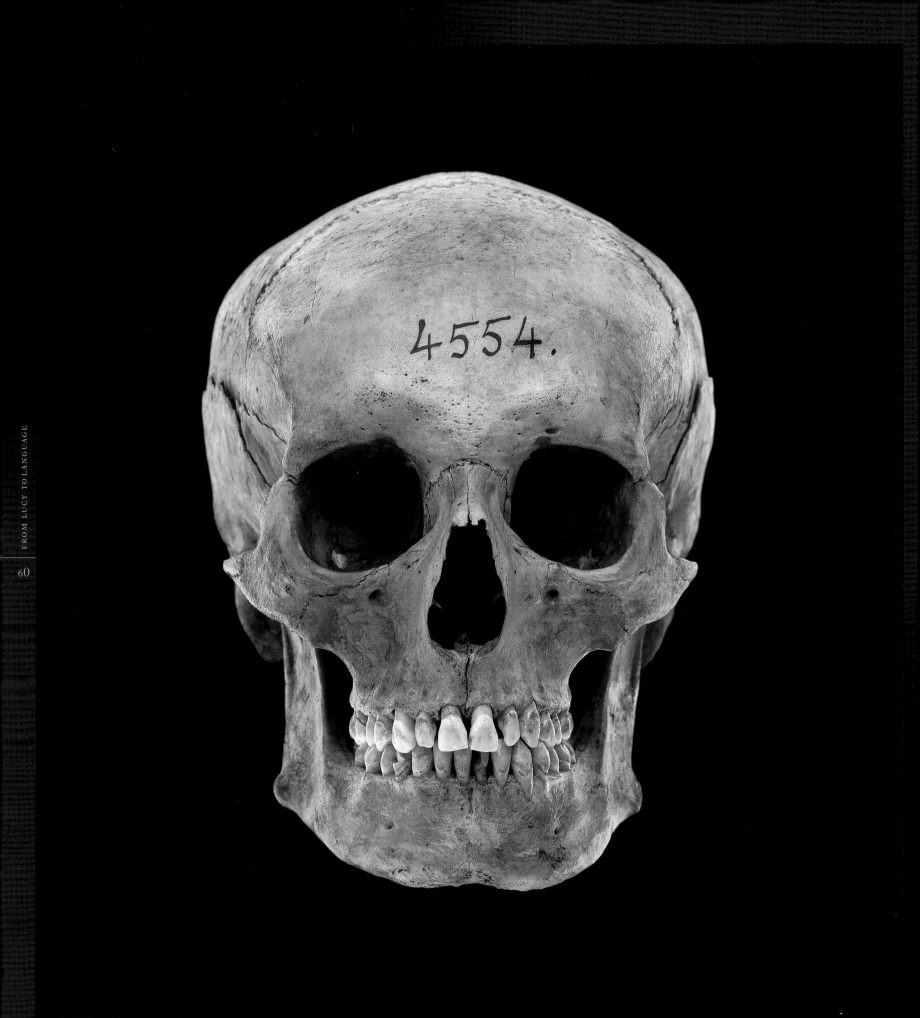

Skull of modern *Homo sapiens* from Germany, Europe. This skull comes from a young male between the ages of 25 and 30. Actual sizes. *Photographs by David L. Brill; courtesy of American Museum of Natural History.*

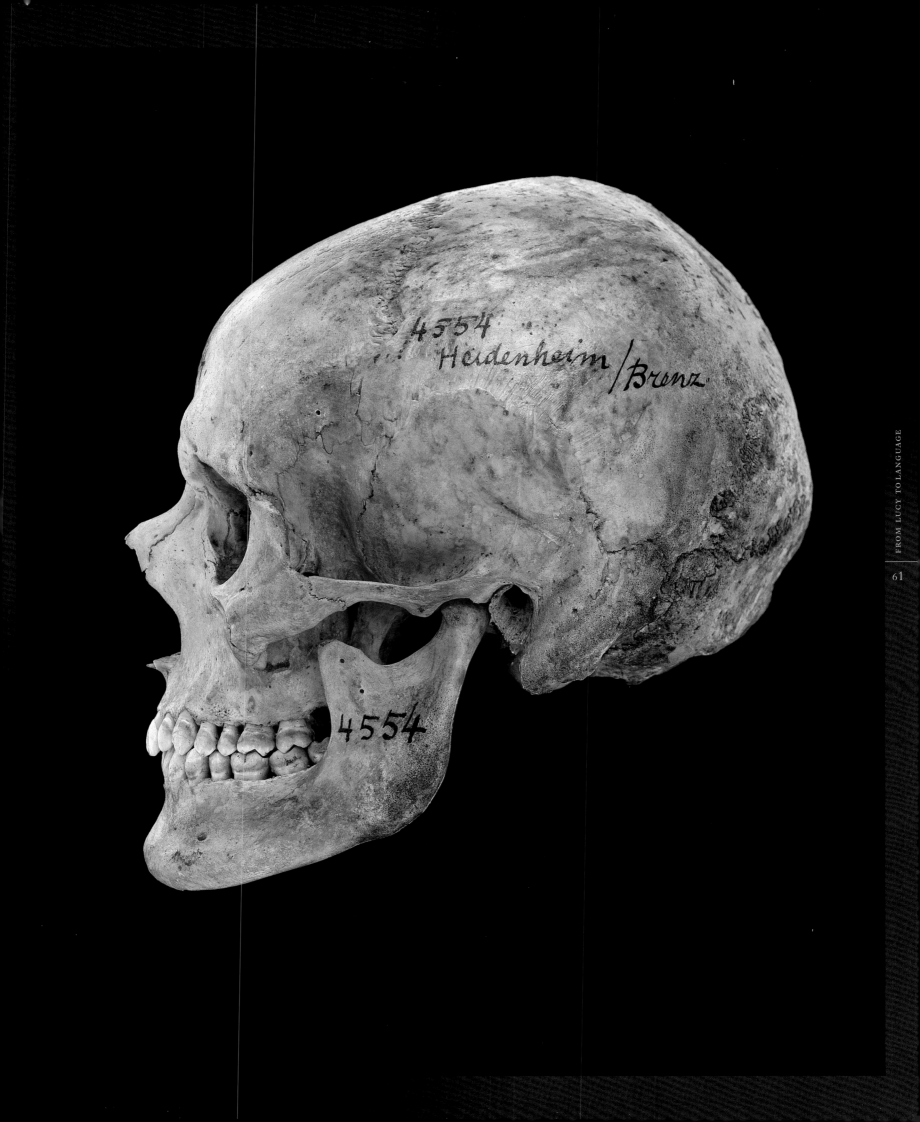

Skull of modern *Homo sapiens* from Borneo, Asia. Although collected in Indonesia,
this skull is that of a Chinese woman between the ages of 25 and 30. Actual sizes.
Photographs by David L. Brill; courtesy of American Museum of Natural History.

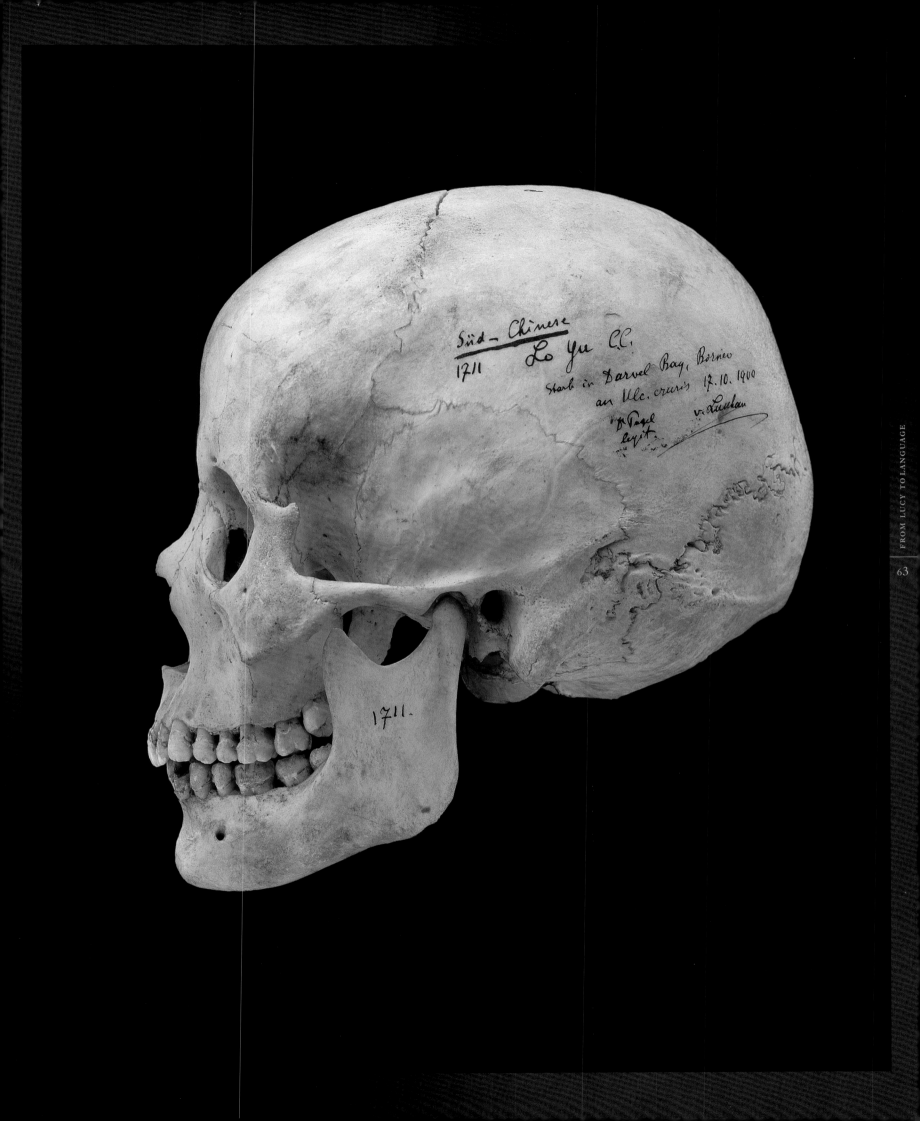

Skull of modern *Homo sapiens* from India, Asia. The teeth of this young Bengali male are stained from eating betel nut. Actual sizes. *Photographs by David L. Brill; courtesy of American Museum of Natural History.*

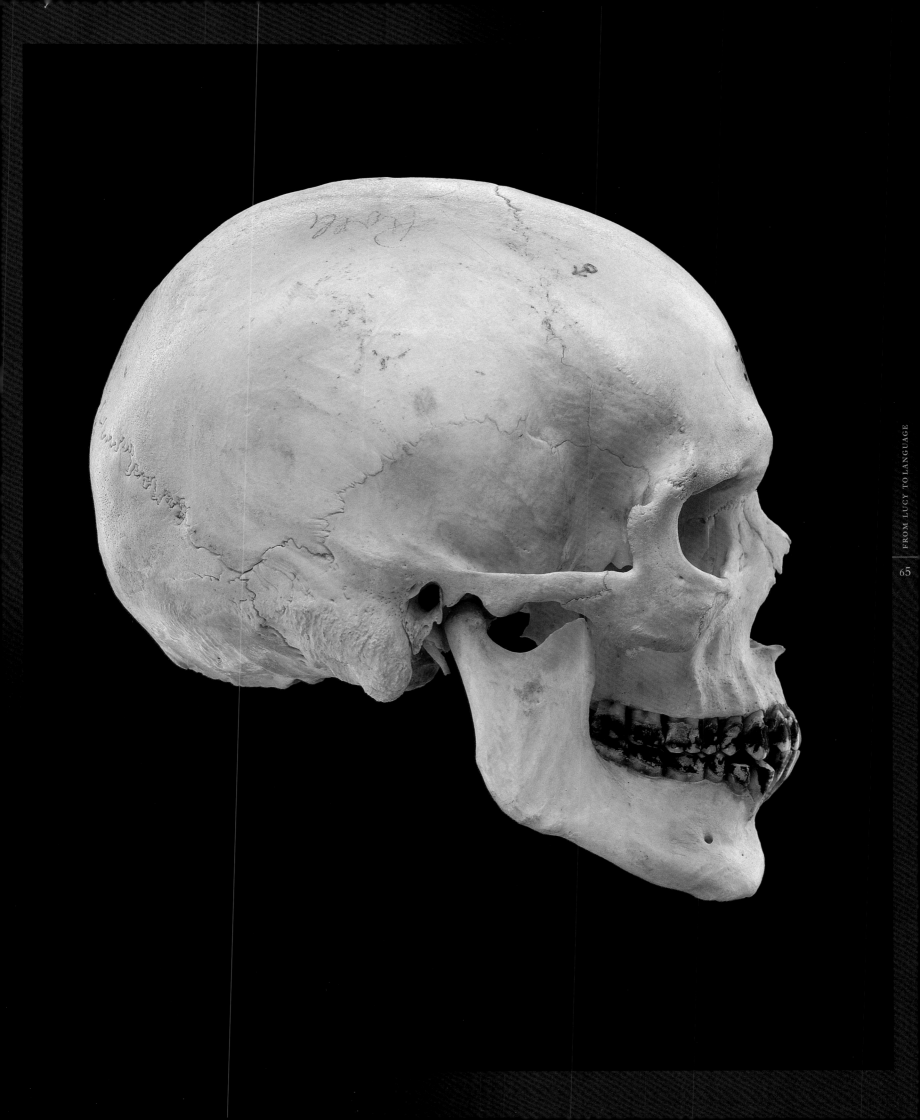

Skull of modern *Homo sapiens* from Solomon Islands, Australasia. This male specimen
was collected in 1893 at Guadalcanal. Actual sizes. *Photographs by David L. Brill; courtesy
of National Museum of Natural History.*

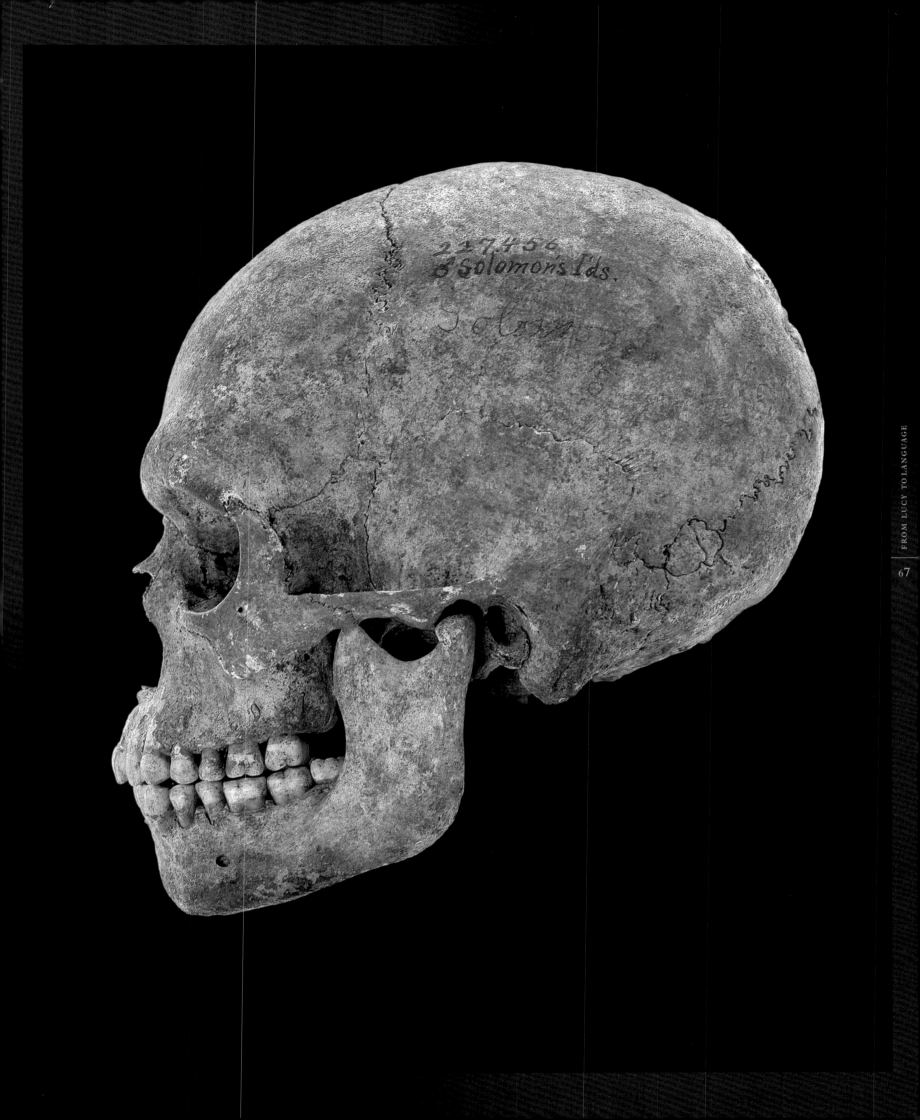

Skull of modern *Homo sapiens* from North America. This male Inuit is between the ages of 35 and 40. Actual sizes. *Photographs by David L. Brill; courtesy of American Museum of Natural History.*

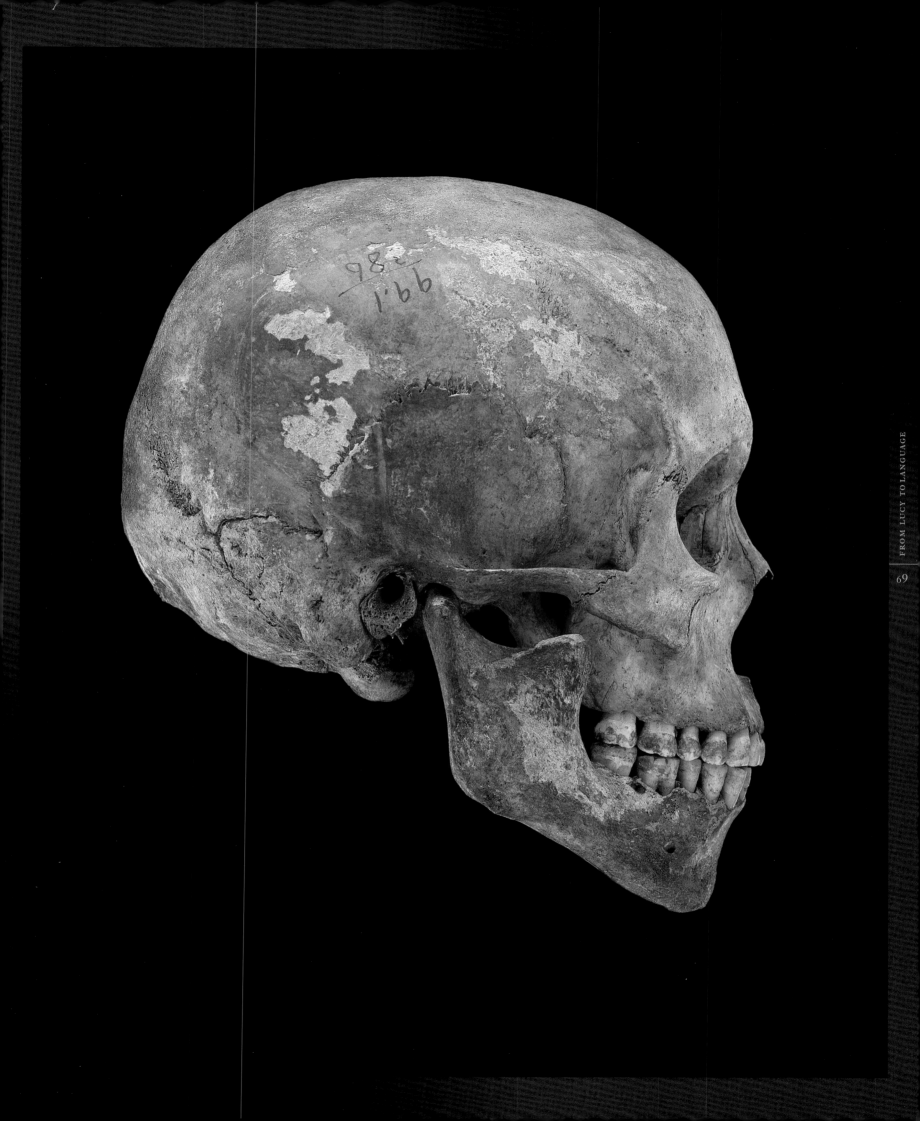

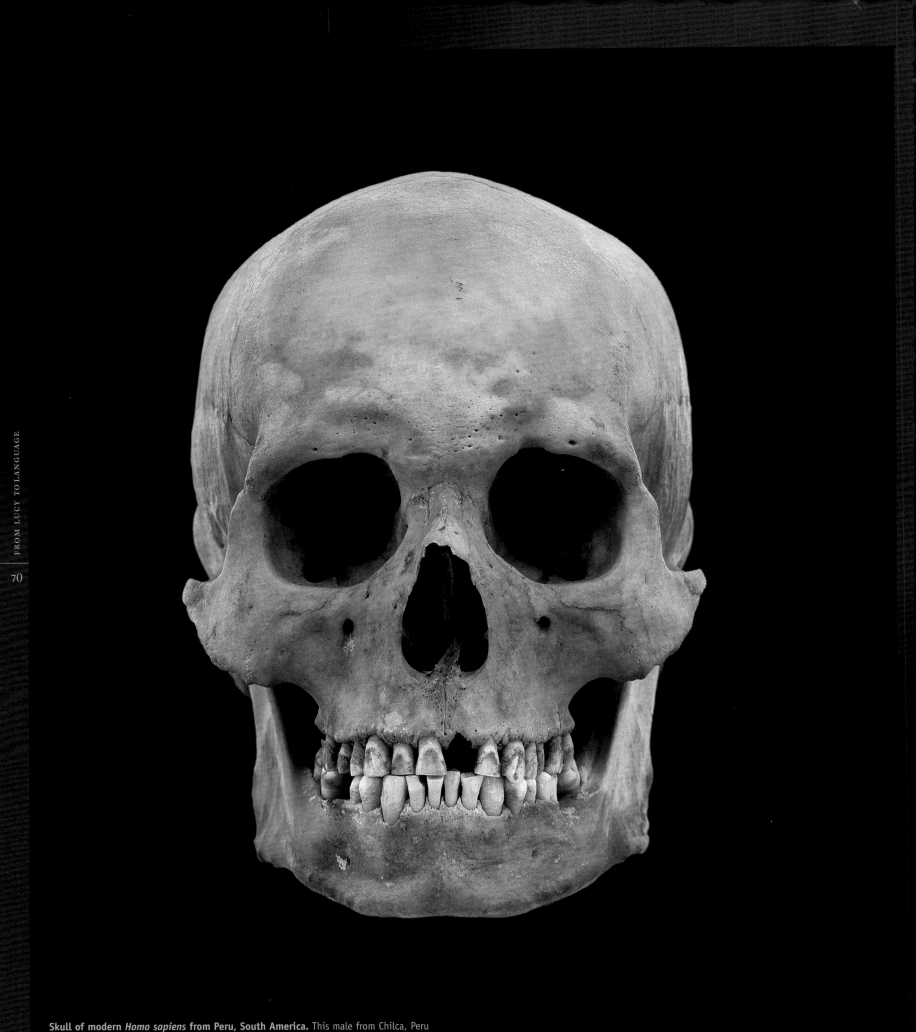

Skull of modern *Homo sapiens* from Peru, South America. This male from Chilca, Peru dates from about the fifteenth century. Actual sizes. *Photographs by David L. Brill; courtesy of National Museum of Natural History.*

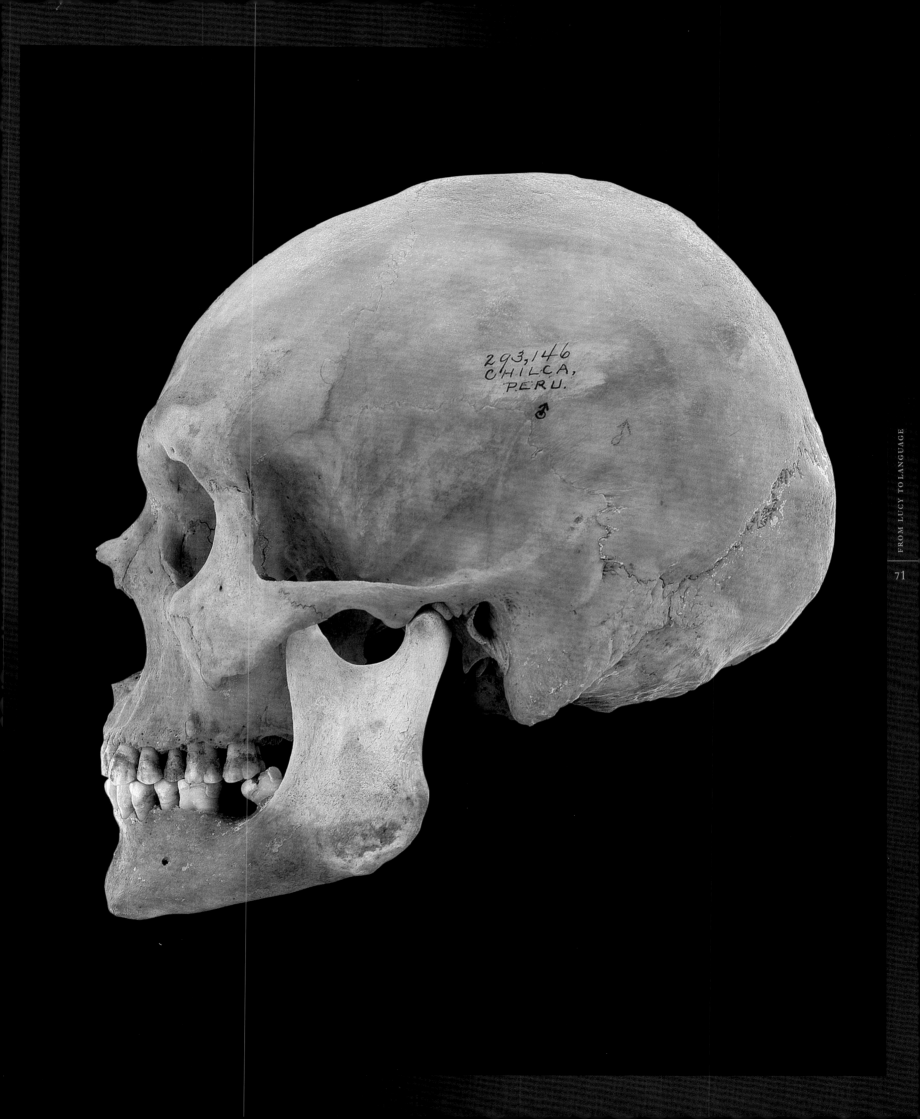

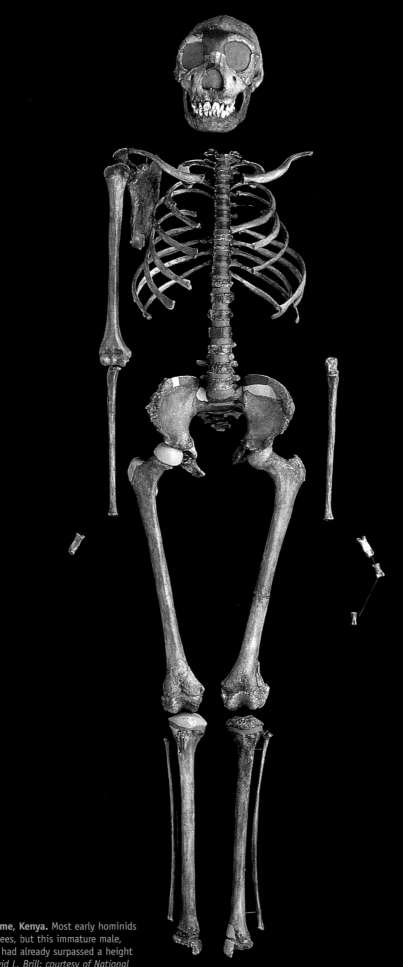

Skeleton of *Homo ergaster* from Nariokotome, Kenya. Most early hominids were not much larger than modern chimpanzees, but this immature male, WT 15000, evolved a tall, thin physique and had already surpassed a height of five feet when he died. *Photograph by David L. Brill; courtesy of National Museums of Kenya.*

A. robustus and *A. boisei* are often referred to as "robust" australopithecines, the adjective applies to their large teeth, not their body size, because they are not much bigger than the other two species, *afarensis* and *africanus*. ◖ It is apparent that during the australopithecine phase of human evolution there was stasis in body size, with males weighing between 40 and 50 kilograms and females weighing between 28 and 34 kilograms. This does not fit with general assumption that there was a trend of increasing body size in hominids through time. Indeed, a specimen from Olduvai Gorge, OH 62, that has been assigned to *Homo habilis* and dated to 1.8 million years ago, has the surprisingly small body size of a mere 24 kilograms. In fact, a significant increase in body size does not seem to have occurred until the appearance of *Homo ergaster*, as represented by the Turkana Boy (KNM-WT 15000). This remarkably complete skeleton is estimated to have weighed close to 67 kilograms and to have stood about 1.6 meters tall—a stunning increase in size at 1.7 million years ago.

28 · Sexual Dimorphism

HUMANS, like virtually all mammals, are characterized by distinct features in size, shape, and behavior that differentiate between males and females—a condition known as sexual dimorphism. For example, human males tend to be larger and stronger than females. Human males and females differ in growth rate, the distribution of subcutaneous fat, and many other features. But the most obvious contrasts are seen in the secondary sexual characteristics such as the genitals, patterns of hair distribution, and breasts. ◖ Following Darwin, it is widely thought that reproductive competition among males leads to selection for physical and behavioral traits that enhance their success in siring more offspring. This increased competition to mate with females may lead to larger male body size and to the development of weapons such as large canine teeth or antlers. The selection for larger body size in males increases the level of sexual dimorphism in a particular species. Darwin first drew attention to this sequence and called it sexual selection—the evolutionary process that favors adaptations that increase mating success. This form of sexual selection often takes the form of male-male mating competition in which the larger, more aggressive males breed more offspring, thus passing on to them those genes responsible for the characteristics that were advantageous for mating. ◖ The degree of sexual dimorphism, therefore, would seem to have some connection with mating systems. As a rough generalization, primates that exhibit little or very reduced dimorphism tend to be monogamous. On the other hand, where sexual dimorphism is extreme and males are twice or more the size of females, the males tend to be polygynous and to live either in groups having a single mature breeding male (such as gorillas) or in multi-male, multi-female groups (such as baboons). Closer study, however, of body size in relation to sexual dimorphism shows a more mixed picture. Some monogamous species are quite dimorphic, and some species with reduced dimorphism are polygynous. ◖ Another generalization is that the overall body size of a species is directly correlated with the degree of sexual dimorphism. Usually cited is the gorilla, which is the largest in size and among the most dimorphic of all primates (see pages 84-85). Again, there are exceptions to this generalization. ◖ The need for males to protect females that are foraging or looking after infants is another factor determining the degree of sexual dimorphism. In an open savanna setting, where potential predator pressure is high, large males possessing formidable canines would have a distinct advantage in defending the troop and driving off predators. ◖ Sexual dimorphism can also be examined from the point of view of its advantages for females, and here we note that in most dimorphic species, the females are smaller than the males. If being large is good, allowing males to challenge competitors and predators, why aren't females also large? Field studies of baboon troops have shown that pregnancy and

lactation place severe energy demands on females. Subsisting on a fairly low-quality vegetarian diet, female baboons are hard-pressed to eat enough to nourish their fetuses or continue to produce sufficient milk to provision their offspring. Therefore, smaller females with less nutritional needs have an advantage over larger females because they are more easily able to eat enough to keep themselves alive and meet other energy demands of child-rearing. ◖ There also seems to be a division of labor by sex that is partly related to body size. In baboons and chimpanzees, where occasional hunting has been observed, it is the males that habitually hunt. Although females have been observed hunting in close proximity to the troop, they tend to concentrate on gathering plant foods. Perhaps this exploitation of different resources functions to reduce feeding competition between the sexes. For example, in the extinct huia birds of New Zealand, the males had short, stout beaks which they used to peck away at branches in search of insects. Females used their long, curved beaks to probe into crevices. In this way, the sexes were not competing for the same food. ◖ The larger samples of fossil hominids from Africa exhibit marked sexual dimorphism, as manifested in the size of teeth, jaws, and other skeletal elements. Estimates of body size (weight) based on lower limb joint size for the best-known early australopithecine, *Australopithecus afarensis*, suggest that males were considerably heavier than females. The average size for six *A. afarensis* males is 45 kilograms, and for three females it is 29 kilograms. The ratio of male to female body weights gives an estimate of the degree of sexual dimorphism. In *A. afarensis* this ratio is 1.52, a value exceeding that of 1.35 for *A. africanus*. For comparison, ratios in modern humans and apes are as follow: humans, 1.22; chimpanzees, 1.37; orangutans, 2.03; and gorillas, 2.09. Clearly, the ratio for *A. afarensis* far exceeds that seen in modern humans and chimps, but it falls short of the large values for gorillas and orangs. ◖ Precisely why *A. afarensis* is so dimorphic is not totally understood, but in part the answer has to do with its terrestriality. In all species of primates that are predominantly terrestrial, we see substantial size dimorphism. What is most surprising about *A. afarensis* is the reduced size of the canines in males, resulting in a low level of canine dimorphism. In most but not all primates there is a correlation between body size dimorphism and canine dimorphism. That is, in primates of low body size dimorphism there is low canine dimorphism and in species with large body size dimorphism there is large canine dimorphism. Furthermore, large canine dimorphism seems to be related to high levels of aggression or male-male competition. ◖ The smaller canines in *A. afarensis* males may therefore suggest that, for whatever reason, there was reduced male-male aggression. Another important consideration is that male primates are largely responsible for warding off predators. In the case of baboons, for example, males are large and have dagger-like canines, both useful features for impeding predator attacks. ◖ In *A. afarensis*, whose males lacked large canines, large body size may have been selected for as part of a strategy for protecting against predators. A presumed reduction in male-male aggression, because of the lack of large canines, might also have led to more cooperation between males to drive off the large predators that occupied the same territory as these early hominids. Chimpanzees, in which males within a group are genetically closely related, have been observed to collectively drive off leopards. ◖ Large predators roamed the landscape at Hadar and Laetoli, and *A. afarensis* would have been subject to predation. In addition, Hadar was more heavily wooded, like the environment in which baboons live today, where incidents of predation are highest. It is in such a habitat, with impaired visibility, that these early hominids would have benefited from larger male body size and male cooperation. Such model building must be viewed cautiously, but it is an interesting example of the importance of features such as sexual dimorphism, body size, and canine size in trying to understand the behavior of our earliest ancestors.

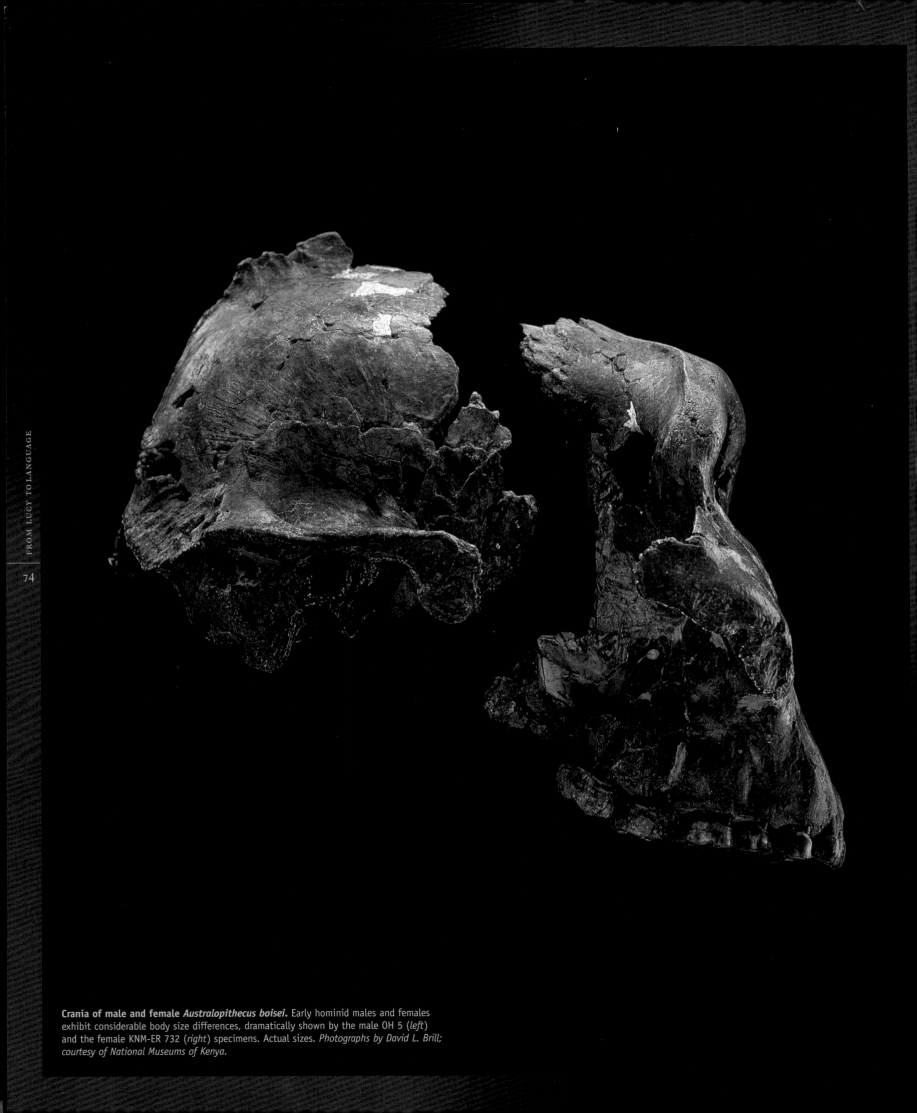

Crania of male and female *Australopithecus boisei*. Early hominid males and females
exhibit considerable body size differences, dramatically shown by the male OH 5 (*left*)
and the female KNM-ER 732 (*right*) specimens. Actual sizes. *Photographs by David L. Brill;
courtesy of National Museums of Kenya.*

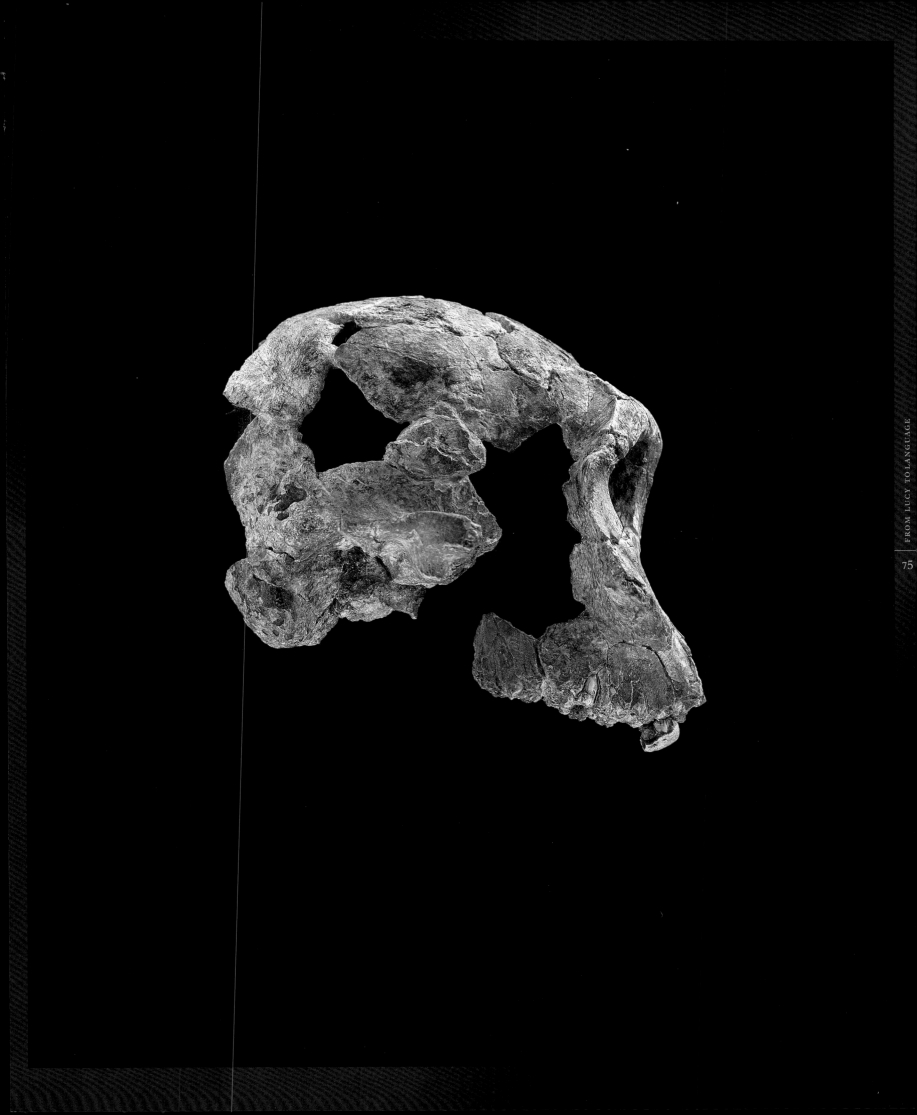

29 · Gestation

FOR our relatively modest mammalian body size, humans have a long period of gestation in the womb, lasting anywhere from 38 to 42 weeks. But we live long lives, and life span does correlate with length of gestation. The longest gestation of any mammal—22 months—occurs in the female elephant, which lives an average of 55 years. Brain size also correlates with gestation length. Mammals with larger brains at birth have a longer gestation, and based on our brain size relative to brain size in other mammals, humans should have an even longer gestation than we already do. ❰ Our gestation fits the general primate pattern of slow fetal body growth rate but rapid fetal brain growth, relative to growth rates in other mammals that give birth to well-developed offspring, such as a zebra foal that soon springs to its feet after a birth on the savanna. Even though human gestation is relatively long and the infant emerges large for its mother's body size, the human brain is born immature. A newborn chimpanzee's brain, for comparison, has already reached half its adult weight, but a newborn human's brain has only a quarter of an adult brain's weight. To compensate, the fetal pattern of rapid brain growth continues after birth for the first year of life, during which the brain more than doubles in size from its newborn dimensions. And as any human parent knows, newborn human babies are extremely dependent in this first year of life. So, as far as our brains are concerned, human gestation lasts 21 months. This growth pattern, termed secondary altriciality, is unique to humans, and it evolved because we have struck a compromise between having big brains and being bipedal. ❰ During delivery the human infant's head is turned sideways as it enters the widest part of the birth canal, flexes chestward, rotates 90 degrees to fit through the narrow middle of the birth canal, which is widest from front to back, and exits facing forward. A newborn human may be the twice the weight of a newborn ape, but the human mother probably weighs only half as much as the mother ape. That is why the process of human birth is among the most difficult for any animal. For a chimpanzee, birth comes easily; the infant's head passes through the birth canal without much constriction. The human pelvis, however, has been extensively reconfigured for two-legged locomotion, and in order to accommodate a large-brained baby, it has expanded from front to back and contracted slightly from the sides, creating a rounder birth canal. Interestingly 95 percent of adult brain size is reached between 5 and 6 years, which coincides with the emergence of the first permanent molar. ❰ Australopithecines apparently came much closer to apes than humans in gestation length and difficulty of birth. From the shape of two fossil pelvis specimens of *Australopithecus afarensis* (A.L. 288-1, or Lucy) and *Australopithecus africanus* (Sts 14) (see pages 133 and 148), it is possible to infer that these hominids gave birth to infants with heads close in size to those of newborn chimpanzees. Birth was probably pretty quick and easy, although Lucy's pelvis is constricted from front to back and the infant's head most likely entered the birth canal sideways. Some recent radiographs of monkey and baboon births show that these primates also rotate in the birth canal, so it may be that, like humans, early hominid infants went through a similar rotation, perhaps to let their shoulders pass through the narrowest section of the canal. ❰ But if australopithecines generally had an apelike gestation and birth, when did the human gestation pattern appear? One recent hypothesis holds that Neandertals had a longer gestation period than ours. This argument relies on evidence from a particular part of the pelvis. Although most of the pelvic bones—the ilium, ischium, and sacrum—are similar in shape and size in Neandertals and modern humans, the two groups differ in the anatomy of the pubis, the bone that forms the front of the pelvis. Neandertals possess an extremely long, slender, and flat superior pubic ramus, a bar along the top of the bone, whereas the pubis of early modern humans and their descendants has a short, thick ramus. (In this detail of anatomy, at least,

Pelvis of *Homo neanderthalensis* from Kebara, Israel. The elongated superior pubic ramus at the top of a Neandertal pelvis was cited as evidence that this species had a longer gestation and birthed larger, better developed babies than modern humans. Recovery of this pelvis from the Kebara 2 skeleton showed that the birth canal was not significantly larger than a human's, however, so Neandertal gestation was probably not any longer. Actual size. *Photograph by David L. Brill; courtesy of Sackler School of Medicine, Tel Aviv University.*

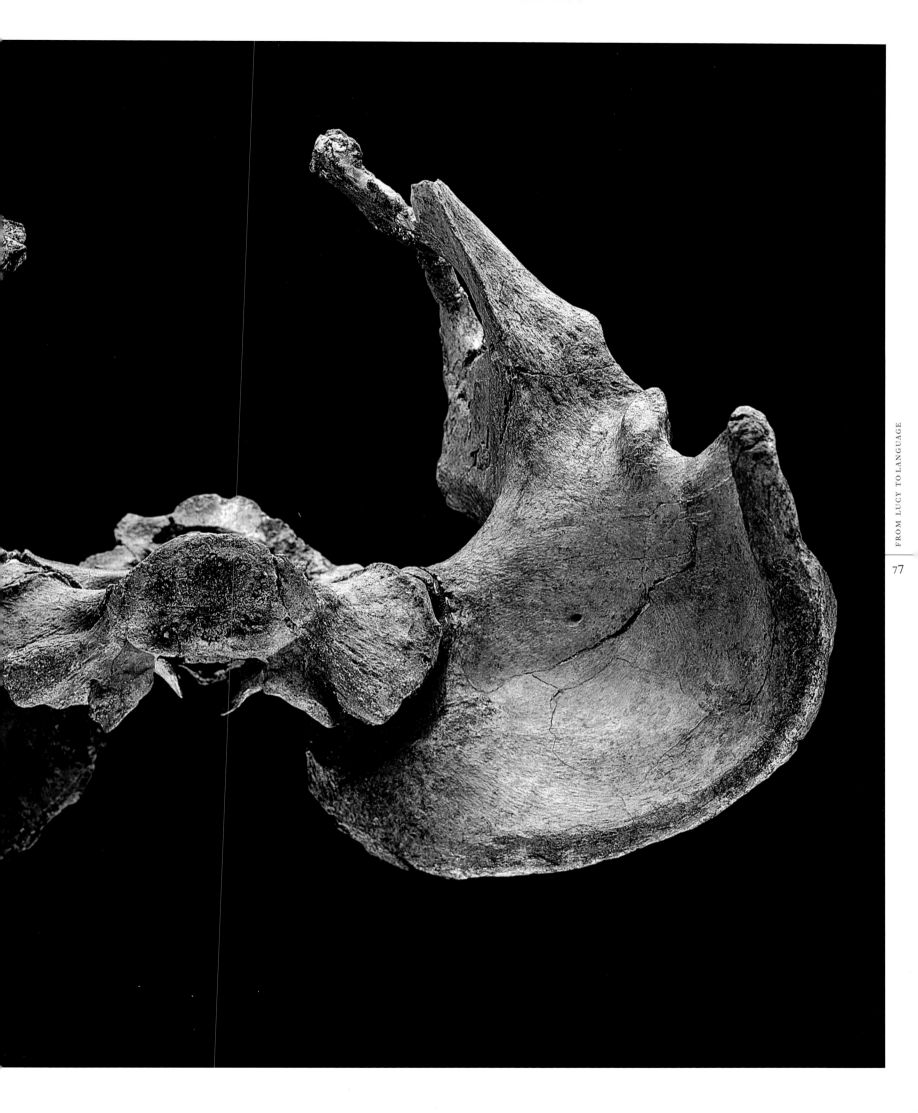

modern humans appear to be more strongly built than the burly, barrel-chested Neandertals.) ❧ After studying the seven Neandertal pubic bones then available, Washington University paleoanthropologist Erick Trinkaus proposed that the long, thin ramus would have broadened the pelvis and thus enlarged the birth canal of Neandertal mothers by up to 25 percent, allowing mothers to give birth to bigger, more developed, less dependent babies. Neandertals might therefore have had a longer gestation than modern humans, perhaps as long as 12 months. ❧ This idea generated lots of speculation about whether Neandertal babies were born more mature and whether the shorter gestations of modern humans required the evolution of greater social support to care for and protect the underdeveloped infants. A shorter gestation would have led to reduced time between births and might have given modern humans a reproductive edge that eventually allowed them to out produce and out compete Neandertals. Further evidence soon forced a revision of this whole story. ❧ The 1983 discovery of the Neandertal skeleton from Kebara Cave in Israel single-handedly refuted the idea of longer gestation periods in Neandertals. This specimen, Kebara 2, included the most complete Neandertal pelvis ever found. The specimen's superior pubic ramus measures about 90 millimeters long, longer than that any other Neandertal and nearly a third longer than the ramus of modern human males. Yet the pelvic inlet width is only 13 percent wider in the Neandertal. So the pelvis was wider from side to side but not deeper from front to back, and the volume of the birth canal did not change. Although Kebara 2 is a male, we can assume that a female Neandertal's pelvis would have similar proportions, and that the greater length of the pubic ramus still did not contribute to a larger birth canal. ❧ Evidence from another complete skeleton, the 1.6-million-year-old *Homo ergaster* specimen KNM WT 15000 (see page 196), suggests that the human pattern of postnatal brain gestation may have arisen in this species. This skeleton belongs to an adolescent boy, but its pelvic width is so narrow that it is hard to imagine an adult female of the species being able to birth a baby with the presumed brain weight—perhaps 200 grams—of this hominid. Unless, that is, this hominid had a pattern of brain growth more like that of modern humans, with brain weight at birth relatively small but continued rapid growth outside the womb to build an absolutely bigger brain. The implication is that *Homo ergaster* also gave birth to infants that required long-term constant care after birth, and perhaps in this strange twist on primate gestation lie the seeds of the elaborate socialization and prolonged learning that came to characterize our species.

30 · Maturation

ALONG with bipedalism, relatively large brains, and symbolic language, one of the unusual aspects of human life history is our delayed development and growth. Humans take twice as long as living apes to reach maturity. For instance, chimpanzees mature after eight or nine years, but human females become sexually mature at around age 13 and continue to grow for five more years. Humans, unlike any other primate, also have inserted a spurt of rapid growth, between puberty and maturity, known as adolescence, when a significant amount of total growth occurs throughout the skeleton. Our extended childhood and adolescence has often been considered a means of enhancing our capacity for learning and thus may be a critical trait of our species. We clearly differ from apes in this respect, but what was the case for our hominid ancestors? ❧ Although hard data about the rates of hominid development are somewhat elusive, intriguing techniques have been developed to answer this question. It is possible to estimate the age of death for a skeleton by careful scrutiny of the bones. Joints offer one set of clues. The growth plates, or epiphyses, at the ends of limb bones fuse to the bone shafts at different rates. In humans, the joints tend to fuse completely in this order, beginning at an

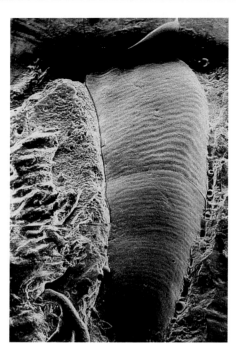

Growth lines on tooth of juvenile *Australopithecus robustus*. Growth lines, or perikymata, on teeth can help determine the age at death of an individual. This tongue-side view of the SK 62 incisor from Swartkrans, South Africa shows several parallel perikymata. *Courtesy of Christopher Dean, University College London.*

average age of 18 years: elbow, hip, ankle, knee, wrist, and shoulder. ❧ Teeth provide an excellent record of growth and development as revealed by their timing of eruption. In humans, the three molars on one side of each jaw erupt about six years apart, beginning with the first molar at age six. The second molar's appearance corresponds with the onset of puberty and adolescence, while the third molar erupts as an adult height is reached. Other teeth can be used to pin-point similar periods of early development. ❧ A more innovative aging technique considers the incremental growth lines that form within the tooth enamel, like tree rings on teeth. Such lines inside the enamel are called the striae of Retzius. As the tooth grows these lines become visible on the outer surface of the enamel and are called perikymata. Although the actual cause of the lines remains uncertain, it is thought that a new line appears every six to nine days as the tooth forms and grows, so the lines provide a ruler of the individual's age. Under a microscope, perikymata can be seen on human and hominid incisors. Because they are among the first teeth to form, incisors become particularly useful for making age estimations of very young individuals. Similar studies have been made of incremental growth in molar crowns of late juvenile or adult specimens. ❧ In order to make accurate estimates of the age at death of hominid specimens, it is necessary to know whether their growth rates were more like apes or closer to our own. Initial studies by paleoanthropologist Alan Mann in the 1970s of juvenile hominid teeth from South African cave sites such as Swartkrans suggested that early hominids had maturation rates similar to that of modern humans. Mann based this conclusion on a collection of juvenile australopithecine teeth from South African cave sites such as Swartkrans. Mann's estimated ages for isolated teeth in the sample led him to conclude that australopithecines had a pace of tooth eruption widely spaced over time, and hence a long period of immaturity. ❧ More recently, the detailed analysis of perikymata and a broader understanding of ape dental development have caused paleoanthropologists such as Holly Smith and Christopher Dean to rethink this idea and conclude instead that early hominids had shorter growth periods and matured at rates more like those of modern apes. This rethinking forces a revision in the age estimates for juvenile fossils. By shifting from a human to an ape growth model for australopithecines, the estimated age at death for a specimen like the *A. africanus* skull from Taung, for example, gets halved from seven years to slightly more than three. ❧ Judging from the rate and sequence of enamel deposition on their molars and the formation of incisors, robust australopithecines apparently had a unique growth pattern different from that of apes, modern humans, and other australopithecines. In robust australopithecine molars, enamel was applied quickly and mostly to the tops of molar crowns, creating a pattern similar to that of human baby teeth. It is as though the robusts retained this juvenile trait in response to some evolutionary pressure—perhaps their probable diet of

Skull of *Australopithecus africanus* from Taung, South Africa (see also page 154). If
the famous Taung child matured at a rate similar to modern apes, then its estimated age
at death was three years. An eagle may have preyed upon this individual. Actual size.
Photograph by David L. Brill; courtesy of University of the Witwatersrand.

hard nuts and seeds—to produce teeth quickly and erupt them early. ❧ It may be that even early members of genus *Homo*, including *Homo habilis*, had apelike rates of growth. Data for two juvenile dental specimens from Koobi Fora, KNM-ER 820 and KNM-ER 1590, suggest that early *Homo* kept the primitive pattern of growth and had not yet evolved delayed maturation. If an apelike pattern of growth characterized human evolution for the first several millions of years, when did this pattern change? ❧ The discovery of the 1.6-million-year-old KNM-WT 15000 skeleton (see page 196) at Nariokotome, Kenya, in 1984 provided a unique opportunity to study a nearly complete juvenile skeleton with its associated skull and teeth. Estimating WT 15000's age at death depends on just what skeletal evidence is used and whether the specimen is compared to an ape or a human growth model. Judging from tooth eruption, the Nariokotome boy had not reached adulthood. His permanent first and second molars had come in, but the third molars (wisdom teeth) had not, and there are still milk canines in the upper jaw. Fewer than half of his permanent teeth had completely formed, suggesting that he died while in the first half of adolescence, perhaps around age eight, which is supported by the growth markers in the teeth. ❧ Evidence from other parts of the skeleton, including the degree of fusion in the upper arm bone epiphyses and the presence of cartilage between the pelvic bones, puts the specimen at younger than 15 years and probably close to 13 years. But based on overall stature—he stood 1.73 meters tall—the skeleton matches a modern human of at least 15 years. By any human standards, this was a big boy. ❧ Based on a human scale of growth, WT 15000 was somewhere between 11 and 15 years old when he died. If an ape scale is used instead, the age estimates drop to between seven and nine years. The ape growth model seems to fit WT 15000's molar development, but the anterior teeth have already reached a more advanced stage of growth when compared to apes. The correct answer to the skeleton's age may lie somewhere in the middle of the estimates derived from ape and human standards, say nine or ten years, and it is reasonable to conclude that maturation in this species occurred at some intermediate rate. ❧ Dental growth markers suggest that our recent contemporaries the Neandertals, matured 15 percent faster than modern humans. The perikymata of an unerupted incisor from a Neandertal child skull found at Devil's Tower on Gibraltar suggested that this child had died at three or four years of age. Yet the skull already had well-developed molars for its age and a brain size of 1,400 cc, approaching the average for modern humans. So Neandertals, in contrast to us, may have grown up quickly, and perhaps this accelerated growth contributed to their bigger heads and brains. ❧ Since Alan Mann's pioneering work two decades ago, we have assembled techniques to help us arrive at an understanding of how other hominids developed and grew. While the answers should become clearer with the addition of new juvenile fossils and more sophisticated analyses, we have already learned that the rate of maturation was as important an adaptation in the lives of our ancestors as it is for ourselves.

31 · Evolution of the Human Brain

THE SPONGY, 1.3-kilogram organ inside our skulls consists mostly of water, but it contains the key to the past 2 million years of our evolutionary story. With its estimated 100 billion neurons—as many communicating nerve cells as there are stars in the Milky Way—and a trillion glial cells to support those neurons, the brain regulates our behavior, our motions and emotions, our instincts and ideas—all that distinguishes the human experience. ❧ In absolute size, the human brain breaks no records. Elephant brains exceed ours by a factor of four, and some whale brains are even bigger. Apparently, bigger bodies require larger brains in order to operate. Brain weight increases about two thirds as fast as body weight over a range of small to large mammals, so that the biggest mammals have absolutely larger but relatively smaller brains. ❧ Regardless of body size, however, primates tend to have relatively large brains. Primates are visually oriented animals, and color stereoscopic vision must require some complicated neural connections. Monkeys, apes, and humans possess the biggest brains relative to body weight of any terrestrial mammals. So, from one perspective, the human brain is just a highly elaborated ape brain. ❧ Yet there is still something different, something unique, about the size of the human brain. Our brain is three times larger than the predicted size for a hypothetical non-human primate of average human body size. Almost all of this added size has evolved within the past 2 million years. But size isn't everything. ❧ Our brain also differs significantly from those of apes in the proportion of various parts. It did not just uniformly grow bigger and thus endow us with intelligence. Some parts of the human brain changed little, while others became barely recognizable. The pons and medulla, primitive parts of the hindbrain, have not been especially enlarged in humans. We show some enlargement of the midbrain, but the real difference lies in the cortex or gray matter of the forebrain. Human brains display an amazing inflation of the cerebellum and neocortex (which covers much of the cerebrum), areas that play crucial roles in many aspects of learning. The cerebral cortex constitutes 80 percent of our brain mass. The primitive forebrain functioned primarily in the sense of smell. Although our forebrain remains connected to the olfactory lobes, these have become reduced, and our forebrain now serves a new function. ❧ That function is reflected in the role of the prefrontal cortex of the left hemisphere. An average primate with a brain blown up to the size of ours would still have a prefrontal cortex that is more than 200 percent smaller than that of the human brain! The prefrontal cortex includes those regions of the brain, such as Broca's and Wernicke's areas, that are strongly associated with the production and understanding of language. Compared to apes, we have greatly enhanced the size and neural connections in that part of the brain. In contrast, the part of our brain devoted to controlling motor skills takes up only about a third of the space it does in a monkey's brain. (We won't be leaping agilely through forest canopies anytime soon.) As a result, we sacrificed some capability for seeing and smelling, but gained new capacity for symbolic and computational thought. ❧ In addition, the human brain is a sponge that soaks up sensations and observations, and it is a masterful organ for storing, retrieving, and processing a wide range of detailed and complicated information. To borrow a metaphor from biologist Christopher Wills, "We differ from other animals not in the ability to juggle but in the number of balls that we can keep in the air simultaneously." Our brains have permitted us to evolve culture and enhance it to an unprecedented degree in the history of life on Earth (see page 21). ❧ So, size alone does not explain our unusual mental abilities. What counts is what's inside the package and how it is all arranged. Cognitive neuroscience is an exciting and rapidly expanding field of inquiry, but we are still just scratching the surface of figuring out how the human brain works. We should not expect to find a simple relationship between brain size, complexity, and intelligence. ❧ We cannot answer exactly why we evolved our large brains, but fossil evidence tells us when we evolved them. During the Lower and early Middle Pleistocene (between 2 million and around 700,000 years ago), hominid brain size doubled, from about 440 cc to more than 900 cc. This change affected mainly the side-to-side and back-to-front dimensions of the cranium, especially the frontal and occipital bones and the cranial base. In part, this doubling occurred because of a general increase in body size, but its rate and extent suggest that natural selection was also

Brain endocasts and cranium of *Australopithecus africanus* from South Africa. A trend toward larger brains occurred in the course of human evolution. Australopithecines like these had brains less in 500 cc in volume, while brain volume in genus *Homo* ranges from 600 cc to about 2,000 cc. Actual sizes. *Photograph by John Reader, Science Source/Photo Researchers.*

82

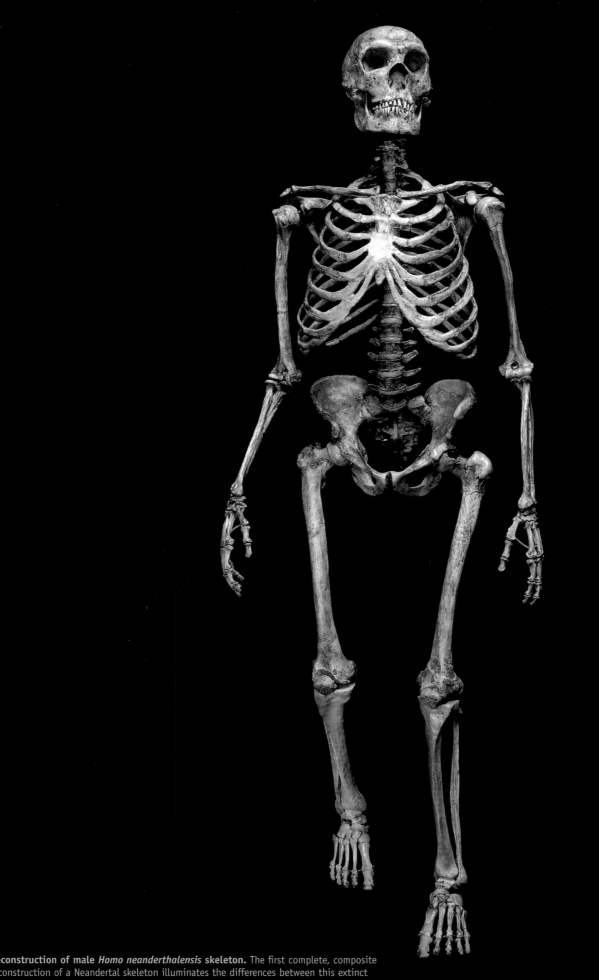

Reconstruction of male *Homo neanderthalensis* skeleton. The first complete, composite reconstruction of a Neandertal skeleton illuminates the differences between this extinct human and modern humans. *Photograph by Denis Finin; courtesy of G.J. Sawyer and Blaine Maley.*

focused specifically on adding cranial capacity. ❧ The biggest burst in brain size increase, however, occurred during the Middle to Upper Pleistocene (around 500,000 to 100,000 years ago), the span of time from late *Homo erectus* to early *Homo sapiens*. This change in both size and shape represents one of the most remarkable morphological shifts that has been observed in the evolutionary history of any mammal, for it entailed both an enhanced cranial capacity and a radical reorganization of brain proportions. And it affected the vertical dimensions of the frontal and occipital bones, allowing for growth of the neurocranium. ❧ Subsequent changes in our cranial capacity, as seen in the shape of the skull, were minor by comparison and included more upward expansion of the frontal bone, a higher and broader biparietal arch, and a rounding of the occipital bone. Thereafter, from the Upper Paleolithic through the present, our skulls and brains have undergone no significant change. Modern human brains weigh an average of 1.3 kilograms and have an average volume of at least 1,350 cc. ❧ People often wonder why Neandertal brains were bigger than ours. Were they smarter, and if so, why are they extinct? The reason for their bigger brains has to do with the relationship between body size and ambient temperature. The "classic" Neandertals of western Europe were adapted to a late Pleistocene cold climate, and part of that adaptation was a relatively larger brain corresponding to their stout stature. Male classic Neandertals had a mean cranial capacity of 1,582 cc. A similar pattern of reduced limb length, stouter body proportions, and larger cranial capacity occurs in modern populations living at high latitudes, including the Lapps or Sääme people and the Greenland Inuit.

IF THERE is a single topic in paleoanthropology that excites more general interest than who our ancestors were, it is what early humans looked like. Piecing together dozens or hundreds of delicate fragments into skulls and limb bones requires both scientific knowledge and puzzle-solving skills, but taking the next steps reassembling these bone fragments and then adding muscle and flesh to the bones constitutes a marriage of science and art. The challenge, essentially, is to perform a dissection in reverse, and few people are up to the task. Three who have made this task their calling are physical anthropologist Gary Sawyer and paleoartists and anatomists Jay Matternes and John Gurche. ❧ Sawyer, at the American Museum of Natural History, for the first time constructed a fully articulated Neandertal skeleton, using casts from real Neandertal bones. The foundation of the reconstruction was the La Ferrassie 1 skeleton (see page 240). The absence of complete ribs, vertebral column, and pelvis from La Ferrassie meant that these missing elements had to be obtained from other individuals, primarily from Kebara 2 (see page 232) with selected elements from Feldhofer 1 (see page 242) and other specimens. The reconstructed skeleton has made it clearer than ever that Neandertals were not a subspecies of modern humans but a separate hominid species. The Neandertal's shoulders were wider than a human's. The pelvis is also wider, even in males. The Neandertal had shorter forearms and shins, a broader trunk, and virtually no waist. The rib cage, instead of tapering off, as in modern humans, was bell-shaped. As Juan Luis Arsuaga of Madrid's Complutense University observed about the reconstructed Neandertal skeleton, "To come across a Neandertal, even a reconstructed one, is a thrilling experience. It was not doubt even more thrilling to our ancestors, who met them in the flesh." ❧ Matternes is best known for carefully reasoned depictions of early hominids in the flesh in his paintings. In addition to representing a group of some hominid species in exacting anatomical detail, these works include a carefully

32 · Reconstructing the Appearance of Early Humans

researched reconstruction of the habitat. His painting of *Australopithecus afarensis* for the November 1985 *National Geographic* magazine shows a troop of 11 individuals, from infant to adult, foraging for figs on the ground of a dense montane forest. Matternes's murals of various extinct primates and hominid ancestors in their environmental context can be seen in the Spitzer Hall of Origins at the American Museum of Natural History in New York. ❧ To make a graphic reconstruction of a specific specimen, Matternes may photograph the fossil—taking care to avoid any parallax distortion—and projects its image on a wall for tracing. He orients the skull projection and subsequent drawings to be in Frankfurt Horizontal (plane), a standard anatomical position defined by connecting right and left craniometric points just above the ear canal opening, called the porion, and the orbitale on the lower rim of an eye socket. This position replicates the natural angle of a human head. (All of David L. Brill's fossil portraits in this book were made in Frankfurt Horizontal.) Matternes uses a series of tissue overlays to draw in muscles and ultimately skin and hair over the bones. For his reconstruction of a male *A. afarensis* that was reproduced in *Lucy: The Beginnings of Humankind*, Matternes opted for a hairy, dark-skinned body, although he decided against adding much facial hair or a beard, surmising that beards had yet to evolve as secondary sex characteristics, as well as to show better the actual facial features. ❧ John Gurche has recreated the heads of several hominids for the National Museum of Natural History, in Washington D.C. He has also created a reconstruction of a male *A. afarensis*, but in three dimensions. Unlike Matternes, who in 1981 relied on a composite skull comprised of bones from several *afarensis* individuals, Gurche had the advantage of basing his head on the AL-444 cranium (see page 137). The result of Gurche's 700 hours of work appeared in the March 1996 *National Geographic*. ❧ Although all three collaborate closely with paleoanthropologists in the course of a commission, such work relies on informed speculation and assumption as well as on science. We cannot be certain of the color of a certain hominid's eyes or skin, the exact shape of the nose, or how much hair covered the head or body. But by using their artistic and interpretive talents to flesh out hominid fossil finds, they have brought our ancestors much closer to us. When we look at these creations, we see both how far we have come and how near we remain.

LONG-TERM field studies in recent decades have shed an immense amount of light on the behavior of our closest living relatives, the African apes. As more primatologists observe more populations in more places, it has emerged, for instance, that chimpanzees in different forests possess separate traditions and learned behaviors. These differences are expressed in many aspects of life, such as the making and using of stone, leaf, or twig tools, types of grooming, diet and attitudes toward prey animals, and the extent of cooperation in hunting. In other words, chimps have distinctive cultures. They may be the most behaviorally variable animal of all, after humans. But the apparent diversity and complexity of chimp behavior make it more difficult to tease out just which behaviors might have been shared by the earliest hominids or the last common ancestor of humans and apes. One approach to this problem taken by primatologist Richard Wrangham was to compare and contrast salient behaviors among chimpanzees, bonobos, gorillas, and humans to determine what behaviors are shared by all, because those behaviors are more likely to have evolved in a common ancestor. If each living species differs in a given behavior, then nothing can be confidently concluded about the likely behavior of the common ancestor. ❧ Wrangham selected several important elements of

33 · Primate Societies and Early Human Social Behavior

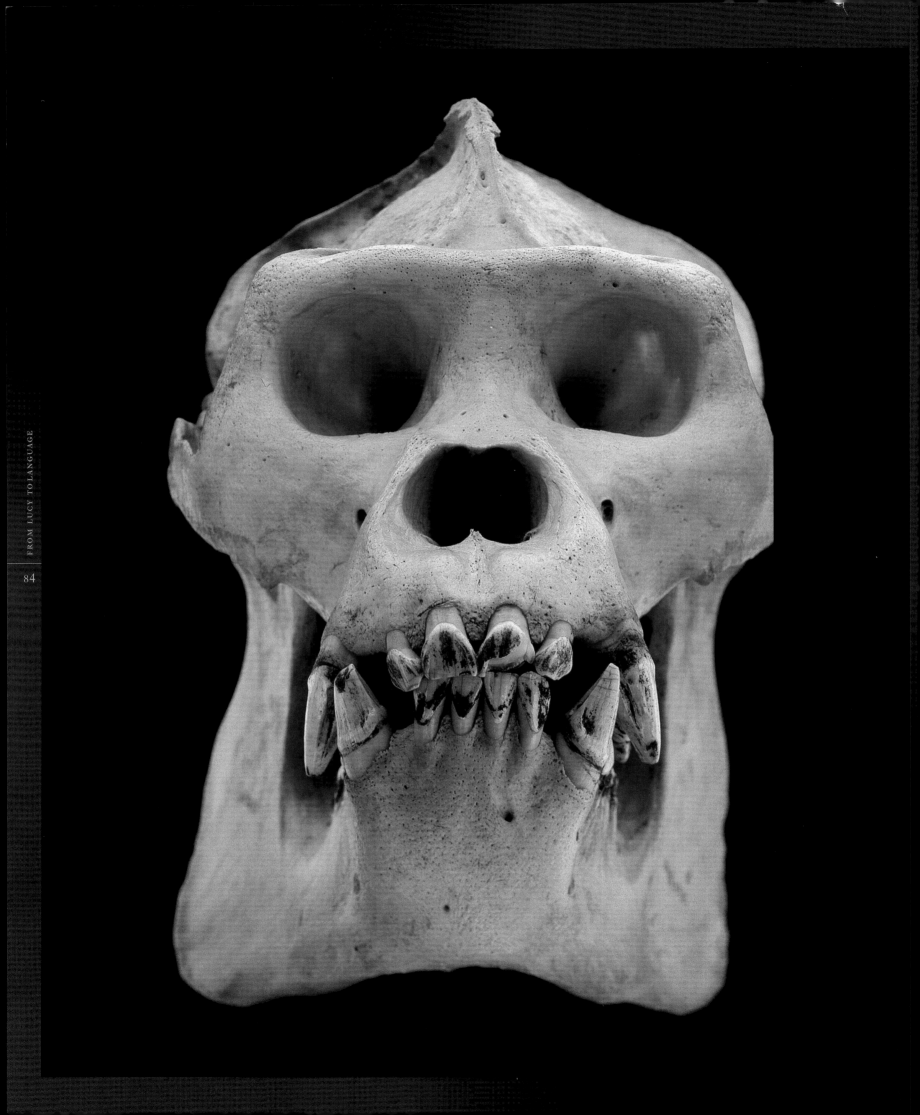

Skulls of male (*left*) and female (*right*) mountain gorilla, *Gorilla gorilla beringei*.
These specimens come from the Virunga Mountains of Rwanda, site of Dian Fossey's long-
term study of this most endangered ape. Gorillas form social groups in which only a single
dominant male mates with females. Actual sizes. *Photographs by David L. Brill; courtesy of
National Museum of Natural History.*

social organization, including adult group patterns, relationships among males and females, sexual relationships, and relationships between unrelated groups. He then examined how each takes shape in various human cultures and in groups of each of the African apes to find a pattern of shared conserved features. For example, many human societies and all African apes have a tendency toward female exogamy, or the forced migration of female members from the group in which they were born, which means that mothers generally associate with adult females who are not close relatives. The unity of this feature among living species suggests it was present also in the common ancestor. ❧ When, however, Wrangham looked at sexual behavior, no clear pattern emerged. Gorillas are polygynous, meaning that one male mates exclusively with several females. Both chimpanzees and bonobos are promiscuous, so neither sex mates exclusively. As for humans, our sexual tastes and traditions vary within and between social groups. So, little can be said about the sexual behavior of a common ancestor. Even so, the fact that male African apes normally mate with more than one female and the prevalence of polygyny in many human societies suggest that similar behavior occurred in our mutual ancestor. ❧ Wrangham's hypothetical common ancestor took form as a primate whose females migrated out of their birth group and formed tolerant, possibly friendly relationships with other females, though did not forge coalitions to gain a social advantage; whose males spent some amount of time living and traveling alone and did not form tight social bonds with other males; whose males mated with more than one female, whether in short-term or long-term relationships; who lived in a closed social network or community; and who had hostile relationships with other groups of its kind that gave rise to attacks and fighting among males. ❧ Aspects of the common ancestor's behavior that remained elusive in Wrangham's study because of variation among apes and humans included the stability of groups gathered to feed or travel; the degree of territoriality; the duration of sexual relationships and whether they were polygynous or just promiscuous; whether males ever mated with close relatives; and whether males formed alliances with each other. ❧ Another researcher, Michael Ghiglieri, followed Wrangham's lead and refined a behavioral model for the most recent common ancestor. He argued, based on genetic evidence, that humans, chimpanzees, and bonobos could be considered apart from gorillas as a more closely linked group (see page 32), and so certain behaviors shared by these three primates, but not found in gorillas, would have presumably characterized the common ancestor. ❧ So, the social structure of Ghiglieri's hypothetical common ancestor included closed, stable groups with many males and females; strict retention of male offspring in their birth group (male endogamy) and female exogamy; the establishment of a male kin-group community system with cooperative behavior among related males to form alliances against rivals, who might have been stalked and attacked; weaker bonds between females; polygynous mating; males active in defending group territory; fusion-fission sociality (individuals stay in large groups even during environmentally stressful periods but also temporarily travel alone); and moderate sexual dimorphism (body size differences due to sex). Male endogamy led to the creation of male kin groups that provided communal defense of females and territory. Ghiglieri further surmises that males of the hominid ancestor formed stable and exclusive mating bonds with one or more females and made a greater investment in the care of offspring, which increased the odds of successful reproduction. ❧ Of course, such models that seek to characterize a common ancestor should be seen simply as guidelines to exploring questions about early hominid behavior. Our ancestors were each unique biologically, ecologically, and behaviorally, with specific adaptations that shaped behavior. Bipedalism, for instance, a form of locomotion not practiced habitually by any ape, would certainly have influenced the behavior of early hominids considerably in terms of how food was obtained and processed, what size territory was occupied or explored, and perhaps even provided the foundation for much of hominid mating and social strategy (see page 88).

QUADRUPEDALISM has served untold numbers of primate species for tens of millions of years. Natural selection, the process that molds living organisms, however, focused on upright, two-footed walking in our ancestry as the primal behavioral adaptation that launched our evolutionary journey. The

34 · Evidence for Bipedalism

major anatomical features associated with bipedalism are seen in the extensive reorganization of the trunk, pelvis, lower limb, and foot. The ability to walk upright reflects an interdependent set of alterations in the soft (muscle) anatomy and, most significantly for paleoanthropologists, in the bony anatomy. ❧ The pelvis of a bipedal hominid differs from all other primate pelves in having broad, vertically short blades that are rotated inward to form a pelvic basin, which supports the viscera in an upright creature. Such an arrangement repositions our large hip muscles to the side, making them useful in balancing our trunk over our lower limbs when walking. Enhanced balance also stems from repositioning the center of gravity within the basin-like pelvis, due in part to the S-curvature of our spine. ❧ The thighbone, or femur, is critical since it supports our entire body weight during locomotion. The relatively long, stout femur fits into the hip socket via a large head. The knee end, due to the oblique orientation of the femoral shaft, is positioned directly beneath the body, adding to the fluidness of our walk by diminishing side-to-side tilting. Features in the knee are designed to facilitate maximal weight transfer and to prevent kneecap dislocation. ❧ It is our foot that ultimately contacts the ground and has become especially well adapted to terrestrial bipedalism. Unlike all other primates, which have a grasping great toe, we have a great toe that is aligned with the other toes, is not opposable, and forms the focal point for forward propulsion-toe-off. Two arches formed on the sole of the foot by strong ligaments act as shock absorbers during locomotion. ❧ Considerable debate has focused on the capacity of the earliest species in our family tree, the australopithecines, for bipedality. Fortunately for the paleoanthropologist, the fossilized remains of our ancestors exhibit an extensive suite of skeletal landmarks that bespeak bipedalism. The skulls, teeth, and isolated skeletal fragments that until recently constituted most of the hominid finds were not able to show the anatomical arrangements of pelvis and lower limb that would resolve the issue of early bipedality, but one skull feature was almost as good. In hominids the foramen magnum, the hole at the base of the skull through which the spinal cord emerges, faces downward, allowing the skull to sit at the top of the vertebral column and spinal cord. In quadrupedal primates, by contrast, the foramen magnum is positioned toward the back of the skull. The position of the foramen magnum eventually proved diagnostic for bipedalism, and when Raymond Dart in 1925 described the skull of the Taung Child, from South Africa, and used it to erect a new taxon, *Australopithecus africanus*, he confidently asserted that it belonged to a bipedal hominid ancestor on the basis of the position of the foramen magnum. ❧ Although portions of the pelvis and lower limb are known for the South African australopithecines, the most comprehensive case for bipedalism is made by the extensive postcranial elements of East African *A. afarensis*. The pelvis and lower limb bones found at Hadar, Ethiopia, and dating to between 3.0 and 3.4 million years ago, provide more than ample evidence that *A. afarensis* was an accomplished biped. The bones of Lucy play a vital role here, but other lower limb bones from Hadar further attest to bipedal behavior in this early hominid. ❧ Restoration of

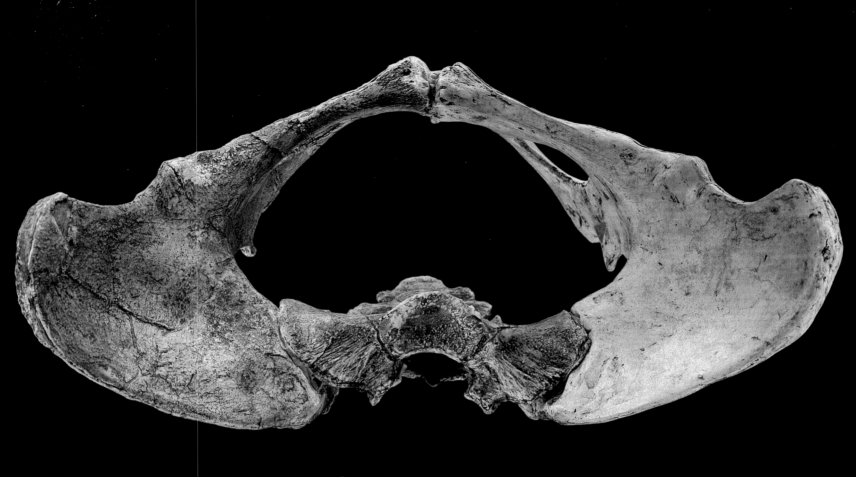

Pelvis of *Australopithecus afarensis* from Hadar, Ethiopia. The partial skeleton of A.L. 288-1, or Lucy, included the left half of her pelvis and a complete sacrum, the five fused vertebrate that make up the rear of the pelvis. The remarkably humanlike form of the pelvis reveals that Lucy had been bipedal. Actual size. *Photography by David L. Brill; courtesy Owen Lovejoy, Kent State University.*

Lucy's complete pelvis from an intact sacrum (tail bone) and left innominate (pelvic blade) revealed all of the major landmarks of a modern human bipedal pelvis. Her pelvis is short, with the blades rotated inward. Slight differences from a modern pelvis are to be expected, such as the flare of the blades, which may have added increased stability to the pelvis during locomotion. Adjacent to a relatively large femoral head, the femoral neck has thickened bone on its lower margin, to prevent breakage from the bending forces generated during walking. ❰ The shaft of Lucy's femur is oblique to the horizontal, an arrangement that causes weight to be transferred primarily through the lateral condyle, the outside knob at the bottom end of this bone. This condyle is also flattened, enhancing contact with the tibia (shin bone) for maximum weight transfer. The broad notch on the front of the femur has a raised ridge that prevents lateral displacement of the kneecap. ❰ Anatomical details of the ankle, especially the articulation of the tibia and the talus (ankle bone), support the bipedal nature of *A. afarensis*. A careful look at the articulation of the large toe with the main portion of the foot shows that this important joint was incapable of the rotation and divergence seen in an apelike grasping foot. ❰ The locomotor skeleton of *A. afarensis* is not identical with our own, an outcome we should expect because of the skeleton's antiquity. In fact, a number of primitive features, such as a short femur, slightly curved finger and foot bones, narrow rather than broad fingertips, a highly mobile wrist, powerful arms, and so on, are unlike those found in modern humans. The intriguing mosaic of specialized bipedal features and primitive features in *A. afarensis* is not in dispute. It is the interpretation of what this amalgam of primitive and advanced traits means for locomotor behavior that is open to debate. ❰ Some scientists stress the bipedal features of *A. afarensis* and conclude that this hominid did not participate in any arboreal activities but was a committed terrestrial biped. The primitive features are interpreted as evolutionary baggage left over from an arboreal ancestor. Other, equally powerful voices have spoken out for the probability that *A. afarensis* spent a considerable amount of time in the trees. Those scientists believe that the primitive features are consistent with the idea that *afarensis* was foraging for food, escaping from predators, or even sleeping at night in the trees. The critical anatomical adaptations, however, seen in the hip, knee, and foot for bipedalism suggest that climbing behavior was not adaptively important and that bipedalism had been under strong selection for some time. ❰ Substantial insights into the locomotor capabilities of *A. afarensis* come from the remarkable discovery of a 28-meter-long trail of 3.6 million-year-old hominid footprint impressions in a volcanic ash at Laetoli, Tanzania (see page 143). The discovery of this hominid trackway in 1978 permitted an unprecedented evaluation of the soft anatomy of a very ancient human ancestor. The hominid footprints found at Laetoli must have been made by members of *A. afarensis*, because the only fossil hominids recovered from Laetoli belong to this species. ❰ Painstaking excavations by Mary Leakey and her team revealed footprints astonishingly like those made by *Homo sapiens* in wet beach sand. The Laetoli prints show a strong heel strike, followed by transfer of weight to the outside of the foot, then across to the ball of the foot, and finally concentrated on the great toe. The great toe is not divergent; there is only a slight gap between it and the lateral toes, similar to that seen among people today who do not wear foot coverings. The impression left in the ancient volcanic ash also reveals an energy-absorbing arch to the foot. ❰ Although other hominid prints from cave sites in Europe date to the Upper Paleolithic, they are from our own species, *Homo sapiens*. The Laetoli prints, made by another hominid species, *A. afarensis*, provide conclusive evidence that hominid bipedality reaches back 3.6 million years in time and undoubtedly stretches even deeper into the past.

35 · The Origins of Bipedalism

EARLY in the twentieth century, substantial debate centered on the sequence of events in human evolution. One school was committed to the view that our brains grew big first and then we became bipedal, while another school saw bipedalism as a precursor to the big brain. To many, bipedalism was a locomotor response in our ancestors to a terrestrial habitat after they left the familiar arboreal one behind. Those who favored the brain first scenario speculated that human intelligence was a necessary precursor for making the decision to walk upright and move out of the forests onto the grasslands. In more elaborate forms, flight from the forest was consistent with the notion of expulsion from the Garden of Eden. Those who saw bipedalism as the initial adaptation claimed that it was the first step toward freeing our forelimbs to manufacture and use tools. This set up a classic feedback mechanism—bigger brains, better tools; better tools, bigger brains. ◀ Resolution of the sequence of events derives from the recovery of a significant storehouse of ancient fossil hominids in Africa. These fossils have provided definitive evidence for bipedalism extending back to roughly 4 million years ago, and a growing consensus postulates that the acquisition of bipedalism may have occurred somewhere between 5 and 8 million years ago. Substantial brain expansion is not seen until roughly 2 million years ago, thus firmly establishing the sequence of development for these two diagnostic features of human evolution. ◀ It is also necessary to abandon the view that our ancestors became bipedal to make and use tools, for early bipeds apparently did not manufacture stone tools. Lithic artifacts first appear in the geological record about 2.6 million years ago, perhaps 2 million years after our ancestors became bipedal. Hence the connection between big brains, tool use, and bipedalism has been effectively uncoupled. ◀ Any explanation for why we became bipedal must also take into consideration that, compared with quadrupedalism, bipedal locomotion is slow, clumsy, and fraught with opportunities for injury. And contrary to much that has been written, walking on two legs is not energetically more advantageous than getting around on four legs.

Some explanations point out that upright posture permits hominids to reach up and pick fruit from a tree or use their hands in special social displays. But nothing prevents a quadruped from doing these things as well, just as chimpanzees and gorillas do. ◀ An often quoted but overly simplistic explanation for bipedalism contends that it enabled our ancestors to stand up and see over the tall savanna grass when they left the forest. From a practical point of view, it is difficult to think of a more vulnerable time to try out a major behavioral change than when moving into an unfamiliar habitat, especially one containing successful predators. Imagine how easily large cats could have taken a slow, lumbering hominid that announced its presence by standing up. Chimps and baboons do look over tall grass to assess the surrounding area, but in case of danger they can resort to four legs and quickly escape. From this perspective, it makes more sense to develop a scenario in which we became bipedal while still in the familiar surroundings of the forest. ◀ Another line of thinking about early upright bipedalism has focused on bodily thermal regulation—the heat load sustained on the open savannas, and how evolution might have helped humans cope with it. The English physiologist Peter Wheeler has postulated that the upright stance of bipedalism would have reduced the amount of body surface area exposed to the sun's rays. An added benefit, he suggests, would have come from our bodies being elevated above the hot ground and exposed to cooling breezes. The underlying idea is that faster dissipation of the heat load would extend the daily foraging times for hominids. Other studies, however, have pointed out that the added foraging time would have been minimal and that hominids, like other animals on the savanna, would have derived greater benefit by resting during the hottest part of the day. Upright bipedalism does reduce the amount of incoming heat due to direct sun exposure, but reduction in the heat load was not the driving force for bipedalism. ◀ Bipedalism probably developed when our ancestors were still in the forest and, as they

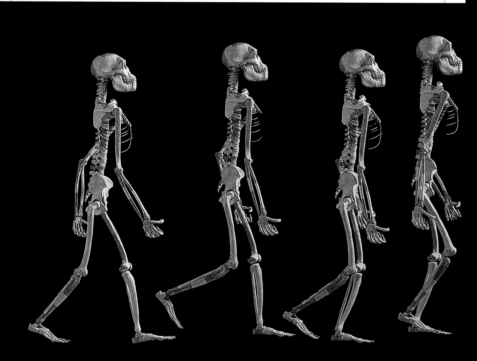

Articulated reconstruction of *Australopithecus afarensis*. The plaster skeleton created by anthropologist C. Owen Lovejoy and his students depicts this hominid as fully adapted to habitual bipedal locomotion. *Photograph by David L. Brill; courtesy C. Owen Lovejoy, Kent State University.*

became active in more open areas, gave them the additional advantage of heat dissipation. Our hairlessness, on the other hand, may be a direct result of living on the savanna because it substantially enhanced cooling through the evaporation of sweat. ◖ Bipedalism is a major evolutionary innovation, and separate from the issue of its numerous advantages for hominids, its source must be sought in behavioral change that enhanced the reproductive success of early hominids. One of the most comprehensive scenarios explaining bipedalism has been offered by Kent State University paleoanthropologist C. Owen Lovejoy. Lovejoy has placed his model within the framework of natural selection—upright walking was selected for in response to some behavioral advantage. Lovejoy has focused on what he calls the "fundamental selective triad," the three areas in which an organism consumes most energy: reproduction, feeding, and safety. The goal of any species is to stay alive, to eat, and to reproduce. Lovejoy points out that beyond a certain point, there is no advantage in expending more energy for feeding and safety, but additional energy spent on reproduction would have a significant evolutionary impact. If a female could dedicate more time to reproduction by reducing birth intervals and caring for more than one dependent offspring at a time, there would be an evolutionary payoff in terms of a greater likelihood that each generation would survive long enough to produce the next. But a faster reproduction rate would require that the female not have to expend as much energy foraging for food or worrying about the safety of herself and her offspring. ◖ One manner in which this could happen would be if males provided high-quality food to females, allowing the females to invest more energy in infant care. Nourishment provided by males would reduce females' foraging time and reduce their exposure to accidents and predation. Thus, the development of longer-term relationships between males and females, the beginnings of pair-bonding, could have had evolutionary significance for hominids. ◖ On the one hand is the notion that males, now bipedal, used their upper limbs for carrying food, to provision females who did not mate with other males, thus securing known paternity. On the other hand, females who selected mates based on their dependability in providing food would gain higher odds of increased survivorship of their young. Following upon these we can look at two possible consequences. ◖ Lovejoy notes that since human females lack external signs of ovulation (estrus) such as sexual swelling and olfactory clues, frequent copulation is necessary for conception to occur. Lovejoy suggests that males were attracted to females who did not display outward symptoms of ovulation, and as a result the males could have avoided competition with other males for those females in obvious estrus. ◖ Lovejoy points out that male and female humans are distinguished from one another by features that serve to attract the opposite sex. For example, males and females have distinctive scalp hair, distinctive voices, and specialized scent glands in their armpits and pubic region. Furthermore, males are some 20 percent larger than females, exhibit facial hair, and have a large penis, the largest among all primates. Human females have permanently enlarged breasts (chimps have enlarged breasts only during lactation) and a particular distribution of body fat. Perhaps penis and breast size were important sexual signaling devices that served to attract potential mates. ◖ An important feature of the unique mating structure seen in hominids is that each sex has something to offer the other. The male provides a reliable source of food as well as added protection for the female and the young. The female guarantees that the male's genes will be passed on to the next generation; and it is precisely the increased survivorship of the offspring in Lovejoy's model that is the key. Because both ape and human infants have extended periods of childhood learning and take a long time to mature, their reproductive rate is low. So, the opportunity to conceive and raise multiple overlapping offspring would have a significant impact on the reproductive rate of a particular species. ◖ In late Miocene times, 5 to 10 million years ago,

cooler climates brought about a diminution of the tropical forests, long the stronghold of the apes. The apes were forced to hang on in more widely dispersed clumps of forest. In contrast, the fast-breeding monkeys proliferated and even began to adapt to the savannas. Thus began the steady decline, which continues today, in the diversity and number of apes. The only ape that flourished was the hominid one, which developed a unique and successful breeding package, an important element of which was bipedalism.

A BONOBO named Kanzi, living at the Language Research Center in Atlanta, Georgia, was given an incentive to make stone tools. Fruit was placed in boxes that were tied shut with cord. After Kanzi was shown how

36 · The Oldest Stone Tools

the sharp edge of a stone flake could be used to cut the cord, he was offered a choice of several flakes for cutting open a box. He chose well, easily cutting the boxes open, and even began occasionally to strike one rock against another to produce his own flakes. ◖ This experiment was conducted to see what kinds of stone tools might be made by a chimpanzee, our closest living relative, and to compare those with the oldest known stone tools in the archaeological record. At first glance, there was an overall similarity between Kanzi's tools and the earliest stone tools, usually referred to as the Oldowan industrial complex because they were first recognized at Olduvai Gorge, in Tanzania. Closer inspection, however, by Indiana University archaeologist Nick Toth, who specialized in stone tool technology, indicated that Kanzi's work, even after several months, did not show the level of cognitive complexity that is seen in the Oldowan material. Kanzi's tools looked much like the broken stones that occur naturally in riverbeds, glacial deposits, and other geological situations where rocks might be haphazardly knocked about. These geofacts lack a distinct pattern of manufacture. Their random pattern of breakage and flaking is easily discerned by experienced archaeologists. ◖ What diagnostic features would allow us to recognize the earliest stone artifacts? Chimps are not alone in the experimental manufacture of stone tools; some archaeologists have become accomplished stone knappers and as a result have come to know a great deal about stone tool technology. Direct percussion—the striking of one rock, called a hammerstone, against another, called a core—produces a flake. The manufacture of high-quality flakes incorporates knowledge of where and at what angle to strike a rock. Even more fundamental is the selection of the raw material for tool manufacture. Fine-grained rocks like chert and obsidian produce finer tools than coarser-grained rocks like basalts. ◖ Flakes made through the purposeful behavior of a hominid show a number of features referred to as conchoidal fracture. This kind of fracture is normally absent from naturally cracked stones, which usually break along natural fractures. Artifact flakes bear a striking platform where the hammerstone knocked off a flake. The inner side of the flake exhibits a bulb of percussion immediately below the striking platform and a convex, rippled surface resulting from the shock waves moving through the stone from a hammerstone blow. The core from which the flake was struck bears a flake scar, the negative impression of the bulb of percussion on the flake. ◖ When Louis Leakey first visited Olduvai Gorge, in 1931, he observed an abundance of stone flakes and cores scattered throughout the gorge that showed signs of purposeful and repeated patterns of manufacture. This archaeological assemblage consisted of rock types such as lavas and quartzite that must have been transported in from several kilometers away. Moreover, in certain areas concentrations of these stone artifacts were associated with broken or butchered animal bones. ◖ The earliest stone tools from Olduvai come from the lowest strata at the site, which are close to 2 million years old. Other East African Rift Valley sites stretch further back into time, per-

Savaging technique. After a few deliberate blows with a hammerstone to crack this impala legbone and a quick twist, fat-rich marrow is available for consumption. *Courtesy of Donald Johanson, Institute of Human Origins.*

mitting exploration of the possibility of even earlier traces of stone tool technology. The Omo region of southern Ethiopia, immediately north of Kenya's Lake Turkana, has been investigated since the 1930s because of the fossil-rich strata known as the Shungura Formation. Thanks to the unique series of volcanic ashes, stone artifacts consisting of quartz pebbles associated with abundant flakes and chips have been reliably dated to 2.3 to 2.4 million years ago in the Shungura Formation. ◖ Oldowan stone tools are known from localities around Lake Turkana and are especially numerous at Koobi Fora, east of the lake, in deposits roughly 1.9 million years old. Considerably more ancient stone artifacts are known from the site of Lokalalei situated west of Lake Turkana where they are found in strata dated to 2.34 million years ago. Cores, flakes, and hammerstones, made predominately of lavas, were excavated *in situ.* Researchers were able to refit flakes onto the original core from which they were struck, indicating that the stone tools were made on the spot. ◖ Many claims have been made for the "oldest" stone tools, but the best documented and most firmly dated ones, thanks to Argon dating, derive from 2.6 million-year-old geological horizons at Gona, Ethiopia, situated immediately west of Hadar. Initially discovered in the late 1970s, recent excavations have recovered numerous Oldowan-type tools consisting of flakes and cores made from chert and volcanic rock. There are also some indications of cut marked bone, presumably processed using the stone tools. ◖ Slightly younger artifacts, although not in the concentrations known at Gona, have been recovered from the site of Bouri (see page 131). These 2.5 million-year-old stone tools appear associated with large mammalian bones bearing cut marks and hammerstone impacts, a result of food processing. ◖ A.L. 666, a locality at Hadar, has produced a very large number of flakes and cores fashioned from volcanic raw materials. The ability to refit flakes onto cores attests again to the notion that tool making occurred precisely at the place of excavation, some 2.33 million years ago. The association of a *Homo* maxilla with the stone tools at A.L. 666 provides the oldest known association of stone tools and a hominid.

Oldowan flake tool from Hadar, Ethiopia. The A.L. 666 site has yielded a fossil upper jaw of *Homo* as well as several stone cores and flakes. With an age of 2.3 million years, these are the oldest known stone tools in direct association with a hominid fossil. Actual size. *Courtesy of Donald Johanson, Institute of Human Origins.*

◖ The overall evidence from East Africa suggests that by roughly 2.6 million years ago, the Oldowan industry tool tradition was well established. Perhaps even older stone tools will be found, but unless their makers returned to a specific spot on the landscape where concentrations of artifacts would be obvious, these sites may be difficult to find. The use of naturally occurring rocks to break open nuts, a behavior that

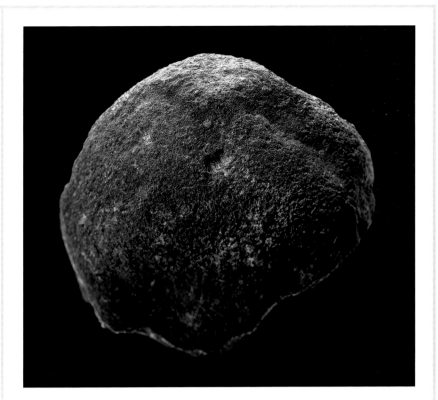

Hammerstone from Olduvai Gorge, Tanzania. This 1.8 million-year-old artifact may closely resemble the earliest type of stone technology. A rounded, unmodified cobble can be used to strike sharp flakes from another rock, creating an effective cutting tool. Actual size. *Photograph by David L. Brill; courtesy of National Museum of Tanzania.*

has been observed in chimpanzees in West Africa (first reported by Darwin in 1871), probably extends far back into our past, but recognition of such behavior from the archaeological record may be impossible. ◖ Although there is no question that the stone tools of the Oldowan industrial complex were fashioned by a hominid hand, there is equally no consensus as to who made the tools. While australopithecines existed throughout the range of time during which Oldowan tools have been found—roughly between 2.6 and 1.5 million years ago—most paleoanthropologists credit the larger-brained *Homo habilis* or *Homo rudolfensis.* ◖ A strong case for *Homo* as the toolmaker is the simple observation that stone tools do not become part of the geological record until *Homo* appears on the scene. Even more convincing is the fact that tools of the Acheulean industrial tradition, mostly hand axes, very quickly replaced the Oldowan tools about 1.5 million years ago, just after the appearance of *H. ergaster.* If robust australopithecines had been making Oldowan tools, one would have expected the tradition to have continued until the extinction of australopithecines, around 1 million years ago, but this was not the case.

OUR earliest ancestors' subsistence behavior is the focus of tremendous attention, but because behavior does not fossilize, teasing details out of the archaeological record has proved arduous. Some attempts to understand

37 · Hunters, Gatherers, or Scavengers?

early hominid lifeways have involved nothing more than projecting into the past an impoverished version of modern hunter-gatherers' lifeways. Although hunter-gatherers provide a comparative base, we must not lose sight of the fact that early hominids undoubtedly had their own unique and highly successful set of lifeways that were probably behaviorally distinct from any living model. The most pervasive single explanation for the survival strategy of our ancestors has been the "hunting hypothesis." Hunting, as understood by predominantly male anthropologists, seemed to be the key activity, one that explained nearly every aspect of humanity. Hunting took cognitive skills, prompting brain expansion; it took cooperation, possible only in more complex societies; and perhaps it even took language to plan and conduct the hunt. ◀ The late archaeologist Glynn Isaac erected a larger framework of behavior that still embraced hunting as a core activity. The more elaborate scenario became known as the "home base hypothesis" and could, Isaac believed, explain the emergence of humanness. Hunting depended on cooperation and planning, which demanded enhanced intelligence and communication skills. Food was brought to a specific place, a home base, where it was shared. This served to establish strong social bonds and stimulated reciprocal behavior, a sort of economic interdependence between males and females based on a division of labor. Males hunted for food and protected the group, females collected plant foods and cared for the children. Support for this hypothesis was sought in the excavations at Olduvai Gorge, Tanzania, where numerous stone tools were found associated with broken-up animal bones, some of which even showed stone tool cut marks. At this locality, known as FLK I, dated to roughly 1.8 million years, remains of *Homo habilis* were also recovered. The association of tools, bones, and hominids was taken as proof of the home base hypothesis, and FLK I was designated a home base where our ancestors left remains of their meals. ◀ Lewis Binford, an archaeologist at Southern Methodist University, has objected strenuously to interpreting the assemblage at FLK I as evidence for hunting by early hominids. If we look into the past with a preconceived notion, such as hunting, the remains at Olduvai will appear to offer proof of that notion. Binford stressed that analogy with modern hunter-gatherer societies grossly biases interpretation of the actual archaeological evidence by projecting what we know about the present into the past when ancient behavior might have been quite different. All that can be inferred from FLK I, according to Binford, is that early hominids collected stones and used them to break open bones. He proposed that at best, our ancestors at FLK I were processing bones they had scavenged from carnivore kills. In other words, our early ancestors were scavengers, not hunters. ◀ To test Binford's theory, a critical study was undertaken in the 1980s by Rob Blumenschine, an archaeologist at Rutgers University, who studied the composition of modern bone accumulations on the Serengeti. He observed numerous feedings by lions, hyenas, cheetahs, and jackals and noted that after these predators had dined on the flesh and crunched up the ribs, the leftovers consisted mostly of limb bones and the skull. He further noticed that roughly 15 percent of those bones showed carnivore tooth marks. ◀ Blumenschine collected the bones left behind by the carnivores and, using an unmodified stone cobble, broke open the bone to extract the marrow, carefully noting the pattern of bone breakage. He then compared his observations on the modern sample to the evidence from FLK I and concluded that the Olduvai bone collection, like the ones on the Serengeti, was probably the result of scavenging behavior. The Olduvai assemblage consisted mostly of limb bones, many of which showed carnivore tooth marks and stone tool damage. The implication was that our ancestors scavenged carnivore kills made by lions, leopards, saber-toothed cats, and even hyenas, collecting predominantly long limb bones and breaking them open to benefit from the bone marrow, a fatty, high-energy food source (see page 182). Most carnivores, including hyenas, would have been unable to crack the thick-walled bones, but a hominid with a simple stone tool would have had no problem doing so. The emerging view of "man the scavenger" did not have quite the heroic proportions of "man the hunter." *Homo habilis* may have benefited from bone marrow, but they were doubtless also fairly flexible in their subsistence behaviors, relying on different food resources for which they developed special foraging strategies. This was especially true for seasonally available vegetable food sources. We can also assume that because hominids must drink every day, our ancestors remained close to permanent sources of water.

Cut marks on a bone from Olduvai Gorge, Tanzania. These parallel incisions were made by a sharp-edged stone flake in the hand of a hominid. *Photograph by David L. Brill; courtesy of National Museum of Tanzania.*

◀ Blumenschine's work further suggests that scavenging would have put an evolutionary premium on intelligence. To be successful at scavenging, early hominids would have had to detect and locate carcasses. They needed to know quite a bit about the behavior of predators—where they hunt, and when. Most important, they needed to know when not to interfere with predators that might have turned on them for a meal. Perhaps some of the skills acquired during scavenging were a prelude to including more meat in the hominid diet. ◀ The time when hunting did become a major aspect of hominid subsistence behavior continues to be the center of much debate in paleoanthropology. Many archaeologists contend that hunting is an ancient human subsistence strategy going back hundreds of thousands of years, perhaps even a million years, to the time of *Homo erectus*. Often cited are the associated remains of large mammals like elephants with hand axes and cleavers. Lewis Binford believes that hunting large animals did not become an important element in human subsistence behavior until the Upper Paleolithic, beginning some 40,000 years ago. The discovery, however, in old coal deposits in Schöningen, Germany of three complete javelin-like spears provides plausible evidence that early inhabitants of northern Europe were big game hunters some 500,000 years ago. The 1.8-2.25 meter-long spears are made from the trunks of spruce trees and are associated with stone tools and butchered remains of thousands of horse bones. Manufacture of these spears demanded great technological sophistication and planning. A perforated rhinoceros's scapula at Boxgrove, England dated to the same time may also indicate use of spears in hunting.

IN THE the United States, the struggle against obesity supports a billion-dollar dieting industry. The same was not true for our ancestors, who struggled to get enough of the right kinds of food to survive. To

38 · Diet

reproduce—the Darwinian goal of the individual and the species—an animal must stay alive, which means it must eat. Our ancestors were successful in getting enough to eat, for we are here today. What remains conjectural is the type of diet they had and how the foodstuffs were processed. ◀ The first discoveries of australopithecines in South Africa immediately elicited some imaginative speculation about their dietary preferences. For example, *Australopithecus africanus* was quickly pictured as a carnivorous, bloodthirsty ape. The distinctive cranial and dental anatomy of *A. robustus* was thought to be a response to an herbivorous diet that demanded extensive mastication of low-quality vegetable foods. These early inferences about the diet of australopithecines rested on a combination of preconceived ideas about the importance of meat eating to early humans and observations of gross

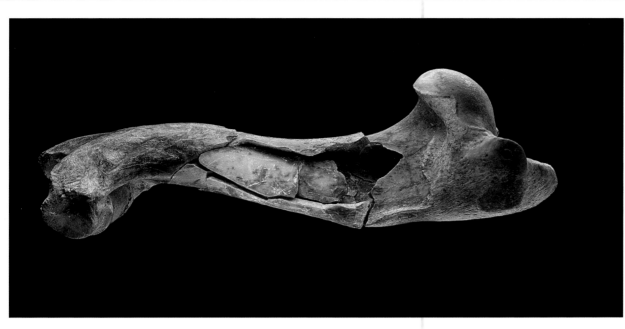

Antelope humerus from Olduvai Gorge, Tanzania. Its shaft shattered to expose the marrow within, this bone from the FLK locality displays both stone tool cut marks and carnivore tooth marks. The array of tools and bones from this site has been used to construct differing scenarios of hunting or scavenging by early humans nearly 2 million years ago. Actual size. *Photograph by David L. Brill; courtesy of National Museum of Tanzania.*

dental wear and anatomy. More recently a number of strategies have been employed to more precisely assess the diets of early hominids. Most of these strategies have been applied to teeth, which are shaped differently to accommodate different diets and also bear the signatures of the types of food consumed. ◀ Comparisons of dental anatomy between groups of animals that subsist on varying types of foods provide some insight into specific dietary adaptations. For example, carnivores, which are specialized meat eaters, have a set of scissors-like back teeth that are used to cut through flesh. Grass eaters, such as horses, have tall teeth with chewing surfaces composed of alternating strips of enamel and dentine. Primates, by contrast, are in large part generalized in their feeding adaptations and therefore lack the specialized dental configurations seen in grazers and carnivores. ◀ Within the order Primates, hominids have teeth that are remarkably generalized, suggesting a wide-ranging omnivorous diet. Unlike, however, the African apes, which have thin dental enamel, hominid teeth have thick enamel. It is thought that thick enamel is the primitive condition—that is, the condition that has not been altered through evolutionary change. This conjecture is in keeping with the different evolutionary histories of apes and hominids. Apes, with a more complex evolutionary history, would have had opportunity to evolve a thin enamel, whereas hominids, whose evolutionary history is sparser in its branching, may have retained the thick enamel from the last common ancestors of apes and humans. ◀ Body size also bears a relationship to dietary preference. For example, insectivores, with a high metabolic rate, expend a lot of energy capturing their prey and tend to be small-bodied. Leaf eaters, with lower metabolic rates (the gorilla is an extreme example here), have larger bodies. Interestingly, frugivores, or fruit eaters, tend to have relatively larger brains than folivores, or leaf eaters. Presumably, because fruits are seasonal and more regionally distributed than leaves, fruit eaters would need larger brains to process the more complex seasonal and geographic information about their environment. ◀ Another line of evidence comes from strontium-calcium ratios in hominid remains. Within the food web there are distinct differences between strontium-calcium ratios in the bodies of different species, which ingest these mineral isotopes with their food and incorporate them into their bones. Plants, a primary source on the food chain, have a high ratio, herbivores a lower one, and carnivores,

which eat the herbivores, the lowest ratio. Attempts have been made to determine the value of these ratios in early hominid fossils as an aid to determining their food preferences. Although the effects of fossilization and burial may limit measurement of the strontium-calcium ratio, some interesting results have been obtained for *A. robustus*. Ratios determined for hominids from Swartkrans fall between those of leopards and baboons and suggest a more omnivorous diet rather than the strictly vegetarian one that was first supposed. These considerations paint in broad strokes the dietary preferences of early hominids, namely, they were opportunistic omnivores. Detailed examinations of the chewing surfaces of teeth, sometimes with the aid of a scanning electron microscope, has refined the picture by permitting quantitative analysis of dental microwear. A very high incidence of enamel pitting is documented in the teeth of *A. robustus* and *A. boisei*. Moreover, the pits and scratches on these teeth are significantly larger than those seen on *A. africanus* teeth. The frequency and size of enamel pits in *A. robustus* are reminiscent of the condition seen in mangabeys, organ-grinder monkeys, and in the orangutan, which consume very hard foods such as nuts, seeds, and fruits with tough rinds. The wear on occlusal or the biting or grining surfaces of *A. africanus* teeth, characterized by fewer pits and narrower scratches, suggests a different dietary preference. Similarities in enamel wear to patterns seen in primates like the spider monkey, which eats more mature, fleshy, softer fruits as well as leaves, suggest that *A. africanus* probably consumed less nuts and seeds than *A. robustus* or *A. boisei*. ◀ Dental wear in the South African australopithecines contrasts strongly with that seen in East African *A. afarensis*. In the South African *A. africanus* and the robust australopithecines, teeth were quickly worn flat after eruption, exposing large islands of dentine. In fact, the canine teeth are so heavily worn that they must have participated in some of the crushing and grinding functions of the postcanine teeth. In *A. afarensis*, not only do the canines exhibit limited occlusal wear, the posterior teeth show minimal wear even when the incisors are heavily worn. This suggests an emphasis on the anterior teeth for food preparation, perhaps of soft fruits, as is the case in chimpanzees. Microscopic examination reveals dental chippage in *A. afarensis*, probably caused by biting down on hard, small food objects like nuts. Close examination of the edges of the incisors show striations from front to back. It may be that some form of vegetation was stripped between the incisors in *A. afarensis*. ◀ Meat consumption in early humans (pre-archaeological hominids) must have been very limited. Dentally they did not have slicing and cutting teeth necessary for eating meat, and they lacked the necessary stone tools for processing it. Chimpanzees have been observed eating the meat of monkeys, small bush pigs, small antelopes, and even the young of their own kind. Chimpanzees are successful hunters in large part because of their strength and speed in the trees. Australopithecines were bipedal, relatively slow on the ground, and limited in their ability to pursue potential game into the trees because they lacked a grasping foot. ◀ It is highly likely that early hominids incorporated a wide variety of foodstuffs into their diets. Yet many items that may have been eaten, such as insects, which at times are eaten in great quantity by chimpanzees, would not have left any diagnostic signature on the very hard enamel. This would also be true for honey, eggs, worms, and small vertebrates such as rodents and reptiles. ❡ At the moment, details of dental wear, and by inference their diets, in more ancient hominids such as *A. anamensis* and *Ardipithecus ramidus* are not available. The molars of the former taxon are thick-enameled and, from photographs, appear to have wear similar to that seen in *A. afarensis*, except that canines tips show heavy wear. The condition in *Ard. ramidus* is somewhat distinct, since it is the only putative hominid that has thin dental enamel, a feature more closely shared with the African apes. ❡ Conjectures about early hominid diet rely on inferences from dental evidence, studies of modern primate diets, remains of animal bones and associated tools at archaeological sites, and information on what modern hunter-gatherers eat. Although the exact details of our ancestors' diets may never be known, they clearly developed effective strategies for feeding, and equally effective strategies for not being eaten themselves.

CHARGES of cannibalism among our ancestors have been made at different times throughout the history of paleoanthropology, with an accusing finger pointed at several species. From our

39 · Cannibalism

modern, culturally civilized perspective, we tend to view cannibalism as an inhuman—perhaps the most inhuman—of acts, repellent and repugnant. Yet, if there is evidence for cannibalism deep in our past, we need to ponder it objectively. It sheds a light, however harsh, on one aspect of human nature. ❡ Among non-human primates, fifteen species have been documented engaging in cannibalism, and the behavior is fairly widespread among mammals. Males and females of our close ape relative, the chimpanzee, cannibalize infants. But without the benefits of direct observation, we have only enigmatic evidence from which to infer such behavior on the part of our ancestors. ❡ For instance, a bashed and broken adolescent *Australopithecus africanus* lower jaw, missing its front teeth, that was found at Makapansgat, South Africa, in 1948 prompted anatomist Raymond Dart to investigate this species' apparently violent nature. Dart declared: "*Australopithecus* lived a grim life. He ruthlessly killed other australopithecines and fed upon them as he would upon any other beast, young or old." But the pattern of broken bones examined by Dart proved on later analysis to have been made by hyenas and other carnivores seeking fat-rich marrow, not by these predominantly herbivorous hominids. ❡ When Franz Weidenreich studied the *Homo erectus* fossils from Zhoukoudian, China, the cave home of Peking Man, in the 1930s, he noticed that there seemed to be too many skulls for the number of limb bones. The faces on some of the skulls had been broken off and the foramen magnum, the hole at the base of the skull through which the spinal cord passes, had been enlarged, as though someone or something had torn open the skull to consume the brain. The cave also preserved evidence of controlled fire, and many bones had been burned, so Weidenreich suggested that Peking Man had been a cannibal. Louis Leakey provided his own interpretation of Zhoukoudian, with Peking Man as the meal of modern humans who, he believed, lived at the site simultaneously. But, as at Makapansgat, it seems that hyenas were the culprits here, too. ❡ These examples illustrate the critical need to distinguish between patterns of bone breakage made by carnivores or other animals and those made by humans. Each leaves a distinctive signature on bone for the careful investigator to find. If it can be determined that early humans were breaking open human bones, it doesn't necessarily mean that cannibalism occurred. Cut marks may indicate ritual defleshing, such as scalping, and broken or burned bones may be part of a mortuary practice. What fossil or archaeological evidence could reveal whether early humans ate other humans? ❡ A cranium from Bodo, Ethiopia (see page 208), belonging to *Homo heidelbergensis* displays diagnostic stone tool cut

marks around the eye sockets, on the cheekbone, on the forehead, and on the top and back of the cranium. It shows no sign of chewing by carnivores. Because the bone had not begun to heal around the grooves sliced into it and because the marks pass under the rock matrix sticking to the skull, the cut marks were made either while this hominid lived or just after he died. Around 600,000 years ago, this individual was intentionally defleshed, in the earliest such incident known. But whether the butcher ate any of this flesh cannot be answered. ❡ A proposed case for early cannibalism comes from Gran Dolina, Spain, where ancient remains of *H. antecessor*, about 780,000 years old, have been recovered (see page 218). The excavators noticed striations on a skull fragment of hominid temporal bone and later found similar markings on a pair of toe bones. An examination of casts of the striations under a scanning electron microscope revealed the telltale V-shaped cross-section of stone tool cut marks like those observed on the Bodo specimen. The dozen cut marks on the temporal fragment occur on the mastoid crest, an attachment point for the sternocleidomastoid muscle, indicating that flesh had likely been removed from the bone. Other animal bones at Gran Dolina have identical defleshing and dismembering marks as well as impact fractures from hammerstones. In contrast, carnivore tooth marks occur on only a small percentage of bones from the cave, so it appears that hominids were mainly responsible for butchering these bones. ❡ Some specimens of European Neandertals provided contradictory conclusions about cannibalism in this species. Two specimens that have been cited as victims or cannibalism or ritualized defleshing are the Circeo I cranium from Grotta Guattari, Italy, and the child's cranium from Engis, Belgium. (Although not recognized on its discovery in 1829, the Engis specimen was in fact the first Neandertal, and thus the first hominid fossil, ever found.) Circeo I was missing the base of the skull when it was found in 1939, lying upside down in an alleged circle of stones. The story spread that this Neandertal's brain had been consumed in a cannibal feast. But the skull bears no cut marks, percussion fractures, or other indications of human induced damage. It does, however, have several marks made by gnawing teeth. And the cave in which it lay was an ancient hyena den. ❡ As for Engis, the suspected cut marks on this cranium were made by humans, but not by any intending to eat the child. The marks, also seen on other Belgian Neandertal skulls, correspond to cranial tracing and measurement tools, the striations left by the sandpaper that was used to smooth the plaster during the reconstruction of missing bone, and scalpel damage inflicted as the molds were being removed to make casts of the fossil. They were all made by overeager nineteenth-century anthropologists. ❡ Recent analysis of bones from the Croatian cave site of Krapina provides a more convincing case for cannibalism (see page 225). The skeletons of several Neandertals were shattered and scattered throughout the site. Collected between 1895 and 1905, unfortunately without modern archaeological techniques, the 1,000 human bones and 1,000 tools tell a chilling story. Some 800 of the bones reveal cut marks and fractures made by hammerstones, while none show any carnivore gnawing. The sides of the skulls were smashed in, and many meat-rich bones are missing, while fragile bones containing little nutritious marrow, such as the fibula of the lower leg, remain intact. It's a perplexing pattern, but Karl Gorjanovic-Kramberger, the excavator of Krapina, may have interpreted it correctly when he wrote in 1918: "These men ate their fellow tribesmen, and what's more, they cracked open the hollow bones and sucked out the marrow. . . ." A similar story may be made for the nearby Neandertal cave site of Vindija, where the skulls were also smashed in and upper arm bones bear cut marks. ❡ If Neandertals occasionally resorted to cannibalism, perhaps to avoid starvation, did this behavior distinguish them from modern humans? Apparently not. A case in point is the number of *H. sapiens* fossils from Klasies River Mouth, South Africa, including a handful of lower jaw fragments, a dozen cranial

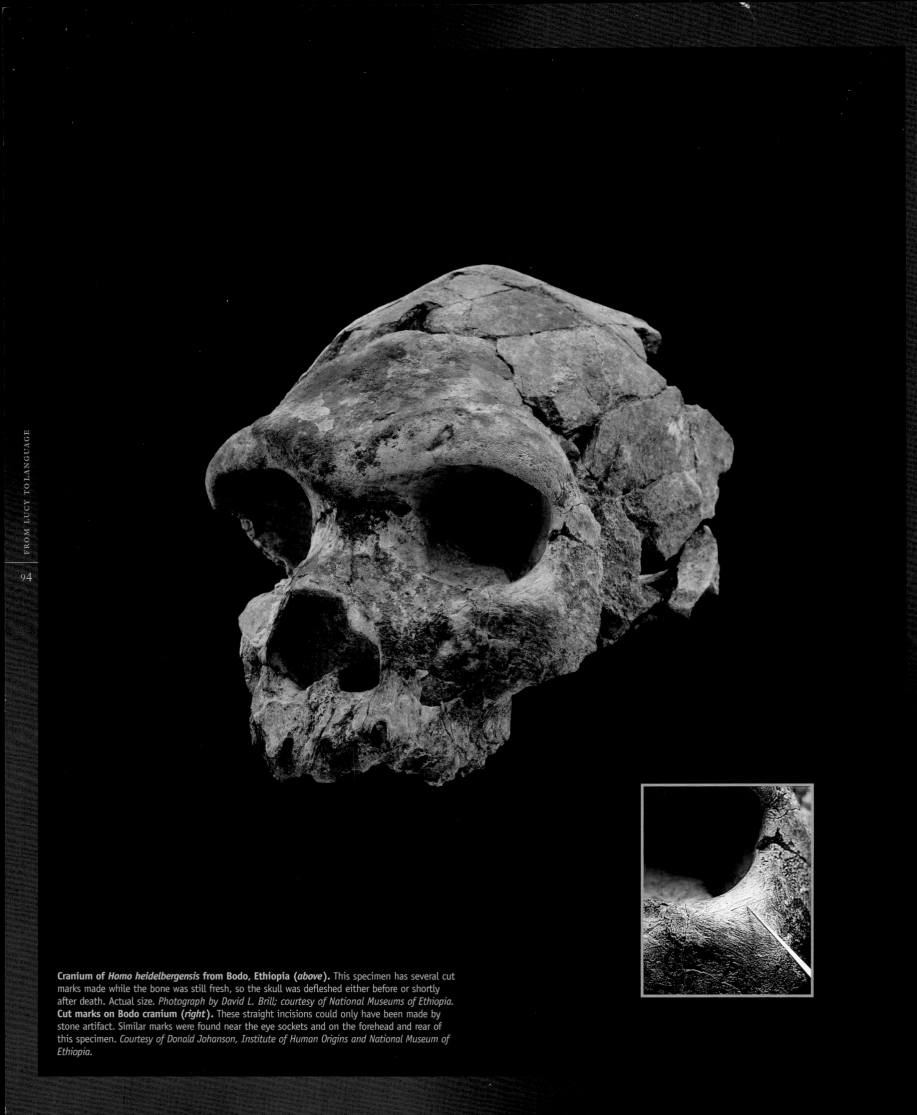

Cranium of *Homo heidelbergensis* from Bodo, Ethiopia (*above*). This specimen has several cut marks made while the bone was still fresh, so the skull was defleshed either before or shortly after death. Actual size. *Photograph by David L. Brill; courtesy of National Museums of Ethiopia.* **Cut marks on Bodo cranium (*right*).** These straight incisions could only have been made by stone artifact. Similar marks were found near the eye sockets and on the forehead and rear of this specimen. *Courtesy of Donald Johanson, Institute of Human Origins and National Museum of Ethiopia.*

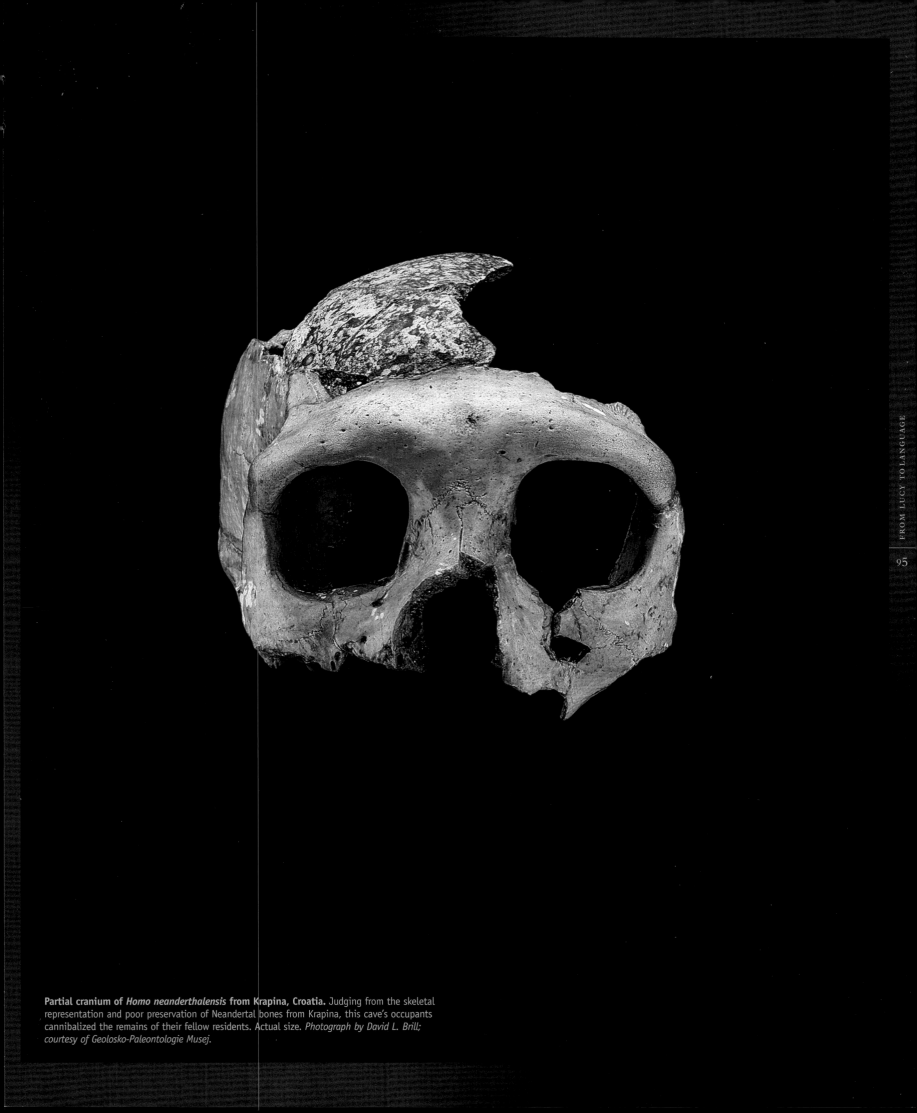

Partial cranium of *Homo neanderthalensis* from Krapina, Croatia. Judging from the skeletal representation and poor preservation of Neandertal bones from Krapina, this cave's occupants cannibalized the remains of their fellow residents. Actual size. *Photograph by David L. Brill; courtesy of Geolosko-Paleontologie Musej.*

fragments, and assorted bits of bones (see page 42). Many of the bones were burned, presumably while resting in one of the fire pits used by the early humans here. A piece of frontal bone from a skull shows the same distinctive cut marks above the eye socket as on the Bodo skull, and evidence that the bone was broken apart while still fresh. The 155,000 year old *H. sapiens* crania from Herto show numerous cutmarks inflicted by a stone tool that was used to deflesh the specimens shortly after death (see page 42). Furthermore, a child's cranium from the same site shows a great deal of polishing as if it were handled repeatedly, perhaps as part of some mortuary ritual. It is not possible to determine if the brains or flesh of these individuals were eaten. ◀ A much more recent case of prehistoric cannibalism comes from the Neolithic cave site of Fontbregoua, France, which was excavated in the 1980s. About 6,000 years ago, broken bones from at least thirteen humans were disposed of in three piles. In ten separate piles were bones from domestic sheep and wild animals that had been broken and butchered in the same way, with cut marks in similar places and frequencies. The bones from animals consumed for food and those of humans were treated identically. Certain bones, such as meaty limb bone shafts and the braincase, are conspicuously absent from both human and animal piles or show up only as discarded scraps. The bones show no sign of having been cooked, but all other evidence indicates that filleted human meat and extracted marrow were consumed. ◀ Since cannibalism is such a controversial issue, anthropologists need to examine the evidence for it with care and caution. A set of criteria can be established, including cut marks, broken bones, the presence or absence of certain nutritiously valuable bones, and burning or other signs of cooking, as contextual clues that should be present in some combination for one to infer that human bones were being processed for nutrients. Even if careful analysis leads to the conclusion that humans ate their cohorts on occasion in the past, we will probably never know exactly why.

HUMANS became adept manipulators of the environment in the course of our evolution, and one of the key steps in that process was the ability

40 · Fire

first to control, and then create fire. As a source of heat and light, fire provided protection from predators and shelter from inclement weather. Eventually, fire expanded the diet of early humans. Seeds and plant foods could be heated to break down toxins. Meat could be roasted to kill parasites and preserve the spoils of hunting or scavenging forays. Mastery of fire ultimately enabled our species to occupy the coldest reaches of the continents and may even have allowed humans to exploit a previously unused food supply by the thawing of frozen animal carcasses. Certainly, fire permitted meat to be smoked and dried to eat when other foods could not be had in harsh northern regions. For our ancestors, fire was the best thing since stone tools (sliced bread was still many millennia away). ◀ Despite its obvious importance as an implement of human culture, the origin of fire has been difficult to trace. As archaeologists dig deeper into time, the evidence for fire becomes both scarcer and more ambiguous, and more time elapsed in which to erase what traces may have once existed. Typically, *Homo erectus* is credited with the first use of fire, in part because the earliest evidence falls within the time period when this species lived and because fire was considered essential for the widespread dispersal of *Homo erectus* outside of Africa. ◀ Claims for early human fire at around 1.4 million years ago at Chesowanja, Kenya, were based on the discovery of about 40 flecks and lumps of burned clay at a site with animal bones and lava tools similar to the oldest stone tools from Olduvai Gorge and Koobi Fora. The excavators ruled out a natural brushfire as an explanation, although other observations suggest that tree stumps consumed by African bush fires could create a pattern like that found at Chesowanja. Lightning and volcanic heating could be alternative explanations for the

Burned bone from Swartkrans, South Africa. A total of 270 burned bones excavated from Member 3 of this cave provides direct evidence that early humans tended fire nearly a million years ago. Actual sizes. *Courtesy of C.K. Brain, Transvaal Museum.*

baked clay. ◀ Burned wood and grass seeds from Gesher Benot Ya'aqov, Israel indicate that fire was used some 790,000 years ago. It is unclear, however, what hominid species was associated with the site and whether the use of fire indicates a controlled or opportunistic use of naturally occurring fire. ◀ More compelling evidence for ancient fire comes from the South African cave site of Swartkrans where the youngest layers contain a small percentage of burned bones of antelope and zebra (and even a few of *Australopithecus robustus*). The few hundred burned bones, darkened by carbon residue, came from the same layer of sediment in a single corner of the cave deposit. Microscopic comparisons with bones heated experimentally to a series of different temperatures determined that the fossils had been subjected to prolonged temperatures of between 600 and 900 degrees Fahrenheit or even hotter, such as would have occurred in a campfire made from the local firewood. Based on the relative abundance of hominid bones throughout the site, an australopithecine may have made the fires at Swartkrans, but *Homo* also appears in the deposits (see page 198), so it is reasonable to conclude that the new hominid in the area, *Homo*, was responsible. ◀ Although the evidence from Swartkrans is intriguing, it is still an isolated occurrence.

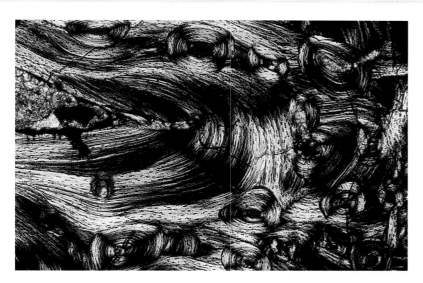

Photomicrograph of burned fossilized bone. In polarized light this thin-section through one of the Swartkrans burned bones reveals how the lamellar structure has been darkened by free carbon that formed while the fresh bone was heated. The same pattern occurred in experimental bones heated to between 300 and 400 degrees Celsius, a typical temperature in a campfire. *Courtesy of C.K. Brain, Transvaal Museum.*

Perhaps fire had been controlled by hominids but not created, at least not as an everyday event. Further archaeological evidence for fire does not appear until around 500,000 years ago. At the site of Zhoukoudian, China, where the famous Peking Man remains (see page 202) were found, *Homo erectus* apparently controlled fire throughout a long period of occupancy during a cold, temperate climate similar to northern China's today. Four ash layers at the site range in thickness from about 1 to 6 meters and date to periods between 460,000 and 230,000 years ago. Charred Chinese hackberry seeds also occur in the cave deposits, along with possibly burned bones, artifacts, and charcoal, although some archaeologists question the presence of actual hearths at this site. And it is structures like hearths that ultimately provide unequivocal evidence for fire later in the archaeological record. ◖ In Europe, for example, evidence for such a fire pit was excavated at Menez-Dregan on the southern coast of Brittany. Here, a thick concentration of charcoal and burnt bones is enclosed by a ring of flat stone blocks. This has been interpreted as a hearth where the occupants may have roasted rhino meat over an open flame. Electron spin resonance dating of burnt quartz places the age of the site between 380,000 and 465,000 years. If the date holds up, this could become the earliest evidence for controlled fire. To the south, on the French Riviera coast, the open-air site of Terra Amata contained hearthlike depressions that were lined on one side with rocks as if to make a windbreak for a fire that burned sometime between 400,000 and 200,000 years ago. Charcoal was found throughout the deposits at the site, and several burned mussel shells were recovered. This site has been interpreted as having several huts, and fires may have burned within each of them. ◖ Hearths did not become common until the Middle Paleolithic, or within the past 100,000 years. By that time, humans had mastered fire, and it must have achieved its important social role of bringing families and groups together and facilitating the exchange of ideas and information. Human use of fire in the Middle Paleolithic can be demonstrated in Africa, Europe, and the Near East. The Israeli cave of Kebara, for instance, was inhabited by Neandertals at least 60,000 years ago, and careful excavation revealed what the excavators called an impressive accumulation of flat and bowl-shaped hearths. The base of the hearths contained blackened, organic-rich silt with charcoal from local oak and carbonized seeds of wild peas. In addition to parching peas, the occupants of Kebara may have been roasting meat, judging from burned bone fragments found in the cave. After fires were put out, the ashes may have spread to warm the ground on which the Neandertals slept. ◖ By the Upper Paleolithic, around 40,000 years ago, fire was in common use for heat-

ing caves and dwellings. At sites such as Abri Pataud in Les Eyzies, France, hearths incorporated draft channels and warming stones for more efficient cooking. ◖ Whether or not fire first crept into our culture around a million years ago, it certainly became a force in human evolution within the past 500,000 years. Use of fire certainly became more frequent within the past 200,000 years, and after 100,000 years ago fire began to serve several specialized functions in daily life. As a complement to our evolving paleolithic technology, fire was a critical cultural tool for our survival and spread as a species, and most importantly, for our social development. The fireside provided the place to share food and share thoughts, a role that it continues to fulfill on a winter's night.

41 · Shelter

THE IDEA of a home or a shelter where we retreat for peace, protection, or privacy is an essential part of our daily lives. Few of us can conceive of being homeless, of having no dwelling to return to after a journey of a single day or one of many months. Yet our ancestors lived without building themselves shelters for most of our evolutionary history. Although structures have been inferred from evidence found at a few sites with associated Acheulean tools—such as pavements of closely packed stones at Olorgesailie, Kenya, and Isimila, Tanzania, and a stone wall supporting a possible lean-to shelter against the cave wall at Lazaret, France, atop a stone—only in the later part of the Middle Paleolithic, around 60,000 years ago, does archaeological evidence for the construction of shelters really become apparent. Even then, the evidence remains sparse and enigmatic. For instance, a single posthole was discovered from a Mousterian level at the long occupied cave of Combe Grenal, France, and excavators made a cast of the 21 centimeter-long wooden stake that once filled it. Nothing can be concluded, though, about what sort of structure this artifact supported or secured. Low stone walls have been reported from the sites of Cueva Morin, Spain, and Pech de l'Aze, France. In La Grotte du Renne at Arcy-sur-Cure, France, a stone wall and 11 postholes delineated the perimeter of an apparent hut covering more than 9 square meters. The postholes once held up mammoth tusks and limb bones, which may have been the supports for a roof. ◖ Despite these few examples, Middle Paleolithic people did not alter their environments to the extent that their Upper Paleolithic successors did, and that includes building unambiguous shelters at cave and open-air sites, often inferred from square and rectangular pavements of river cobbles that formed the base of structures. As modern humans moved out of environments in Europe rich in natural caves and began to occupy open-air sites at higher latitude during the past 40,000 years, shelters became critical for survival. ◖ At Dolni Vestonice in Moravia, Czech Republic, humans lived for five periods between 28,000 and 24,000 years ago. The site is famous both for an unusual triple burial discovered in 1986 and for the remains of five dwellings-the largest covering an area 9 meters by 15 meters—demarcated by postholes, mammoth bones, and blocks of limestone. The remains of two kilns reveal that these people were the first anywhere to experiment with ceramics, creating human and animal figurines that may have been purposefully exploded inside the ovens. ◖ Spectacular evidence of past shelters also comes from further east in Europe, specifically from sites in Ukraine on the central Russian plain between the Carpathian foothills and the Urals. Modern humans most likely moved into this environment to hunt the abundant herds of large herbivores—woolly mammoth, reindeer, horse, and bison—and they set up short- and long-term camps in prominent river valleys. This was an extreme environment for humans, one marked by a frigid climate and harsh weather. Large mammals provided a vital source of food, certainly, but animal skins and furs must have been important for clothing, too. In view of the scarcity of trees at some sites, mammoth bones and reindeer antlers also provided handy construction materials. Ten Russian sites possess the

Cast of stake from Combe-Grenal, France. This cast was made from a posthole in a Mousterian layer of the cave site. The blunt tip indicates that the original wooden stake was pounded into hard ground to support a tent or windbreak. Actual size. *Photograph by David L. Brill; courtesy of Denise de Sonneville-Bordes, Université de Bordeaux I.*

ruins of mammoth bone structures, and eight other sites have the remains of structures made with additional materials. ❦ One of these sites, Molodova, on the Dnestr River, consists of several occupation levels containing numerous hearths, which are associated with mammoth bones arranged to create windbreaks. At the Middle Paleolithic site Molodova I, an oval arrangement of mammoth bones once stood as a shelter. The floor space inside had been cluttered with 29,000 flint flakes and fragments, hundreds of broken animal bone pieces, 15 hearths, and a piece of red ochre. A slightly smaller and less clearly patterned arrangement of mammoth bones occurs at Molodova V, and this site has also preserved a dwelling roofed by reindeer antlers, which were presumably covered with animal skins. ❦ Reindeer antlers and mammoth bones were also incorporated into structures at Mezin, a site on the Desna River. Five bone concentrations found here have been interpreted as the remains of conical huts constructed with animal skin. Because bones from 116 mammoths of different geological ages were incorporated into the structures at Mezin, it is likely that the occupants mostly scavenged bones from mammoths that had died naturally, adding bones as needed from recently hunted animals. ❦ Another site, Yudinovo, had dwellings of a very different design: one was built around a base of two concentric rings of mammoth bones, with possible postholes drilled into some of the bones in the inner ring; elsewhere at the site, 56 mammoth skulls surrounded a large depression dug a half meter below ground level and divided by lines of bones into six partitions or "rooms" of a long house. ❦ The most impressive mammoth bone dwellings come from Mezhirich, a site southwest of Mezin on the Dnepr River where human occupation dates to between 18,000 and 14,000 years ago. This winter camp contained at least five bone houses. One of these had been built from 385 mammoth bones. A semicircle of skulls, topped by a second layer of skulls, pelves, and scapulae, formed the interior wall, and 95 mandibles, stacked chin down in rows, fortified the outer wall. Ivory tusks arched across the perimeter to support an animal skin roof. Beneath the bones of Dwelling Number One lay 4,600 flint tools, plus bone artifacts, charred bones, and scattered pieces of red ochre and amber for beads, all surrounding a central circular fire pit. Some bones placed upright around the pit may have been part of a barbecue. The entire shelter had been about 4.5 meters across. ❦ At each dwelling, the shapes of the bones were put to best advantage to create a sound structure. That each had a unique arrangement of bones implies an aesthetic sense among the builders, who attempted a variety of architectural styles. The builders of one dwelling adapted elements of construction from two other huts and added some individual touches. ❦ Building these houses was no easy task. A single mammoth skull with a small pair of tusks weighed more than 90 kilograms, and Mezhirich's Dwelling Number One contained about 20,700 kilograms (23 tons) of bones. Each house may have taken ten people at least four or five days to build. Such effort, plus the amount of accumulated debris from their daily lives, suggests that people stayed here, throughout long winters, for several years. The site's population at any time cannot be determined for certain, but a population of 25 has been estimated. Mezhirich was a full-fledged settlement, and in its buildings and material items we see a foreshadowing of the sedentary societies that became a more universal part of human life only 10,000 years ago, with the advent of agriculture.

42 · Clothing

AS WITH SHELTER, early humans had no need for clothing during most of their evolutionary history. They occupied tropical and subtropical regions and, at least until *Homo ergaster* and *Homo erectus* appeared, probably had a healthy coat of body hair. Judging from the proportions of the boy's skeleton from Nariokotome (see page 196), *Homo ergaster* had a tall, thin build similar to that of modern equatorial Africans; it may have been more energy efficient for this hominid to be relatively hairless in the tropical heat, and we have no

Bone needle in matrix from Enlene, France. Evidence of early human clothing has not survived in the archeological record, but tools specifically suited for tailoring have been preserved from as far back in time as 26,000 years ago. Actual size. *Photograph by David L. Brill; courtesy of Société Civile de Domaine de Pujol.*

evidence to conclude that *ergaster*, or even globe-trotting *Homo erectus*, wore clothes. ❦ The earliest hominid that is likely to have had attire is *Homo neanderthalensis*. Although there is no scrap of archaeological evidence for Neandertal clothing, the distinctive sloping wear on Neandertal incisors may derive from their use as a vise to grip leather and sinews, perhaps during the curation of animal skins. Neandertal tools would have been useful for scraping mammal hides, and skins and furs would have helped them survive in often glacial environments. Limited evidence for Neandertal body adornments was uncovered in the Grotte de Renne at Arcy-sur-Cure, France. In addition to Châtelperronian stone artifacts were bone and antler tools and an intriguing set of eight plausible items of personal adornment, including carved and grooved teeth from deer and fox. ❦ Only at modern human archaeological sites, however, do we find artifacts specifically suited for creating and wearing clothes. Bone needles almost indistinguishable from modern sewing needles have been found at 26,000-year-old sites in Central Europe, and at western European sites dating to 23,000 years ago in the Solutrean period. Needles become more common during the succeeding Magdalenian period. The needle shown here comes from Enlène, France, a site best known for its thousands of portable art objects, including engraved, flat stones that were laid on the cave floor like tiles. Of course, the needles have proved more durable than the clothing they stitched, so we can only guess at how these outfits looked. ❦ Buttons made from bone or stone disks date to at least this time as well. At the Late Magdalenian site of Montastruc in the Dordogne region of France, an engraving shows the front of a human figure with a series of small circles running up the middle of the torso and chest. Round objects interpreted as buttons also have engraved animals on the surface, while other buttons were toggle shaped. ❦ Body adornments, such as beads and pendants of bone, mammoth ivory, shell, amber, stone, and other natural objects, appear about 35,000 years ago, at the beginning of the Aurignacian period of the Upper Paleolithic. They occur scattered in large numbers at ancient campsites and in living areas. Beads probably served a variety of uses, including as buttons, necklaces, or bracelets, and sewn directly onto clothing to add decorative texture. Body adornments were frequent features of early modern human graves, including those of Cro-Magnon (see page 260). ❦ Archaeologist Randall White has reconstructed the stages by which humans manufactured ivory beads in southwest France during the Aurignacian period. The procedure began with four-inch stakes of mammoth ivory cut from the soft outer layer of a tusk and fashioned into a cylindrical rod as thick as a pencil or twig. Because discarded tusk fragments are rare at French archaeological sites and intact mammoth tusks are absent, the ivory must have been transported from a distant source. The ivory sticks were incised and broken into shorter segments. The craftsperson whittled away layers of ivory to thin one end of a segment into a stem, then gouged a hole through the ivory at the stem's base using a pointed tool. Further grinding and polishing reduced and rounded the stem and the bulb at the opposite end of the segment until the hole was left near one edge of the finished bead. This technique created a cylindrical bead around a hole, rather than attempting to pierce an already finished bead. ◗ Perhaps the most famous example of early body adornment comes from Sunghir, an archaeological site near Moscow that dates to about 24,000 years ago, where the burials of three individuals contained 10,000 ivory beads and pierced teeth. On the body of an elderly male, beads were arrayed in strands across the forehead and chest and along both legs. This pattern suggests that the beads had been sewn onto a hooded tunic, or perhaps a separate cap and jacket, and tailored pants. The remains of ivory bracelets and pendants and shell and tooth necklaces also decorated each grave. Undoubtedly, some sort of protective clothing would have been a necessity for humans to spread out across Ice Age Europe and as far north as the Russian steppe. A bone figurine found at Buret, Siberia, depicts a human figure wearing a hooded, parkalike outfit. These people may have dressed entirely in fur, because many fur-bearing mammals, including foxes and wolves, were hunted at these open-air sites. ◗ Because so little evidence of human clothing has survived at archaeological sites, it is necessary to move beyond the time range of this book and jump to the end of the Late Stone Age for a glimpse of what earlier humans might have worn. The extraordinary discovery in 1991 of a Late Stone Age hunter-gatherer whose frozen body had been preserved by a glacier in the Tyrolean Alps provided a very rare and informative look at utilitarian clothing from more than 5,000 years ago. Archaeologists carefully collected and reconstructed seven articles of attire preserved in bits and pieces along with the Iceman and his small tool kit. From head to toe, he wore a conical fur hat with a leather chin strap; a sleeveless deerskin tunic; a plaited grass cloak that stretched from shoulders to knees and was fringed to permit free leg movement; loose-fitting leather leggings that hung by straps from a belt and tucked into the shoes; a leather loincloth that protected a belt pouch containing tools and tinder; the pouch itself, which appears to be made of calf leather, possibly from the now extinct aurochs, an ancestor of domesticated cattle; and shoes made of oval pieces of leather bound with straps and stuffed with grass padding, held in place by a net of grass cords. ◗ The leather tunic or cape, probably tanned over a smoky fire, had been scraped on one side. It may have been reversible so that the fur side could be worn in or out, depending on the weather. Sinew and grass held together the several vertical strips of hide, and some seams show signs of a hasty repair with grass thread. In addition to deerskin, bear, ibex, and chamois hide may have been incorporated into the shirt. Although weaving was already an established technique by this time, the Iceman wore no woven clothing.

SIGNS of purposeful burial do not appear until relatively late in the archaeological record. We have widespread and excel-

43 · Burial   lent evidence for burials in excavated graves, often accompanied by grave goods and body adornments, beginning in the Upper Paleolithic, or around 40,000 years ago. The evidence for purposeful burial associated with Neandertal skeletons from the preceding Middle Paleolithic is more ambiguous, for these burials lack obvious signs of attendant ceremony or ritual. Even earlier, the rich trove of 300,000-year-old hominid fossils from the Sima de los Huesos in

the Atapuerca Mountains of Spain (see page 218) may represent a case of deliberate disposal of corpses down a dark cave shaft, but there is no evidence here of individual burial. ◗ Beginning with the Middle Paleolithic, as people began to occupy caves and rockshelters more frequently, the odds increased that complete or nearly complete skeletons would be preserved intact. Thus, to determine if the bodies represent intentional burials, other evidence besides the flexed skeleton, such as excavated grave pits or accompanying grave goods, should be present. Unfortunately, there is ample opportunity for such information to be disturbed or lost between the time of burial and excavation, or during and after excavation. ◗ Among the Neandertal skeletons proposed as deliberate burials are specimens from Krapina, La Chapelle-aux-Saints, La Ferrassie, La Quina, Le Moustier, Regourdou, Roc de Marsal, and Spy, in Europe, and from Amud, Kebara, Kiik-Koba, Tabun, Teshik-Tash, and Shanidar in Asia. Some of these — the Krapina specimens and several from Shanidar — are more likely the hapless victims of cave ceiling collapses, while others may have been preserved through natural processes of burial. But there is more convincing evidence that such specimens as La Chapelle I, La Ferrassie 1 and 2 (found lying head to head) and four other individuals found within pits at that site, the young child from Roc de Marsal, the adult and infant from Amud, and the Kebara skeleton discovered in 1983, among others, were indeed deliberately buried by their contemporaries. ◗ Whether the act of burial reflects any religious or spiritual beliefs in Neandertals, an emotional attachment to specific individuals, or simply the fear of scavenging animals and the pragmatic need to remove a decaying body from a campsite cannot be determined. A commonly cited case of Neandertal ritual burial is Shanidar 4, known as the "Old Man," from the Shanidar Cave in Iraq. Soil samples collected from the grave revealed a high concentration of pollen from a broad range of colorful wildflowers, some with medicinal value. The scenario developed that this partially blind and crippled individual, who had obviously survived with the care and attention of others, had been attended by survivors, who cast flowers over his grave. With the 1960s still fresh in the cultural consciousness, the Shanidar Neandertals were called the "First Flower People." Yet it is not really possible to rule out wind as the source for depositing pollen atop the body or even two of the Iraqi excavation crew who wore flowers tucked into their sashes. Indeed, with the exception of Shanidar 4, Middle Paleolithic burials show a lack of special character, and what grave goods that do attend corpses — stone tools and animal bones, but never any art or adornments — could have already been present in the sediment used to cover a body. At La Chapelle-aux-Saints, for instance, the leg and foot bones of a bovid found near the body were interpreted as a haunch of meat that had been given as an offering of food for the deceased, and numerous nearby stone tools in the sediment were taken as suggestive evidence that the tomb had been the site of recurring funeral feasts by other Neandertals. Both conclusions are clearly speculative. ◗ In contrast to Neandertal graves, Upper Paleolithic burials of modern humans appear to be much more common (nearly 100 examples from Eurasia, versus about three dozen for the Middle Paleolithic Neandertals). Upper Paleolithic burials occur in both caves and open-air sites and often include a rich assortment of grave goods, such as stone and bone tools, shells, ivory beads or pendants, animal bones, and red ochre. The bodies in these burials are more often fully extended rather than in the flexed position of the Neandertals and are much more likely to occur as multiple burials, as in the double burial of an infant with the adult female Qafzeh IX (see page 255) or the enigmatic

Cast of *Homo sapiens* burial at Arene Candide, Italy. In contrast to Neandertal burials, Upper Paleolithic graves commonly included artifacts, body adornments and sprinklings of red ochre. This reconstructed cast of the 20,000 year-old burial shows mammoth ivory pendants and a flint blade accompanying the body of a teenage boy. *Courtesy of Giacomo Giacobini, Universitá di Torino.*

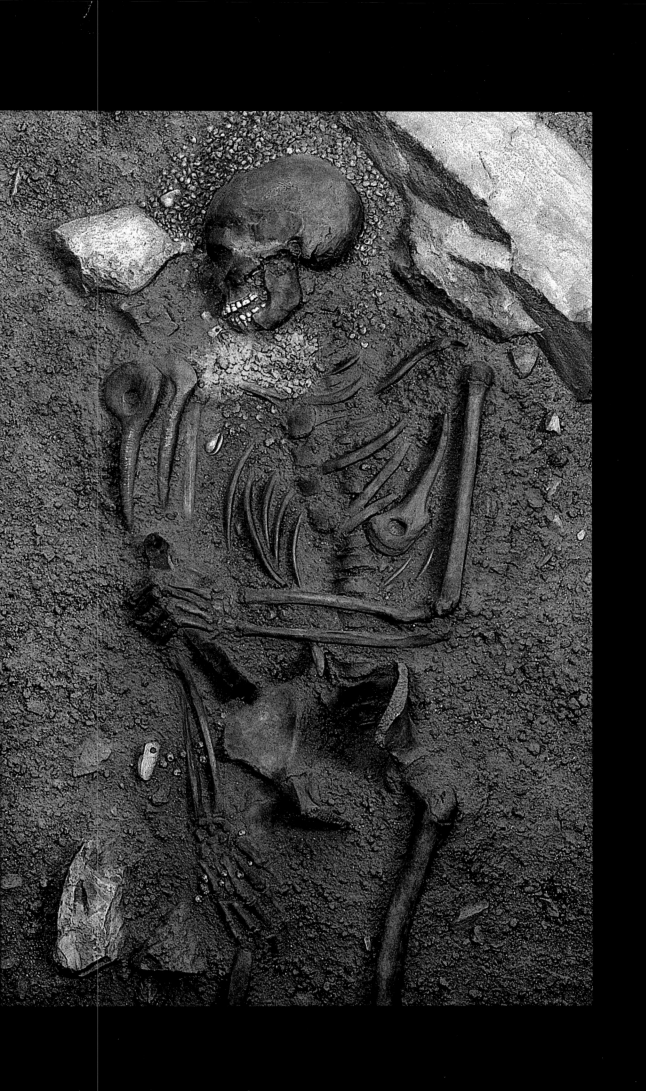

triple burial at Dolni Vestonice of a presumed female and two young males, one with a wooden pole penetrating his hip. ❡ The 20,000-year-old burial of a male known as the "Young Prince" at Arene Candide, Italy, exemplifies the elaborate nature of Upper Paleolithic burials. The body was dusted with red ochre and decorated with mammoth ivory pendants, four *batons de commandement*, shells that may have been linked in a bracelet, and a 23-centimeter flint blade inserted in one hand. Red ochre is a recurring feature in Upper Paleolithic burials, although its symbolic meaning can only be guessed at. Its use was also widespread: the body of a male (see page 263) buried at Lake Mungo, Australia, around 30,000 years ago had been sprinkled with ochre.

IF ART is an attempt to imitate nature, our Upper Paleolithic ancestors

44 · Art

were master artists. It is impossible to visit a cave like Lascaux in southwestern France or Altamira in northern Spain and not be moved by the images of horses, bison, deer, and other prehistoric animals painted on the ancient cave walls. Reaching across eons of time, these lifelike yet hauntingly impressionistic paintings immediately connect us with the artists who rendered their world on cave walls nearly 20,000 years ago. ❡ When the painted cave of Altamira first came to the attention of researchers, in 1880, the immediate reaction was that such sophisticated and well-executed paintings could not have been made by prehistoric people. With the discovery of other decorated caves in France, however, Upper Paleolithic art was confirmed to be genuine. Today more than 300 decorated caves are known in southwestern Europe that date from the Upper Paleolithic and provide a window on the aesthetic life of our forebears. Caves decorated with art are still being found, such as the cave at Chauvet, in the Ardèche, which dates to some 30,000 years ago and was discovered in December 1994. An even more remarkable discovery is the cave of Cosquer, near Marseilles, France. Cutoff by rising sea levels when the glaciers melted, about 12,000 years ago, it was not seen again until found by a scuba diver in 1991, who discovered it at the end of a 150 meter submerged passageway. ❡ The Upper Paleolithic, roughly between 40,000 and 10,000 years ago, saw an explosion of human cultural activities. Our ancestors began to gain some mastery over nature and with it, perhaps, the confidence to celebrate nature. The earlier, survival-oriented existence was gradually replaced by one in which imagery, innovation, ceremony, and ritual were interwoven in highly complex patterns. Humans were beginning to cope with intangible fears, anxieties, and mysteries, perhaps in concert with an increased awareness of their own intellectual powers and consciousness. And as the Upper Paleolithic people spread throughout the world, they created art. Some art in Australia may be as much as 40,000 years old, and since it is still part of the Australian Aboriginal people's way of life, it is the oldest continuous art tradition in the world. Art from Africa dates back to nearly 30,000 years, but little has survived, for the painting was often done in areas exposed to the elements. The discovery, however, of 78,000 year old engraved ochre plaques and possible beads at Blombos, South Africa, as well as perforated ostrich eggshell beds in Tanzania, hint at the possibility that African art significantly predates occurrences in Europe. ❡ The exact meaning of the art from this period may be lost forever, but the sophistication, elaboration, and widespread manifestations of art in Asia, Africa, Australia, and Europe attest to the importance of artistic activity in Upper Paleolithic times. Art was likely the product of professionals, and this suggests that their society enjoyed an economic surplus that freed certain individuals from purely subsistence labor. The images themselves, however, would have been deeply embedded in the broader cultural fabric of the group, representing the interests of all. ❡ One of the best-known and most extensively studied examples of cave art is the cave at Lascaux, in France. The immediate impression one gets on entering the Hall of the Bulls is that the art was

carefully planned perhaps reflecting at least in part the intentions of a single artist. Oil lamps that provided illumination for the artists were found in the cave, and a piece of charcoal from the cave floor has been dated to 17,000 years ago. Holes in the cave walls may have been used to anchor some sort of scaffolding for the artists and their assistants. Some 17,000 years before Michelangelo painted the ceiling of the Sistine Chapel, an Upper Paleolithic artist was supported 6 meters above the cave floor to complete some of the paintings. ❡ The cavalcade of horses and bulls in Lascaux seems to be in motion, especially when viewed by the flickering illumination of an oil lamp. A mythical animal bearing two forward-jutting straight horns appears to command the performance. Some think it is a shaman draped in an animal skin. Caves have their own aura of mystery, and Lascaux may have been chosen because of its unique personality or because it had special significance to the group. ❡ The paintings at Lascaux reveal much naturalistic detail and the use of a variety of techniques and pigments. The brown and red ochre that appear in several hues on the walls could have been taken from the sediments in the floor of the cave. Manganese and charcoal were used for black. It took special skill to paint on the uneven, calcite-covered walls—mistakes could not be hidden. Outlines of even the largest animals were drawn by a sure hand in a space of appropriate size. Lines were drawn using an ochre crayon, and dots were made by stabbing at the wall with a pigment stick. ❡ In areas where the pigment appears diffuse and "soft," it may have been blown on, perhaps through a hollow bird bone; or the pigment may have been chewed in the mouth, then spit on the wall. The techniques of spitting or blowing a pigment onto the wall may have had special significance for the artist. Breath gives life, and French prehistorian Michel Lorblanchet conjectures that using the breath to decorate a wall bestowed "life" on the image. He speculates that, in a manner of speaking, the painter was transformed into the animal being painted. ❡ Although images of humans are rare in Paleolithic cave art, images of animals are plentiful and seem to interweave magical and realistic spheres. Some animals painted on Upper Paleolithic cave walls, such as horses and bison, appear to be speared. Were such depictions an attempt at sympathetic magic, to give the hunters dominion over prey and ensure hunting success? In other paintings, natural bulges on the cave walls were used to accentuate the belly of an animal and make it look pregnant. Perhaps these realistic animal representations were created in the hope that game would always be plentiful. Were they painted in the recesses of subterranean caves as a symbolic gesture to the belief that Earth was the womb of all life? ❡ At one French cave, Les Trois Frères, there is a fanciful creature known as the sorcerer. It wears a helmet of antlers and has owl eyes, a horse's tail, wolf ears, bear paws, and human feet and penis. The melding of natural life forms gives a general impression that this was a place of initiation where, under the spell of a shaman, individuals underwent certain rites of passage. Were some of the symbols and images the result of hallucinations as shamans entered a trancelike state? ❡ Cave wall art, although undoubtedly created for the entire society, nevertheless is of an invitational or ritualistic nature. Public to a degree, its very hiddenness kept it from being public to all. Public art in the form of monuments or generally acclaimed works is common in the world today, often reminding people of their belonging to a particular place or of having a shared history. Few such pieces of public art survive from the Upper Paleolithic, but one, in Cap Blanc, France, is quite impressive. A frieze of horses, carved in soft limestone under a rock overhang, is visible from the valley below. We can only speculate on its function—perhaps it was a visual reminder to the wandering bands of hunter-gatherers that this region belonged to a particular group of people. This kind of public art was for everyone, even those outside the social group of the artist, and it clearly served a different purpose from ritualistic art or private, individual art. ❡ Individual art, if it existed, would have belonged to the category of portable art. Portable art took on myriad forms: carvings made of stone, bone, antler, and ivory in the form of human

Bear head figurine from Isturitz, France (*above*). Human occupants of this site purposefully aligned the bones of cave bears and fashioned this realistic image of an animal that
must have inspired awe and fear. Actual size. *Photograph by David L. Brill; courtesy of
Musée des Antiquités Nationale.*
Bison figurine from La Madeleine, France (*below*). This head of a spear-thrower, which
depicts a bison licking its flank, is one of the classic images of hand-held art and comes
from The type site for the Magdalenian period. Actual size. Photograph by David L. Brill;
courtesy of Musée des Antiquités Nationale.

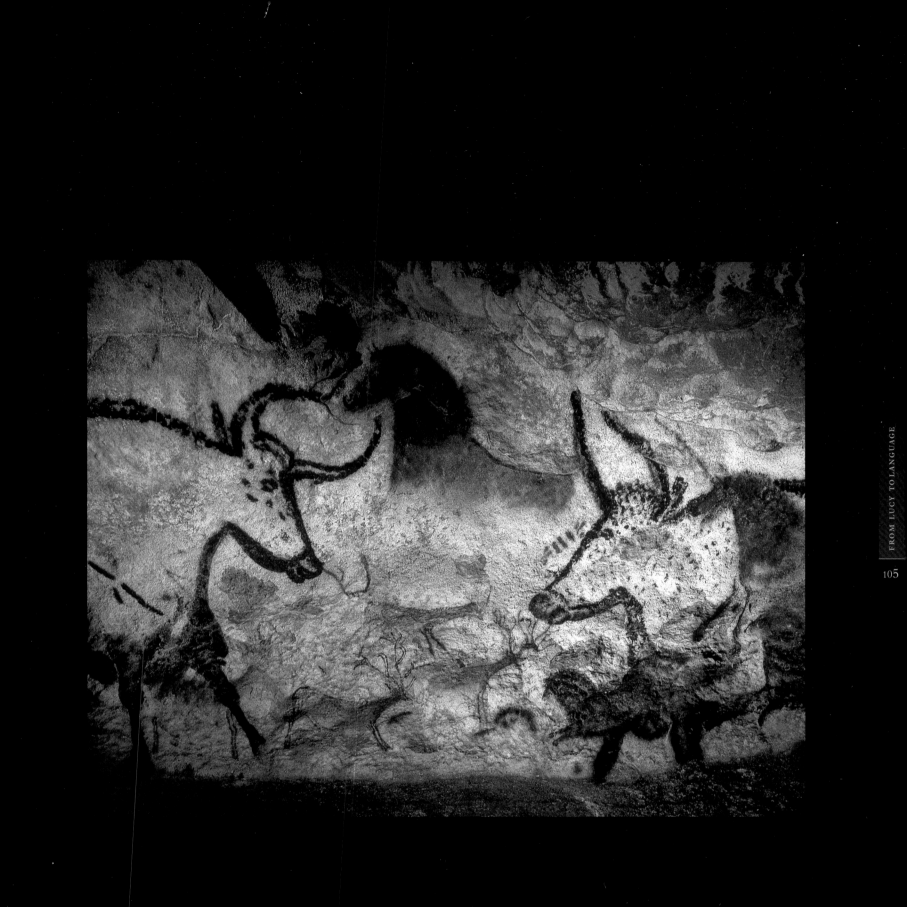

Rock art from southeast Algeria (*left*). This image of a giraffe hunter was created by the
Tuareg people perhaps as long as 3,000 years ago. *Photograph by David Coulson.*
Cave painting from Lascaux, France (*above*). Seventeen thousand years ago, an artist in
the cave's rotunda created this face-off between two aurochs, an extinct cattle, each depicted
with an associated symbol. A small horse appears between the aurochs' horns, and three stags
can be seen below. *Courtesy of Norbert Aujoulat, Centre National de Préhistoire, Ministère de la
Culture.*

heads; carvings of the entire female body with accentuated breasts and vulva, called "Venus" figurines, that served as fertility fetishes; carved animal heads and bodies; perforated animal bone, seashells, and teeth that were strung together as necklaces; an intricately decorated spear thrower or sculpted *baton de commandement*, and so on. Such mobile art may have been very specific pieces that a person carried throughout life. Perhaps a certain object signified belonging to a particular clan, or served as the symbolic bond between two people, in the manner of a modern-day wedding ring. It is also probable that some owners regarded these pieces as talismans and attributed supernatural or protective powers to them. ◀ Some of the most intriguing examples of portable art are pieces that appear to be records of historical events. The changing seasons of the year are likely to be among those events recorded, and from La Vache we find a bone knife with, on one side, the face of a doe, water, and flowering plants. These images probably recorded springtime. On the opposite face a bison in autumn rut, dying flowers, and nuts suggest autumn. ◀ With the end of the Upper Paleolithic in Europe and the dawn of the agricultural revolution, some 10,000 years ago, art became more improvisational. The first great efflorescence of prehistoric art occurred during the height of the last Ice Age. In Europe once the open steppe and grasslands began to disappear with the melting of the ice sheets and forests began to dominate the landscape, the Upper Paleolithic way of life gave way to agriculturalism.

45 · The Origins of Language

LANGUAGE is a distinctly human feature that all of us depend on daily. Language informs and enriches our world and our perceptions. It is the foundation on which all modern human behavior rests. Imagine human culture without language in all its manifestations: books, plays, operas, poetry, newspapers, television, movies, the Internet, casual conversations, and formal debates. Because language provides such a central part of our being, much effort has been spent in tracing its origins. ◀ Some paleoanthropologists have argued that language evolved almost overnight, emerging in modern humans about 50,000 years ago as part of a genetic mutation that suddenly rewiring of the brain and permitted a cultural flowering seen throughout the archaeological record of the Upper Paleolithic. Traditionally, though, language has been considered a more ancient innovation that must have evolved over millions of years, at least since the beginning of genus *Homo*. How can we decide between these disparate hypotheses? Fossil evidence, in the form of cranial endocasts that reveal the peaks and valleys on the surface of hominid brains, has been brought to bear on the question of when language evolved. Because surface morphology of the brain may not strictly correlate with internal function, however, drawing conclusions from endocasts proves difficult. Endocast experts look for asymmetry in the brain hemispheres and the presence of Broca's area, Wernicke's area, or other structures linked to language in modern humans. The presumed presence of certain ridges, grooves, or furrows in these structures on endocasts has then been taken as evidence of language competence in various early hominids, including *Homo habilis* and even *Australopithecus africanus*. ◀ But interpretations of the presence and position of these critical structures differ, as do conclusions about the capacity for language. Wernicke's area proves to be especially tricky to identify in endocasts. As for Broca's area, traditionally viewed as the motor center for speech, recent data indicate that control centers for language reside in several regions of the brain and that Broca's area may more generally serve motor functions. ◀ Others have cited aspects of our primate heritage as evidence that language must have evolved early in our lineage. As social primates, hominids lived in large groups. Perhaps group size created a need to hold these groups together, with language providing the social

glue. As a bonding mechanism, language, or at least an enhanced type of vocal communication, could have fulfilled the task more effectively than grooming does, while permitting the exchange of useful information. But other primates who live in large social troops and who communicate vocally still lack the equivalent of human language. Nevertheless, the extent of social communication in other animals indicates that human language must have likewise emerged in a social setting. ◀ Language evolution is probably intimately linked to brain evolution, and since our brain has been growing and reorganizing over the past 2 million years, it seems unlikely that language suddenly arose from some radical new mutation. Human brains could have been language-competent long before spoken languages appeared. The enlarging brain of early *Homo* no doubt was capable of complicated cognitive coordination and calculation and as such relied on and used skills important to language. Perhaps language evolved in tandem with our enlarging brain or was a cause, rather than a consequence, or brain enlargement during the Pleistocene. ◀ Yet the idea of early language runs into trouble when compared with archaeological evidence. Although the first appearance of tools corresponds with the time to which some paleoanthropologists assign the origin of language, the technology of early *Homo*, the Oldowan and especially the Acheulean, shows no sign of innovation or diffusion of new ideas beyond a standard set of tool types and flaking techniques. More important, if language, which constitutes a symbolic use of signs to communicate and an imposition of symbolic meaning on reality, evolved early, why do we not find obvious archaeological evidence of symbols until very recently, particularly within the past 40,000 to 50,000 years? ◀ The tools of this latter period show extreme variety and innovation in raw material, design, and function, plus regional styles and other features suggesting a role for language in developing and disseminating these new forms. What also appears around this time is art, both paintings and engravings as well as portable carvings. And art, of course, is inherently symbolic. Perhaps the appearance of full-blown phonetic language as we think of it today, with its syntax and symbols, made this possible. ◀ Does this mean that only modern humans among the hominids had the capacity for language? At least our close contemporaries, the Neandertals, probably also possessed language. Analysis of the Neandertal cranial base led to the hypothesis that the Neandertal larynx was too primitive to permit spoken language. In humans the larynx sits low in the throat, which facilitates uttering a wide range of sounds but makes us the only mammals incapable of simultaneously drinking and breathing, lest we run the risk of choking. Judging from their lack of cranial flexion, Neandertals had a high-sitting larynx and a more apelike vocal tract incapable of the full range of human sounds. ◀ Then a compelling piece of evidence for Neandertal language turned up in the burial of Kebara 2 in Israel. A nearly complete hyoid bone — the only one from the Middle Paleolithic until the skeletal discoveries from Atapuerca, Spain — was found beside the specimen's mandible. The only free-floating bone in the human body, the hyoid attaches to soft tissue of the larynx and anchors throat muscles important to speaking. Because the Kebara hyoid has an essentially modern morphology, as opposed to a chimpanzee hyoid, we might infer that this Neandertal had a modern-looking larynx and was capable of producing the modern range of spoken sounds or words that comprise language. ◀ Beyond the question of when language appeared, we must consider the even more elusive but critical question of why it evolved at all. Some hypotheses have considered language to be a secondary occurrence rather than a target of natural selection. As linguist Steven Pinker points out, however, language itself contains clues that lead to the conclusion that language evolved as a specific adaptation. Language, whether spoken or written, exists in all human societies today and existed in all known past societies. All of these languages possess complex rules of grammar, and certain nonfunctional,

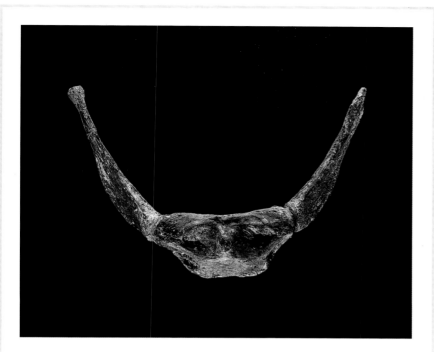

Hyoid of *Homo neanderthalensis* from Kebara, Israel. This rarely preserved throat bone closely resembles the hyoid of a modern human, which suggests that Neandertals shared the human capacity for spoken language. Actual size. *Photograph by David L. Brill; courtesy of Sackler School of Medicine, Tel Aviv University.*

universal rules also hold across these languages. This suggests that the unifying force is not culture but nature, in the form of the human brain. All societies seem to have the equal potential to learn language, as do all individuals within a given society. Most interestingly, between the ages of two and three years, generally, children in all cultures begin to assimilate language naturally, without cues or classes, from the experiences and sensations around them. ◖ Like a good adaptation, language increases our reproductive fitness; it is essential to our strategies for obtaining mates. It allows us to exchange potentially lifesaving information about our physical and social environments. Language has a complex anatomy and can be considered to be as complicated an organ as the vertebrate eye, which serves a specific function; language thus is a product of natural selection. Language also appears to operate independently from the systems behind other cognitive processes; for instance, it can exist in the absence of intelligence, and vice versa. So language is unlikely to have evolved by chance, by random genetic drift, or by some process other than directed natural selection. It has been built into our biology as the most efficient and effective means to communicate our thoughts.

WE have looked at the evolutionary history of many parts of our bodies from the feet to

46 · The Problem of Consciousness

the brain. But what about that apparently extrasomatic structure of the mind, or consciousness? Where does the brain end and the mind begin, or are they one and the same? Is it consciousness—more than bipedalism, language, or evolved culture—that really sets humans apart? ◖ What is consciousness? No single definition may suffice for such an elusive concept, but we can describe consciousness as self-awareness and self-reflection, the ability to feel pain or pleasure, the sensation of being alive and of being us, the sum of whatever passes through the mind. ◖ So, consciousness deals with the subjective experiences of our lives, and as such it differs from the logical cognition conducted by computers. Since the middle of the twentieth century, when mathematician and Enigma code-breaker Alan Turing posed the question of whether machines could think, researchers in artificial intelligence

have devised generations of machines to find an answer. Significant doubt remains, though, as to whether a machine will ever experience subjective feelings or the sorts of emotions emoted by the supercomputer Hal in Stanley Kubrick's film *2001: A Space Odyssey*. As Robert Wright asked recently in *Time* magazine, about the chess-playing computer that challenged (unsuccessfully) world champion Garry Kasparov, "Could Deep Blue ever feel deeply blue?" ◖ Paleoanthropologists probably can make little contribution to solving the problem of consciousness. Feelings don't fossilize. Yet scholars from a surprisingly diverse range of disciplines—including philosophy, neurology, immunology, molecular biology, and physics, to name a few—have attempted to answer the vexing question posed here, usually be delving into their respective sciences for inspiration. As titles from some recent books on the subject show—*Consciousness Explained, The Problem of Consciousness, Consciousness Reconsidered*—opinion is divided on how conclusive these efforts have been. ◖ Two recent hypotheses of consciousness differ in the emphasis placed on feeling and thinking. One view, articulated by Nicholas Humphrey in *A History of the Mind*, holds that consciousness is nothing more or less than immediate sensory experience; sensation-arousing stimuli define who we are, how we feel, what we know. Philosopher Daniel Dennett, author of *Consciousness Explained*, argues that consciousness can be understood from the metaphor of a computer. He views the mind as the software to the brain's hardware, a program that writes a narrative of our experience, edited and compiled from multiple drafts of information streaming into the brain. In this view, the present moment of sensation is insignificant compared to the subsequent mental reflection and contemplation, from which meaning arises. Consciousness—the mind—is simply a product of the brain. ◖ Yet another hypothesis, named neural Darwinism, was posited by immunologist Gerald Edelman in response to the brain-as-computer model. Comparing the brain to the immune system, Edelman argues that shifting alliances of neurons compete in a process similar to natural selection that results in shaping the brain with selected stimuli from outside. A fourth hypothesis, proposed by physicist Roger Penrose, dismisses the computer metaphor for one drawn from quantum mechanics, proposing that the interaction of subatomic forces in the brain sparks our sense of self. ◖ Admittedly, neither of us has fully pondered the merits or shortcomings of these theories, but none of the answers seems particularly satisfactory. Consciousness, being inherently singular and subjective, is a tricky prospect for objective scientific analysis. Indeed, at least one philosopher contends that as products of evolution, our brains are simply too limited cognitively to grasp the concept of consciousness. Unlike bipedalism or language, physical and tangible phenomena, it is not clear how consciousness could act like an adaptation to increase our fitness for survival. Perhaps it doesn't, if consciousness is merely the sum of our sensations; what once may have been a critical component of our ability to survive may now instead add layers of richness to our lives. We need to close the gap between the physical and subjective realms of this topic before we can hope to reach an understanding of consciousness. Until then it remains, according to *Scientific American*, "biology's most profound riddle."

EXTINCTION. It's a weighty word that resounds with the deep tones of finality and inevitability. Extinction is

47 · Will Humans Become Extinct?

as natural a process as speciation and evolution. Just as every individual life has its end, so every species, if we look back through the fossil record, ultimately comes to pass. We know relatively little about how extinction works, about its specific causes and broad consequences, but its influence on the course of evolution has been considerable. ◖ Virtually all species that have ever

Cranium of *Australopithecus robustus* from Swartkrans, South Africa. This species, represented by the SK 48 specimen, appeared about 2.0 million years ago and went extinct roughly 500,000 years ago. Actual size. *Photograph by David L. Brill; courtesy of Transvaal Museum.*

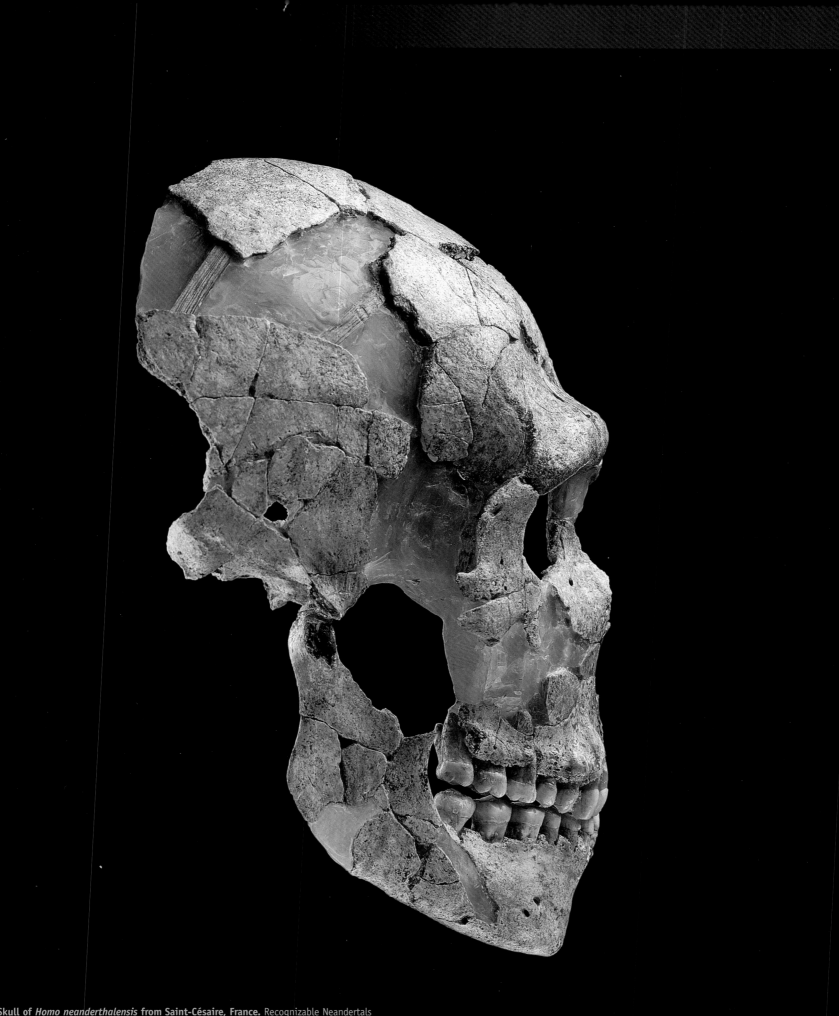

Skull of *Homo neanderthalensis* from Saint-Césaire, France. Recognizable Neandertals
first appeared about 200,000 years ago but were extinct before 30,000 years ago. This
36,000 year-old specimen was one of the last known survivors. Actual size. *Photograph
by David L. Brill; courtesy of Université de Bordeaux I.*

lived have gone extinct. The fossil record contains good data on some 250,000 past species extinctions. We can also recognize in the fossil record a series of mass extinctions that may have recurred every 26 million years for the past 250 million years, leading paleontologists to seek evidence for cosmic catastrophes—such as impacts from periodically passing asteroids—as culprits of major extinctions. Massive volcanic eruptions and lava flows may also have been responsible, particularly for the Permian mass extinction about 250 million years ago, which wiped out between 75 and 95 percent of all life on Earth. ❧ The Permian was by far the largest of five past mass extinctions that left their mark in the fossil record, most recently the Cretaceous-Tertiary extinction 65 million years ago. This event eliminated the dinosaurs and opened new opportunities for the diversification and radiation of mammals, including primates. (Three other mass extinctions occurred during the Ordovician, Devonian, and Triassic periods, respectively 440, 365, and 210 million years ago.) Whatever may have triggered these events, they were devastating for species unable to cope, adapt, and survive. ❧ So, if we strictly consider the odds from the past, then yes, we are likely to go extinct. We are the sole survivors of a relatively small and unusual family of primates, which are a peculiar group of mammals. We have reached a position of unparalleled dominance in the natural world in terms of our use of diverse resources and ability to alter environments, often to detrimental effect. Biologists such as E.O. Wilson believe that by disturbing and removing natural habitats worldwide, we are perpetuating a sixth mass extinction that threatens much of the planet's biodiversity. While there is reason to think that we can have a long tenure on Earth (recall that *Australopithecus afarensis* lasted for around a million years), there are many indications from our behavior that perhaps our end will come, too. Evolution comes with no guarantees. ❧ Extinction has been an increasingly frequent factor in hominid evolution. The last known appearance of *A. afarensis* is in 3 million-year-old sediments at Hadar, Ethiopia, and no subsequent hominid species appears to have lived as long (with the possible exception of *Homo erectus*, if this species is defined to include fossils otherwise placed in the species *Homo ergaster*.) After the extinction of *afarensis*, hominids diversified between 3 and 2 million years ago, with several species existing simultaneously by the end of this period. But after 2 million years ago, the hominid pattern switches from diversification to extinction. *Homo habilis* and *Homo rudolfensis* both appeared and disappeared before 1.5 million years ago. Between 1.5 and 1.0 million years ago, *Australopithecus boisei* and *A. robustus* went extinct. Within the last million years, *Homo erectus, H. heidelbergensis, H. floresiensis,* and *H. neanderthalensis* have all likewise gone extinct. ❧ As for us, the conditions for significant evolutionary change in our species have been greatly diminished, perhaps eliminated. Human populations are no longer isolated, and for the most part we have become one giant global gene pool. There may be little significant morphological change awaiting us under natural selection because we are also operating under the ameliorating influence of cultural evolution. Our culture allows us to be generally adaptable to environments and new conditions. Biologically we are, in many ways, still foraging the savannas, but culturally we are exploring outer space. Culture is our key survival strategy, as long as we can keep the planet intact and alive. If conditions became critical for our survival, would our brains, honed by 2 million years of biological and cultural evolution, be up for the challenge? (Our brains, by the way, are expensive organs to run and maintain, demanding about 20 percent of our body's energy intake. In the face of future environmental stresses and

Cave painting from Cosquer, France. Among the oldest European art, this image was painted about 27,000 years ago. Handprints may signify an artist's signature and indicate that Upper Paleolithic people were conscious of themselves and their environment. Actual size. *Photograph by J. Collina-Girard; courtesy of Jean Clottes, International Committee on Rock Art.*

the continuing march toward global cultural homogenization, we may continue to see a decline in brain size, which has been shrinking, along with our overall stature, for at least the last 10,000 years.) ❧ Evolution, and the future, are inherently unpredictable. If environmental conditions suddenly change, or if new diseases and viruses or political anarchy and terrorism threaten significant numbers of people, who can say whether culture will see us through? But it has brought us to this point, where we are poised to both consider our past and contemplate our future. To quote Czech president Vaclav Havel, "For the first time in the history of man, the planet he inhabits is encompassed by a single global civilization. . . . [T]his makes the modern world an essentially dramatic place, with so many peoples in so many places resisting coexistence with each other. And yet its only chance for survival is precisely such coexistence."

For, after all, what is man in nature? A nothing compared with infinity, all when compared with nothing. A middle point between nothing and all. Infinitely far from understanding the extremes... equally incapable of seeing the nothing when he emerges and the infinity in which he is engulfed. — B. Pascal, *Pensées*

48 · The Place of Humans in Nature

IN THE MAIN, this book is concerned with the fossil evidence for human evolution. But study of the ancient past has yielded more than simply knowledge about our ancestors. Paleoanthropology has brought into the common marketplace of ideas Darwin's theories about evolution and extinction. From that broader perspective we are able to see *Homo sapiens* as one species among many, and to face our own possible extinction. In evolutionary terms, we are almost assured of extinction. As our perspective on our own species through time enlarges, a single question emerges into an ever brighter spotlight as the true core of our quest: what is our place in nature? ❧ There are some who take the extreme position that humans are intruders on the natural order and don't belong on Earth. But we are here, having survived the usual evolutionary processes of elimination, and we have equally as strong a warrant (not more) to be here as any other organism. But unlike any other creature that has ever lived, we have been endowed with a great intelligence, through which we have achieved unprecedented dominion over nature. And with this gift of intelligence comes the responsibility to use it well-as guardians of Earth and its life. ❧ If we hold up the fossil record of our ancestors to modern life, it is evident that we have become highly culturally dependent. Our link to nature has loosened. For example, as individuals, we no longer provision ourselves daily and directly from Earth's resources. Yet this dependence on culture for survival has occurred in only an instant of geological time. By contrast, millions of years went into shaping us physically, and biologically we have changed little from our ancestors millions of years ago. ❧ Perhaps, in some manner of speaking, it is this imbalance in the clocks, an imbalance between our biological and cultural evolution that is at the root of our modern dissatisfaction, our sense of unease and unrootedness, of being out of tune with nature. Cultural and technological evolution advances happen unbelievably quickly, but genetic change is glacially slow. Genetically, we are less than two percent different from chimpanzees, and we have been separated from them for 6 to 8 million years. How much closer must we be to the earliest *Homo*, which appeared more than 2 million years ago! ❧ Ancient hominid fossils like Lucy are a link to our common origin with the African apes; this is no longer contested. Yet the fossils have an even more important message for us: they are a testament to our link to the natural world. The more links we find, the closer to nature we see ourselves. We are, without doubt, part of the continuum of life. By living so thoroughly in a world of culture as we do, we tend to see ourselves as distinct from nature. So, it is terribly important for us to be reminded that we are part of nature. We cannot ignore the natural world from which we evolved. ❦ The human time frame is generally limited to the life span of the reflective individual. We may, perhaps, appreciate the past of our parents and grandparents, and contemplate anxiously the future of our children and grandchildren. But this concern encompasses about a hundred-year span. To be good custodians of Earth, a responsibility our brains and adaptive capabilities impose on us, we must have the foresight to consider the extended future as well as the hindsight to take lessons from our ancient ancestors. ❦ Let us see what our past might indicate about our future. From what we know now, *A. afarensis*, Lucy's species, walked on the African continent for approximately 1 million years—about 50,000 generations. Behaviorally modern humans have been around for roughly 100,000 years—only 5,000 generations, one-tenth the length of time that Lucy's species persisted. To equal the length of time that Lucy's species lived, we would need to survive for another 9,000 centuries, and by all estimates the next century alone will be a very tough one. We are a very recent addition to Earth's biomass. ❦ Knowing our place in nature enriches us beyond comprehension. Nature is familiar, it is in our genes. It is reawakened when we sit around a campfire and exchange stories, just as our ancestors did for hundreds of thousands of years when we were hunter-gatherers, living in tune with the natural world. If we are to continue for the next 9,000 centuries, we must abandon our arrogance and become the introspective species. The choices that we and our descendants will make must be ones that benefit not only us, but all of our fellow travelers in life. ❦ *Homo sapiens* is the most intelligent, the most dangerous, and the most cooperative of all animals that have ever existed, but we must understand that we are not the final product of 3 billion years of evolution on Earth. Our species, like all others, is an evolutionary work in progress. Unlike other species, however, we have some ability to control our fate. We have the know-how to influence how the game is played out.

Rock Painting, Kakadu National Park, Australia. *Courtesy of Donald Johanson, Institute of Human Origins.*

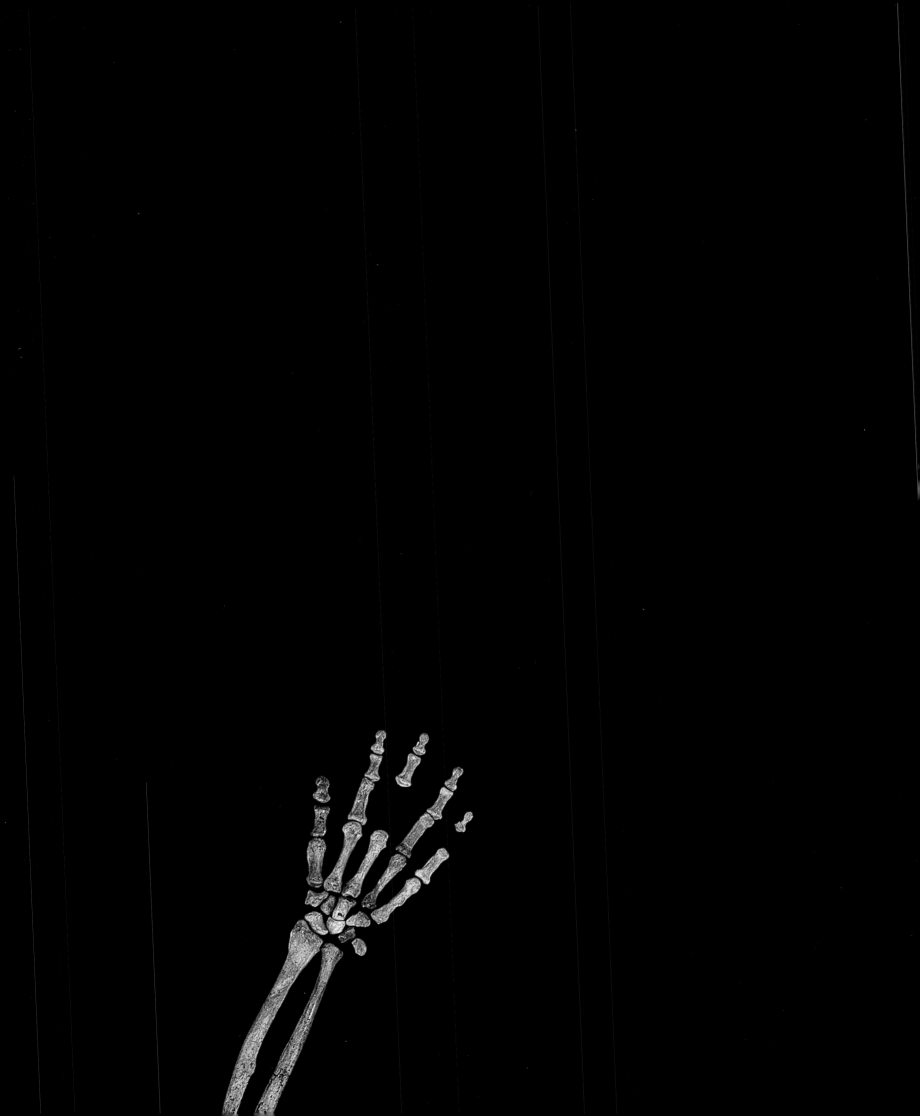

ENCOUNTERING THE EVIDENCE

Part 2

SPECIMEN	LOCALITY	AGE	DISCOVERERS	DATE	PUBLICATION
Cranium	Toros-Manalla 266, Chad	6-7 million years	Ahounta Djimdoumalbaye	2001	Brunet, M., F. Guy, D. Pilbeam, H.T. Mackaye, A. Likius, D. Ahounta, A. Beauvilain, C. Blondel, H. Bocherens, J.-R. Boisserie, L. de Bonis, Y. Coppens, J. Dejax, C. Denys, P. Duringer, V. Eisenmann, G. Fanone, P. Fronty, D. Geraads, T. Lehmann, F. Lihoreau, A. Louchart, A. Mahamat, G. Merceron, G. Mouchelin, O. Otero, P.P. Campomanes, M. Ponce de Leon, J.-C. Rage, M. Sapanet, M. Schuster, J. Sudre, P. Tassy, X. Valentin, P. Vignaud, L. Viriot, A. Zazzo, and C. Zollikofer. 2002. A new hominid from the Upper Miocene of Chad, Central Africa. *Nature* 418: 145-151.

Sahelanthropus tchadensis
TM 266-01-060-1

Our current view of early hominid evolution derives almost exclusively from two geographical areas, the Great Rift Valley of eastern Africa and the karstic cave sites of South Africa. These two windows into the past are so productive because the geological and depositional history of those areas were conducive to not only preserving bones of ancient animals, but also making them accessible to scientists millions of years later. Many anthropologists, however, have harbored the notion that we might be seeing only a portion of the entire portrait of African human evolution, providing us with a greatly limited and inadequate glimpse into our evolutionary past. While some anthropologists surmised that the Great Rift Valley was in fact a geographical barrier, restricting early hominids to eastern Africa, the majority opinion embraced a pan-African distribution for our early ancestors. But where was the evidence?

French anthropologists have long been interested in Chad because of the 1961 discovery there of an enigmatic "hominid" dubbed *Tchadanthropus uxoris*, which was later shown to be a heavily eroded skull of a modern human. Undeterred, Michel Brunet of Poitiers University returned to work in Chad in the early 1990s and published a new species for a 3-3.5 million year old mandible fragment his team (Mission Paléoanthropologique Franco-Tchadienne) recovered from the site of Koro Toro. He saw affinities with *A. afarensis*, but chose to distinguish it as *Australopithecus bahrelghazali* (the river of the gazelles southern ape). Brunet and his colleagues concluded that certainly by this stage in human evolution our ancestors had begun to extend their geographic range well beyond eastern and South Africa.

Then in 2001 Brunet published the discovery of a 6-7 million-year-old, badly crushed and distorted cranium from the Djurab desert in Chad. Working under extremely inhospitable conditions where blinding sandstorms force researchers into tents for days on end, the French-Chadian team recovered a small-brained cranium (estimated 360-370 cc), a bit of a mandible, and a few isolated teeth they consider to be the earliest hominid. As a result of this new discovery, taken together with the

previous finds at Koro Toro, the case for a pan-African arena for hominid evolution gained currency. After all, the Chad fossils were found some 1500 miles west of the Great Rift Valley and more than 3,000 miles north of those from South Africa.

Molecular studies of modern ape and human DNA suggest that the common ancestor for hominids and the African apes existed roughly 5 to 7 million years ago. So, if the Toumaï specimen, as it is called in the local Goran language meaning "hope of life," is actually a hominid, it is situated very close to the bifurcation event when apes and hominids began their separate evolutionary journeys. It was then logical that the cranium should bear features typical of ape anatomy, but also show hints of being a hominid. This is in fact the case, and one scholar noted that the cranium is apelike in back of the brain case, but more humanlike in the structure of its face.

What makes the Toumaï cranium a hominid? One major hallmark of being a hominid is reduction in canine size. Apes have large projecting canines, especially the males, but throughout hominid evolution there has been a trend towards canine reduction. In the Toumaï specimen the canine is not dramatically projecting and is relatively reduced in size, though still within the size range of variation seen in African apes today. An additional feature allying the Chad specimen with hominids is seen on the canine where the tip is worn down, unlike ape canines, which remain pointed throughout life.

One other diagnostic feature of the canine, differentiating apes and hominids, is seen in the manner the upper canine occludes with the lower first premolar. In apes the projecting upper canine overlaps with the lower first premolar, promoting a polished wear facet on the cheek (buccal) side of the lower premolar as the trailing edge of the upper canine is sharpened. This honing complex appears to be absent in the Chad specimen, but no lower first premolar has yet been recovered.

A partial mandible (TM 292-02-01) containing the left canine and P4 conveys a number of features which are shared

with later specimens considered hominid. Wear on the canine is consistent with a lack of an apelike shearing complex, the symphysis (chin area) is vertical, the canine crown is small, and the transverse torus (a bone buttress on the inside of the mandible beneath the incisors) is only weakly developed unlike apes, in which it is often formed into a shelf of bone.

For many anthropologists the true hallmark of being a hominid is bipedalism. Unfortunately no postcranial bones were found associated with the Toumaï skull, so it cannot be unequivocally stated that this creature was an obligate biped. The position of the foramen magnum, however, seems to be placed fairly far forward on the base of the skull (which has a short basioccipital), so that the skull balances over a vertical vertebral column consistent, a condition with upright walking.

The face of the cranium has enormous brow ridges, a bit of a riddle since such development is not really seen until *Homo erectus*, 5 to 6 million years later. Interestingly, australopithecines lack such protuberant brows, which suggests that if Toumaï represents an ancestor, somewhere along the evolutionary journey the brow ridges were lost and then regained in *H. erectus*—an unusual sequence to say the least. The midface of the Chad specimen lacks the significant degree of prognathism typical of apes and is in fact even less projecting than in *A. afarensis*, presenting a more vertical profile.

A high-resolution computed tomography scan was employed to generate a virtual reconstruction of the Toumaï cranium that has removed much of the distortion by carefully reorienting and repositioning many bone fragments in virtual space. This has resulted in a much better appreciation of the morphology of the specimen, reinforcing the initial view of the discoverers that this TM 266 cranium is a hominid. In lateral view the cranium appears elongated front to back with a nearly horizontally oriented braincase with little or no hint of a forehead unlike all subsequent hominids and even some of the African apes.

For the moment the Toumaï specimen presents us with a tantalizing glimpse of potentially the earliest hominid. Some scholars have looked at the evidence and suggested that the cranium may in fact represent the first evidence for gorilla evolution and has little bearing on the origins of hominids. Should it turn out be an ape ancestor it might be even more exciting since up to now we have found no ancestral fossils for the African apes.

The cranium was dubbed *Sahelanthropus tchadensis* ("Sahel Man from Chad") by its discoverers who remain cautious about giving the new species a firm position on the hominid family tree. This reticence is laudable since comparable cranial material is really not available until the time of *A. afarensis*, which means that these two taxa are separated by 3 to 4 million years of time, and drawing even dotted lines from *S. tchadensis* to presumed descendents would be premature.

S. tchadensis might be an ancestor to later hominids, it might not, at the moment it is really impossible to determine. Perhaps it is part of a broader, more complex picture of the earliest stages in hominid evolution when, like among other mammal groups, several different species of apelike creatures were vying for the role as "the" ancestor. Until the fossil record between 7 million years ago and 4 million years ago, when we see incontestable evidence of bipedal hominids (*A. anamensis*, see page 129) becomes better populated with discoveries we have to be satisfied with the knowledge that African hominid evolution was geographically widespread, just as we had always assumed.

Sahelanthropus tchadensis, **TM 266-02-060-1.** This crushed cranium from central Chad, Africa extends the range of early hominids geographically. Actual size. *Photograph courtesy of Michel Brunet.*

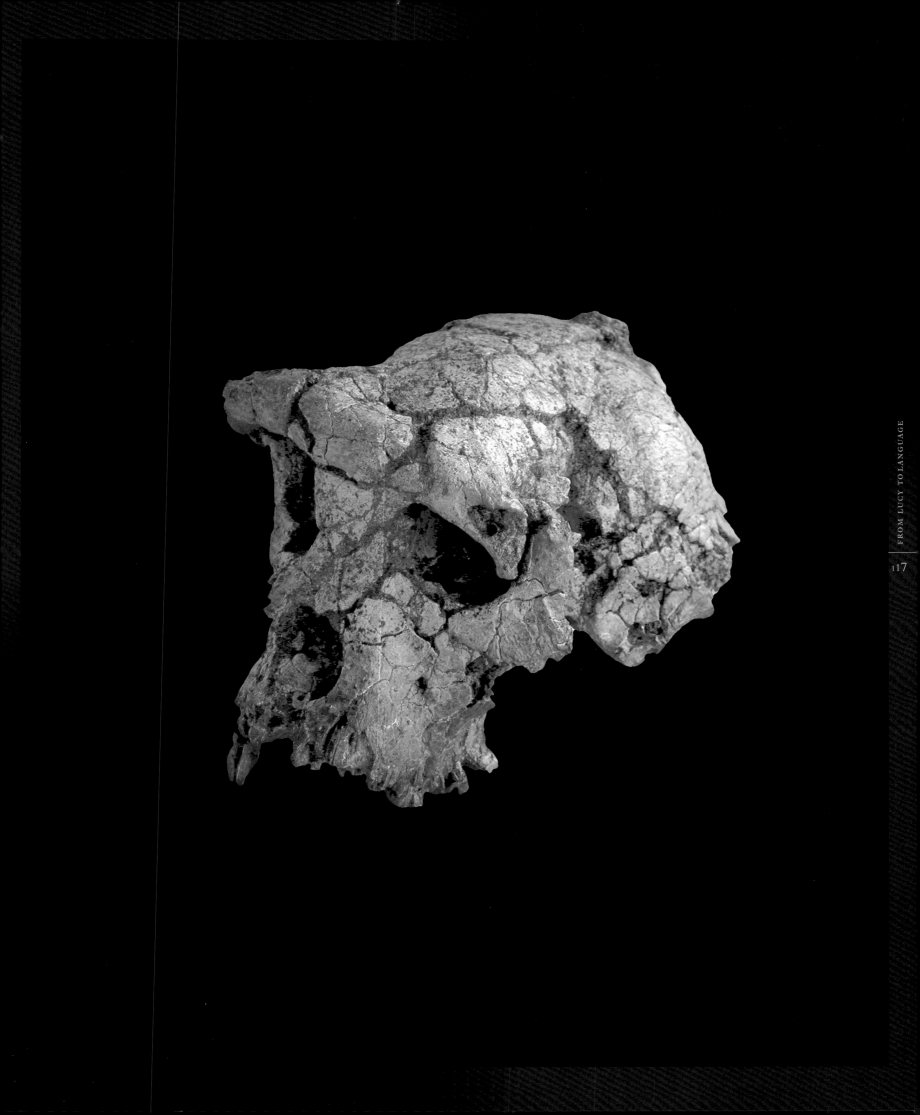

SPECIMEN	LOCALITY	AGE	DISCOVERERS	DATE	PUBLICATION
Mandible	Tugen Hills, Kenya	6 million years	Kiptalam Cheboi	October 25, 2000	Senut, B., M. Pickford, G. Gommery, P. Mein, K. Cheboi, and Y. Coppens. 2001. First hominid from the Miocene (Lukeino Formation, Kenya). *C. R. Acad. Sci. Paris*, série IIa 332: 137-144.

Orrorin tugenensis
BAR 1000'00

Almost like the space race, the race for the earliest hominid has been heating up in recent years with a number of claimants. Primary among these is the assertion by the leaders of the Kenya Palaeotology Expedition, an international collaboration between the Community Museums of Kenya and the Paris based Collège de France under the direction of Martin Pickford and Brigitte Senut, that their discoveries in the Tugen Hills located in northern Kenya stand as the most ancient evidence for bipedalism, and hence hominids. The validity of this assertion has been widely questioned and especially the proclamation that the discoveries are more "humanlike" than the later australopithecines which the discoverers consign to a side branch.

Discovered in 2000 and announced the following year, a collection of some 13 fossil specimens, reliably dated to 6 million years ago, was dubbed "Millennium Ancestor." It was no surprise that the fossils possess an amalgam of ape and human characters. The curved finger bone, humerus (upper arm bone), and a number of features of the teeth demonstrate affinities with a more apelike ancestor that was perhaps somewhat arboreal. Contra these apelike features, the Millennium Ancestor's femurs demonstrate anatomical similarities with the femur of Lucy, an undoubted biped.

To distinguish the Tugen Hills specimens from other hominid taxa, Pickford and co-workers assigned the material to a new genus and species, *Orrorin tugenensis*, selecting a fragmentary mandible as the holotype specimen. The name orrorin is from the local Tugen language meaning "original man," which sounds like the French word "aurora" for dawn or daybreak, thus designating these specimens as "the original man from Tugen."

Recognizing that bipedalism is considered by most paleoanthropologist to be the defining feature of what it means to be placed on the human lineage rather than the ape lineage, intense analysis has focused on the three femoral fragments recovered from the Tugen Hills. BAR 1002'00 is the most complete, preserving the femoral head, neck, and two-thirds

of the shaft, but lacking the diagnostic bottom end that forms part of the knee joint. The greater trochanter, the attachment area at the top of the femur for the lesser gluteal muscles important in bipedal locomotion, was unfortunately chewed away by some ancient carnivore that fed on this individual 6 million years ago. A number of features, such as a somewhat enlarged femoral head (relative to the femoral shaft), presence of a line on the anterior surface of the top end of the femur (intertrochanteric line), a long femoral neck, and presumed broad notch on the top margin of the femoral neck, all demonstrate affinities with later hominids such as *Australopithecus*.

One feature on the back or posterior surface of the *Orrorin* femoral neck, called the obturator externus groove, is a constant feature in bipeds. This groove reflects the position and action of the obturator externus muscle, which at full extension of the lower limb impresses itself on the posterior surface of the femoral neck. The groove can sometimes be present in other primates such as the gorillas but is considered by many as an important marker of bipedal locomotion; it is also seen on the Maka femur (see page 2).

The cross-sectional anatomy of the femoral neck is widely considered to be vital for distinguishing between bipeds and quadrupeds. In bipedal locomotion the head of the femur is the pivotal point where the weight of the upper body is transferred to the lower limb in bipedal walking. Loading the femoral head in this manner generates bending stress in the femoral neck. In bipeds the lower portion of the neck, susceptible to compression and thus breaking, is thickened and in contrast the superior or upper portion of the neck is actually quite thin. Quadrupeds, on the other hand, have thick bone in the upper and lower portions of their rather short femoral neck.

CT scans of the BAR 1002'00 femur are unfortunately not very definitive, but it appears that there is a thickening of the cortical bone on the inferior portion of the femoral neck. With the upper portion eroded or chewed, the thickness of this

cortical bone is less clear. While the morphology of the Tugen Hills femur is largely consistent with upright bipedalism not all scholars are convinced that *O. tugenensis* was upright.

Perhaps too much attention has been centered on the femoral anatomy of *O. tugenensis* at the expense of some of the potentially important dental evidence. Recent evaluation of fragmentary fossil remains now assigned to *Ardipithecus kadabba* targeted upper canine anatomy and function as vital for ascertaining the affinity—ape or hominid—of these 5.2-5.8 million year old fossils from Ethiopia. In spite of some very minor differences between the Tugen Hills upper canine and the one assigned to *Ard. kadabba*, there are a number of similarities. Most specifically, neither of the reduced canines appear to have been part of a canine-premolar honing complex so typical of apes where the upper canine sharpens it trailing edge over the cheek (buccal) side surface of the lower third premolar leaving a long, highly polished facet. The *Ard. kadabba* lower premolar lacks the diagnostic polishing on the buccal face but the condition in *O. tugenensis* is unknown since no lower third premolar as been found for this taxon.

Assertions by Pickford and Senut that *O. tugenensis* is more "advanced" that later australopithecines, specifically *A. afarensis*, do not hold up against scrutiny. They point to the fact that *O. tugenensis* has smaller molars and larger postcranial bones than Lucy, which makes it more likely that we can trace our ancestry back to larger bodied ancestors, such as *O. tugenensis*, rather than to *Australopithecus*. This argument does not take into consideration two things. First, the conditions of field discovery prevent the direct association of the *O. tugenensis* teeth and mandibles with the femora, and secondly, looking at the range of variation for *A. afarensis*, Lucy's species, the presence of femora similar in size to the Tugen Hills femora, attests that there were large bodied individuals within *Australopithecus*.

The expanding discovery of what might be called "basal hominids," which sample a time perhaps not far distant from the

pivotal evolutionary event that sent apes and hominids in different directions, is emphasizing the increasingly apelike nature of these creatures. A detailed evaluation of subtle clues in the dentition and postcranium will be essential for sorting out the evolutionary relationships, diet, and behavior for these tantalizing, but for the moment frustratingly incomplete fossils.

Recognition of these earliest "basal hominids" raises a number of intriguing issues. Evidence for the ancestry of the extant African apes is totally missing from the fossil record presumably because bones are rarely fossilized in forested habitats due to acidic soil. Both the *Ard. kadabba* and *O. tugenensis* specimens, however, derive from paleohabitats that were apparently quite forested. Is it not an option that one of these taxa is an ancestor to modern chimpanzees, an alternative hinted at by Pickford and Senut for *Ardipithecus*? Postcranial bones, especially those from the hands and wrist, would tell us if they were knuckle walkers and a pelvis or a distal femur would add much sought after details of their locomotor system. Another, less satisfying option is that neither *Ard. kadabba* or *O. tugenensis* left any descendents, but were extinct cousins of both apes and hominids.

Orrorin tugenensis, **BAR 1000'00.**
Although fragmentary, these specimens from northern Kenya, dated to 6 million years ago, may represent the earliest evidence for bipedalism. Actual sizes. *Photograph by Martin Pickford; courtesy of Community Museums of Kenya.*

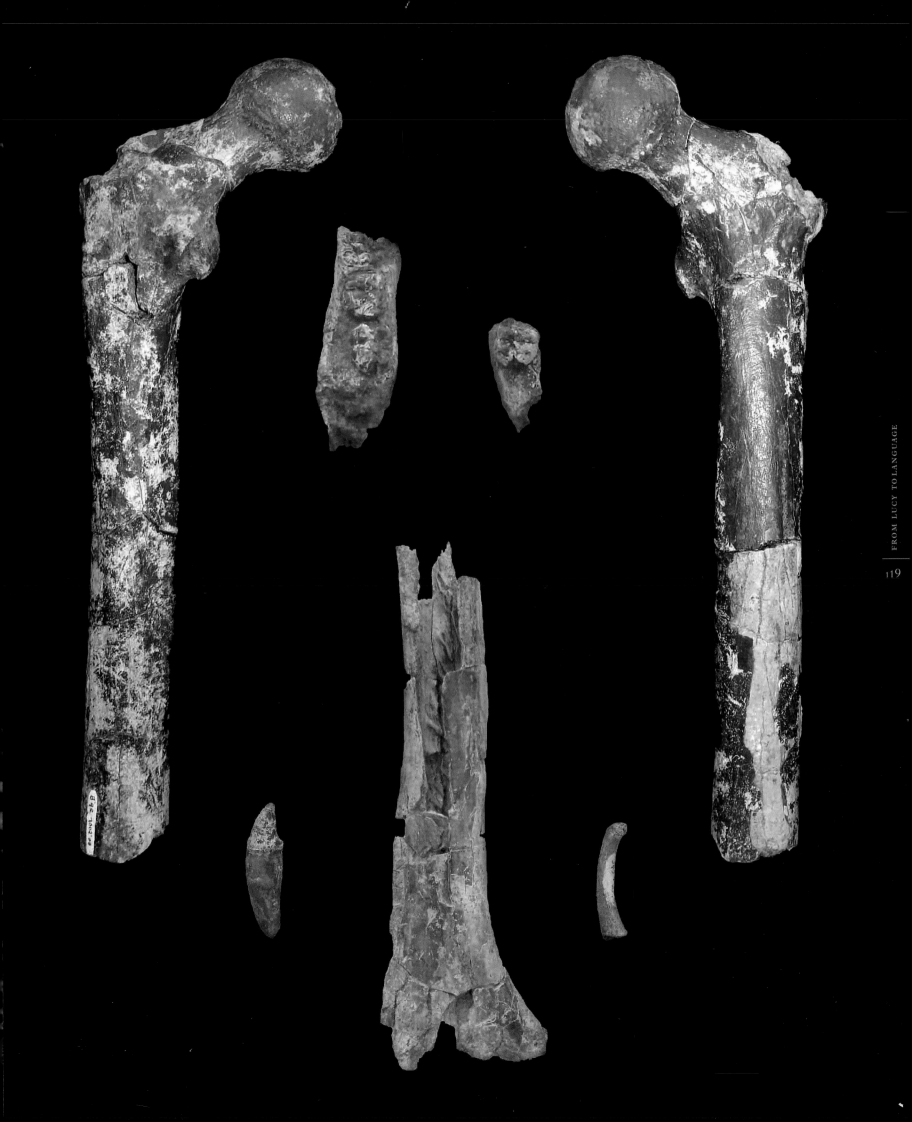

SPECIMEN	LOCALITY	AGE	DISCOVERERS	DATE	PUBLICATION
Juvenile partial mandible	Aramis, Ethiopia	4.4 million years	Alamayehu Asfaw	December 1992	White, T.D., G. Suwa, and B. Asfaw. 1994. *Australopithecus ramidus*, a new species of early hominid from Aramis, Ethiopia. *Nature* 371: 306-312

Ardipithecus ramidus
ARA-VP-1/129

Some 70,000 generations prior to Lucy's birth, an enigmatic but highly provocative and very apelike hominid walked the African landscape in what is now Ethiopia. Some 4.4 million years later its nearly complete skeleton may offer some intriguing new insights into a time when hominids were not that different from the common ancestor to our own zoological family, the Hominidae, and the African apes.

Little information has been made available about the "mystery skeleton" collected in 1994 and 1995 at Aramis in the Middle Awash region of Ethiopia, at locality ARA-VP-6/500, by an international team led by anthropologists Tim White and the late J. Desmond Clark from the University of California and Berhane Asfaw from Addis Ababa, Ethiopia. Its bones are crushed and broken and still, in large part, encased in plaster jackets, awaiting reconstruction. Excavation of the skeleton was unusually demanding because the bones were tightly packed in a hardened clay. To prevent the bone from turning to powder, preservative had to be applied to each bone before the bone was removed from the ground. The slow, painstaking work of excavation often continued late into the night under illumination supplied by the headlights of a Toyota Land Cruiser.

The skeleton is remarkably complete; for example, it has seven out of the eight wrist bones and nearly all of the finger bones, which are long and curved. The pelvis, lower limb, and foot bones may reveal a locomotor mode unlike that of any living or extinct primate. Thus far the finders have been cautious in saying that *ramidus* was a biped, although the foramen magnum is situated fairly far forward, as is seen in a portion of the basicranium (ARA-VP-1/500), a condition usually considered indicative of bipedalism.

The fossil-rich beds of the Middle Awash were first recognized in the late 1960s by French geologist Maurice Taieb, and in the early 1980s a team led by Jon Kalb, an American geologist at the University of Texas, estimated the sediments in the Aramis region to be roughly 4.5 million years old. It was not until December 17, 1992, however, that the first hominid fossil, an upper right third molar, was found at Aramis. The discovery was made by Gen Suwa, an anthropologist at the University of Tokyo, Japan. During the same field season a diagnostic mandible fragment with a deciduous lower first molar, known by the catalogue number ARA-VP-1/129, was discovered by Alemayehu Asfaw of the Ethiopian Ministry of Youth, Sports, and Culture.

Following another highly fruitful field season at Aramis, in September 1994 the Middle Awash research team officially announced a new species of *Australopithecus*—*A. ramidus*. An associated set of adult teeth, ARA-VP-/1, was chosen as the holotype. The species is distinguished from other hominids by a number of anatomical features, including relatively large upper and lower canines, a chimplike lower first deciduous molar, an apelike temporomandibular joint, thin dental enamel, and a markedly asymmetrical lower first premolar. The overall impression gained from the Aramis fossil collection is one of a significantly more primitive (more apelike) stage than that seen in later hominids. In spite of the "apelike" nature of the Aramis finds, *ramidus* is set apart from the apes by having smaller, less pointed and more incisorlike upper canines, a non-sectorial canine-lower premolar complex, and a short cranial base.

Initially assigned to the genus *Australopithecus*, in a correction published in May 1995 in *Nature* the Aramis material was transferred to a new genus, *Ardipithecus ramidus*. In the Afar language, *ardi* means "ground" or "floor" while *-ramid* means "root"; *pithecus* is from the Greek and means ape. The name reflects in its generic and specific parts the idea of a basal species for the Hominidae. In the correction published in *Nature*, the partial skeleton is referred to *Ardipithecus ramidus*.

A potential nomenclatural problem with the Aramis hominids arises from a mandible fragment with two molars from the site of Tabarin (see page 39). This specimen dating to 4.5 million years was previously assigned to *Australopithecus praegens*. If the hominids from Tabarin and Aramis are determined to be the same taxon, the correct name for them would be *Ardipithecus praegens*.

Significant preparation challenges are delaying publication and description of the Aramis fossils. The highly crushed nature of a cranium recovered from the site prompted one of the finders to call it "road kill" and reconstruction is employing new techniques such as reassembling the skull using micro CT scans. Some 13 years after the initial discovery anthropologists are eagerly awaiting details of *ramidus* that will clarify its position on the family tree and further substantiate its distinctiveness from other Pliocene hominid taxa. Remains of *ramidus* reported from 4.51 to 4.32 million-year-old deposits at As Duma in the Gona area of the Afar region in Ethiopia will also add to our knowledge of this intriguing taxon.

A small collection of some 17 specimens, comprised mostly of teeth, from the Late Miocene of the Middle Awash area in Ethiopia have been placed in a new species *Ardipithecus kadabba* (the species name means "basal family ancestor" in the Afar language). Dated by Argon analysis to between 5.2 and 5.8 million years, the initial discoveries were only considered a subspecies *Ardipithecus ramidus kadabba*. Although the 17 specimens were scattered over an extremely wide area and span 600,000 years of time, they have all been assigned to the same species. *Ard. kadabba* is purported to be the first species on the human branch of the family tree following the split between chimps and humans. The upper canine-lower third premolar complex shows a mixture of an apelike honing process and more advanced features normally considered hominid, such as a somewhat reduced canine. A left foot phalanx, some 600,000 years younger than many of the other specimens, and found some 20 kilometers to the east, shows a morphology intermediate between apes and *A. afarensis*, which may indicate incipient bipedalism. Since the toe bone is isolated in time and space from the dental material used to diagnose *Ard. kadabba*, it is far from certain it belongs to the same species.

Ardipithecus ramidus, **ARA-VP-1/129.** The pointed crown of the deciduous molar in this right mandible fragment proved to be a critical feature in distinguishing this new species. Actual size. *Photograph copyright © 1994 by Tim D. White/Brill Atlanta*

SPECIMEN	LOCALITY	AGE	DISCOVERERS	DATE	PUBLICATION
Cranium	Lomekwi, West Turkana, Kenya	3.5 million years	J. Erus	August 1999	Leakey, M.G., F. Spoor, F. H. Brown, P. N. Gathogo, C. Kiarie, L.N. Leakey, and I. McDougall. 2001. New hominin genus from eastern Africa shows diverse middle Pliocene lineages. *Nature* 410: 433-440

Kenyanthropus platyops
KNM-WT 40000

Adorning the March 22, 2001 cover of the of the respected international science journal *Nature* is the photograph of a cracked and distorted 3.5 million year old cranium with the accompanying title: "The Human Family Expands." The featured specimen was recovered in 1999 during fieldwork directed by Meave Leakey along the western shores of Lake Turkana at the site of Lomekwi in northern Kenya.

The cranium exhibits post-mortem compression, causing significant distortion of the vault and face as well as extensive cracking and expansion of small fragments of bone as can be seen in the photograph. The post-mortem damage to the specimen necessitated a challenging and protracted period of fossil preparation and restoration of the cranium. The cranium consists of two major parts, the brain case and the face—unfortunately the cranial base is not preserved.

Following an in-depth analysis of the anatomy of the KNM-WT 40000, the describers concluded that a new genus and species needed to be coined for the specimen, *Kenyanthropus platyops*, which means flat faced Kenyan man (Greek platus=flat; opsis=face). It is particularly the derived feature of having a flat face, as is evidenced in reduced prognathism and a flatness just below the nasal bones and the nasal aperture, as well as vertically oriented cheek bones that the authors feel makes the specimen so distinctive.

Formerly the only known species of hominid in the 3 to 3.5 million year time interval was *Australopithecus afarensis*. With the increasing awareness that during most of hominid evolution, at least until *Homo sapiens* became the sole hominid taxon on the planet, there has been a diversity of hominid species, lending a "bushy" look to the human family tree. For some researchers it was unusual that *A. afarensis* was a lone species.

KNM-WT 40000 lacks the characteristic projecting (prognathic) face of *A. afarensis* in which the area below the nasal opening, rather than being flat, is convex horizontally and vertically very distinct from that seen in the Lomekwi specimen. On the basis of the differ-

ences between *A. afarensis* and 40000, a species distinction seemed appropriate to the describers.

There is no accepted definition of a genus, as there is for a species. In the opinion of the investigators studying the Lomekwi cranium on the basis of a flat face, presumably small postcanine teeth (based on one molar crown), and lack of derived *Homo* features such as an enlarged brain, it was judged prudent to establish a new genus, *Kenyanthropus*. They further suggest, although do not elaborate, that the pattern of dental and facial anatomy signals a shift in diet, although the nature of such a dietary change is not specified.

Some features of WT 40000 are shared with chimpanzees and other early hominids such as *A. anamensis* with a small ear opening (external auditory meatus) and the apparently quite small cranial capacity. The presumed male status for the specimen is based on the slightly raised and closely placed temporal lines that are the attachment areas on the cranial vault for the large temporalis chewing muscles.

Looking more widely within the African storehouse of fossil hominids, Meave Leakey and her colleagues point to similarities between the face of *K. platyops* and the type specimen for *Homo rudolfensis*, KNM-ER 1470. The 1.8 million-year-old 1470 specimen, when discovered, was only assigned to the genus *Homo*, because of its enlarged brain, 775 cc., and only later designed at new species (see page 189).

The morphological similarities between 1470 and 40000 suggest to the authors that there is a phylogenetic relationship between the two, and they have proposed that KNM-ER be transferred to the genus *Kenyanthropus*, becoming *K. rudolfensis*. One of the many shortcomings of such an arrangement is that 1470 is 1.7 million years younger than the Lomekwi specimen with not a single specimen bridging that long temporal gap. This is a significantly long hiatus over which to ascertain an ancestor-descendent connection.

Subsequent to the 2001 announcement of *K. platyops*, Tim White seriously questioned the validity of using such a distorted and

cracked specimen as KNM-WT 40000 to establish a new species, never mind a new genus. He drew attention to the high degree of "expanding matrix distortion" exhibited by the 40000 specimen, which in his view makes a firm taxonomic judgment problematic. In the face alone there are roughly 1,100 separate pieces of bone, which makes its true shape in life difficult, if not impossible, to know. In other words, the flatness of the face may be an artifact of post-morten distortion, and the specimen might be a Kenyan example of *A. afarensis*.

As is so often customary with new discoveries like *K. platyops*, many new questions arise. For example, the cranium has a rather large, flat face with a forward placement of the cheek bones, suggesting powerful chewing, but the cheek teeth are small. Why such an odd combination of traits? More than likely the exact position of *K. platyops* on our family tree will not be resolved until additional, less distorted fossils are recovered that would clarify the anatomy of the presumed new genus, *Kenyanthropus*.

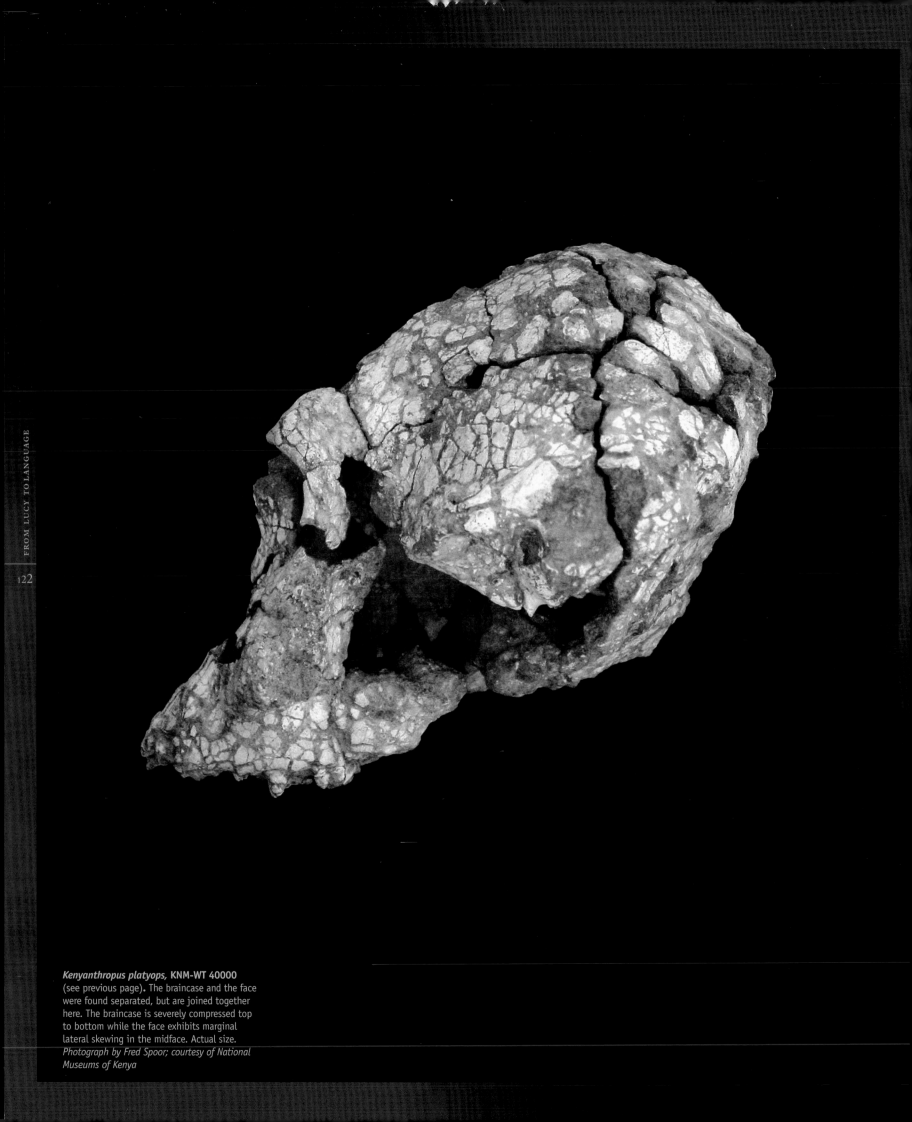

Kenyanthropus platyops, **KNM-WT 40000**
(see previous page)**.** The braincase and the face
were found separated, but are joined together
here. The braincase is severely compressed top
to bottom while the face exhibits marginal
lateral skewing in the midface. Actual size.
*Photograph by Fred Spoor; courtesy of National
Museums of Kenya*

***Kenyanthropus platyops,* KNM-WT 40000.**
This badly crushed 3.5 million-year-old cranium
from west Lake Turkana appears to have a flat
face like *Homo* and not projecting like *Australopithecus*. It may, however, also represent the
distorted face of an *A. afarensis*. Actual size.
*Photograph by Fred Spoor; courtesy of National
Museums of Kenya.*

Australopithecines

Charles Darwin made a brave prediction in 1871 in *The Descent of Man*, when he wrote,

In each great region of the world the living mammals are closely related to the extinct species of the same region. It is, therefore, probable that Africa was formerly inhabited by extinct apes closely allied to the gorilla and chimpanzee; and as these two species are now man's nearest allies, it is somewhat more probable that our early progenitors lived on the African continent than elsewhere.

Vindication of this prediction did not come until nearly half a century after Darwin's death, when Raymond Dart, anatomist at the University of the Witwatersrand in Johannesburg, announced the discovery of a fossilized child's skull from the site of Taung in South Africa. In 1925, he placed the skull in a new taxon, calling it *Australopithecus africanus*. In the specimen he saw a mixture of apelike features such as a projecting face and very small brain (405 cc), but he also surmised from the ventral position of the foramen magnum (the hole at the base of the skull from which the spinal cord emerges) that the creature had walked upright and was therefore a human ancestor (see page 154).

Dart's choice of the name *Australopithecus* is a curious mixture of Latin and Greek: *australo*, Latin for "southern," and *pithecus*, latinized Greek for "ape." These "southern apes" of Africa have turned out not to be confined to southern Africa nor are they apes., Because, however, of the stringent rules of the International Code of Zoological Nomenclature, which govern the naming of taxa, Dart's name cannot be altered. *Australopithecus* is the valid generic designation for a diverse group of early hominids (members of the zoological family the Hominidae, which includes ourselves and our bipedal ancestors), who apparently did not make and use stone implements and were common to both southern and eastern Africa between roughly 4 and 1 million years ago.

Since Dart's coining of the genus, subsequent discoveries have been made which show that *Australopithecus* consist of the seven generally recognized species: *anamensis*, *afarensis* (includes *bahrelghazali*), *garhi*, *africanus*, *robustus*, *aethiopicus*, and *boisei*. These early hominids, collectively called the australopithecines and sometimes referred to as man-apes, reflect the initial radiation of our zoological family and have been found only in Africa. Although they are confined to sites in South and East Africa, it is most probable that they lived over a much broader area of Africa, but their remains did not, unfortunately, become fossilized. Although they possessed the cardinal feature for being placed in the Hominidae, bipedality, they are distinguished from our own genus *Homo* by having relatively small brains. *Australopithecus* brains generally were 500 cc or less, but two specimens have been recorded with slightly larger brain volumes. *Homo* brains, on the other hand, are usually larger than 600 cc (one has been recorded as low as 510 cc; 417 cc, if the skull from Flores, Indonesia is included) and range up to more than 2,000 cc. Modern *Homo sapiens* brains average 1,400 cc.

Beginning in 1936, the famed paleontologist Robert Broom, from the Transvaal Museum in Pretoria, South Africa, began to find additional australopithecine fossils at the site of Sterkfontein. Now numbering more than 500 specimens, these finds from Sterkfontein were first given a new name, *A. transvaalensis* (after the Transvaal region of South Africa), but they are now considered to be of the same species as the Taung Child and are placed in *A. africanus*.

Two other sites in southern Africa, Kromdraai and Swartkrans, yielded remains of a different type of australopithecine. In 1938, Broom recognized a fossil hominid from Kromdraai that differed from *A. africanus* in having a bony crest on the top of the skull, massive postcanine teeth, and a robustly built mandible. With his proclivity for coining new genus and species names for each of his discoveries, and wanting to emphasize the robust nature of the Kromdraai specimen, Broom designated the find *Paranthropus robustus*. *Para* is Greek for "near" or "alongside," and *anthropus* is latinized Greek for "man." The Kromdraai hominid collection is dominated by teeth (50), and although most paleoanthropologists recognize them as a discrete species, they are placed in the same genus as the Taung Child and referred to as *Australopithecus robustus*.

In 1948, Broom and his student John T. Robinson recovered more fossils of robust australopithecines from the site of Swartkrans, situated across an old river valley about 1.5 kilometers from Sterkfontein. True to form, Broom distinguished these finds by a new name: *Paranthropus crassidens* (*crassus* meaning "thick" or "solid" and *dens* meaning "tooth"). Long-term excavation at the site has recovered close to 300 australopithecine specimens from Swartkrans, making it one of the largest samples from a single site anywhere in Africa. Although some paleoanthropologists believe that the Swartkrans hominids are a different species from those at Kromdraai, most researchers place the Swartkrans finds in *A. robustus*.

In the late 1940s, Dart undertook fieldwork at the site of Makapansgat in the northern Transvaal, where several dozen australopithecines fossils have now been found. In spite of the many anatomical similarities shared with the Sterkfontein hominids, Dart ascribed the Makapansgat fossils to a new species, *A. prometheus*. The species name is from the Greek Titan who stole fire from the heavens, and it was chosen because the bones had a dark coloration, presumably caused by being burned in fires made by the australopithecines. Chemical analysis, however, revealed that the dark coloration was due to staining by manganese dioxide. Currently, the Makapansgat hominids are placed in *A. africanus*.

In the nonhominid animal fossils from Makapansgat, Dart saw a high incidence of horns, antelope mandibles, and certain long bones. He interpreted these as having been selectively collected and used by the australopithecines as tools and weapons. The long bones were used as clubs, he speculated, the horns as daggers, and the serrated edges of the teeth in antelope mandibles were employed as saws. He named this assemblage the Osteodontokeratic Culture (*osteo* means bone, *donto* means teeth, and *kerato* means horn). It is now believed that the bone accumulation at Makapansgat is better explained by the scavenging and hunting activities of hyenas, porcupines, and leopards.

Although the South African australopithecine cave sites cannot be precisely dated using radiometric techniques because there are no volcanic ashes or other datable deposits, a comparative chronology has been established using the presence and absence of time-specific mammalian species. The premise is simple: older geological layers have a higher percentage of extinct forms and younger layers have

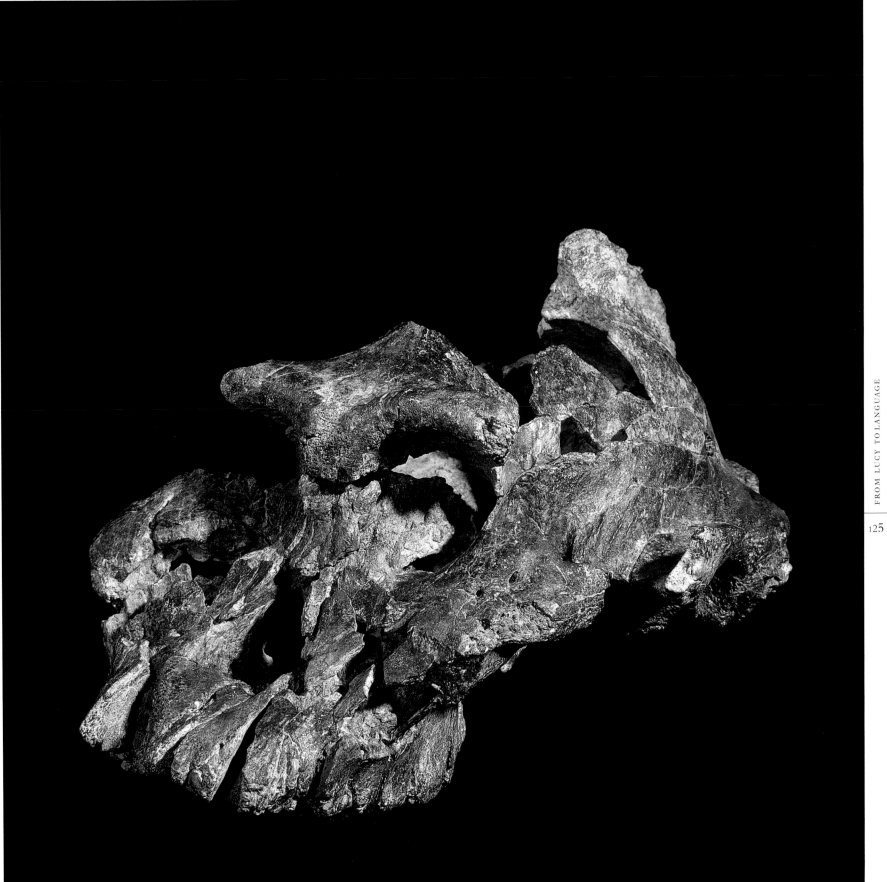

***Australopithecus aethiopicus*, KNM-WT 17000.**
This 2.5 million year-old cranium comes from
the earliest known kind of robust australopithecine, a parallel branch of hominids that went
extinct about a million years ago. Actual size.
*Photograph by Robert I.M. Campbell; courtesy of
National Museums of Kenya*

more species of extant forms. Comparison of extant and extinct species found in the cave sites in South Africa, using this premise, gives the following chronological sequence for the sites, from the oldest to the youngest: Makapansgat, Sterkfontein, Taung, Kromdraai, and Swartkrans.

Because some animal species found in the South African sites are also known from radiometrically dated sites in East Africa, an approximate idea of the ages of the South African sites can be established using biostratigraphy. Some species of antelopes and monkeys, for example, are time diagnostic; that is, they are known only from specific time ranges. With this knowledge, approximate biostratigraphic ages have been suggested for the South African australopithecine sites: Makapansgat, 2.8 million years; Taung, 2.3 million years; Sterkfontein, 2.5-3.0 million years; Kromdraai, 2.0 million years; and Swartkrans, 1.8 to 1.0 million years.

A few other fossil-bearing cave sites in the Transvaal of South Africa have yielded remains of *A. africanus* (Gladysvale) and *A. robustus* (Gondolin and Coopers), and the site of Drimolen promises to be among the most prolific of all. Located less than 10 km distant from the Sterkfontein Valley, Drimolen, biostratigraphically dated to between 1.5 and 2.0 million years, preserves extensive cave breccia remarkably rich in fossil vertebrate remains and to date some 79 hominid specimens have been excavated. The great majority of the specimens are assigned to *A. robustus* while the remainder to the genus *Homo*. Preservation of the specimens is exceptional and further documents the co-occurrence of *A. robustus* and *Homo*, and a significant level of sexual dimorphism in size is apparent in *A. robustus*.

In 1959, at Olduvai Gorge in northern Tanzania, Mary Leakey, a world-famous archaeologist, found the first significant specimen in East Africa to be recognized as an australopithecine. The specimen was nicknamed "Nutcracker Man" because of the enormous cheek teeth. Louis Leakey, a pioneer of paleoanthropology in East Africa, initially dubbed the find *Zinjanthropus boisei* (the species name was chosen to honor a benefactor Charles Boise), but it is now called *A. boisei*. Sometimes referred to as a "hyper-robust" australopithecine, *boisei* is distinguished from its South African counterpart, *robustus*, by its massiveness, enormous cheek teeth, tall face, visor-like cheekbones, and other anatomical features. Numerous specimens of *boisei* have been recovered from sites in Ethiopia, Kenya, and Tanzania from a time interval between 2.3 and 1.4 million years ago.

One of the best-known species of australopithecine, *A. afarensis*, was named in 1978 by Donald Johanson, Tim White, and Yves Coppens, based on fossil finds from Laetoli, Tanzania, and Hadar, Ethiopia. This australopithecine, represented by nearly 400 specimens that include most cranial and postcranial bones in the body, is distinguished by a unique suite of primitive cranial, dental, and mandibular anatomical features. Overall the skull is very apelike in appearance, dominated by a strongly projecting face. The neurocranium is quite small, containing a brain averaging roughly 478 cc in size. The species is best known from the famous 3.2 million-year-old Lucy skeleton, which is clearly bipedal but retains certain apelike features such as relatively long arms and curved finger and toe bones. There is almost universal agreement among paleoanthropologists that *afarensis*, now also known from Kenya and Chad, is the sole hominid species thus far recognized in the 3 to 4 million-year time slice, with possible exception of *Kenyanthropus platyops* (see page 121).

Based on a toothless mandible from the Shungura Formation near the Omo River in southern Ethiopia, *A. aethiopicus* was named in 1968 by two French paleontologists. This taxon is now known from an undistorted and fairly complete cranium from west of Lake Turkana, in northern Kenya, called the Black Skull. This 2.5 million-year-old cranium, which possesses a large, flangelike, posteriorly placed sagittal crest atop the braincase and enormous cheek teeth (inferred from large tooth roots), has obvious affinities with the robust australopithecines. It is distinguished, however, from all other species by having features reminiscent of *A. afarensis*, such as a projecting face, little flexion of the cranial base, and compound temporonuchal crests situated on the occipital (back) of the cranium. Even though less than half a dozen specimens have been identified, *aethiopicus*, dated to between 2.7 and 2.5 million years, is one of the most distinctive of all australopithecine species.

From two fossil sites in northern Kenya, Kanapoi and Allia Bay, 79 australopithecine specimens were assigned to *A. anamensis* by Meave Leakey, a zoologist at the National Museums of Kenya, and her colleagues. Slightly more ancient than *A. afarensis*, dating to between 3.9 and 4.2 million years ago, *anamensis* (*anam* is the Turkana word for lake), they believe, makes an ideal ancestor to *afarensis*. Although *anamensis* shares many anatomical details in the jaws and teeth with *afarensis*, the primitive, very receding mandibular symphysis (front of the

mandible), narrow, parallel tooth rows, and the detailed anatomy of the teeth justify placing these finds in a novel species.

The geographic distribution of australopithecines has always been hampered by lack of favorable geological deposits outside of the Great Rift Valley in East Africa and the cave sites in South Africa. But recently Michel Brunet, a French paleontologist at the University of Poitiers, reported the discovery of the front portion of a mandible with seven teeth from the Central African country of Chad, 2,500 kilometers west of the Great Rift Valley. The associated vertebrate fauna shares similarities with fauna from sites in East Africa dated to between 3.0 and 3.5 million years. Initially assigned to a new species, the Chad specimen exhibits dental and mandibular anatomy similar to *A. afarensis* that make its assignment to this species a logical choice.

Although it is clear that there is diversity in australopithecines, not everyone is agreed on the phylogenetic relationships between them or even on their taxonomy. For some, the robusts are best placed in the genus *Paranthropus*, which includes *robustus*, *boisei*, and *aethiopicus*, and sometimes *crassidens*. The nonrobust

australopithecines are sometimes referred to as gracile australopithecines, probably a misnomer, since their bodies do not appear to be significantly more lightly built than those of the robusts. Using Dart's original appellation, the genus *Australopithecus* currently includes *anamensis*, *afarensis*, *garhi*, and *africanus*.

The rationale for maintaining a generic distinction for these early hominids was clearly articulated by John Robinson in 1972 when he considered the specialized cranial anatomy of *Paranthropus* (the robusts) as representing a specific dietary specialization for processing hard, tough, fibrous foods. The more generalized cranial morphology of *Australopithecus*, for Robinson, reflected a more omnivorous diet that incorporated vegetation as well as meat. This view, dubbed the dietary hypothesis, has been revived by paleoanthropologist Ron Clarke at the University of the Witwatersrand in Johannesburg. For these researchers a genus reflects a specific way of life or "adaptive zone." The species within each genus, *Paranthropus* and *Australopithecus*, simply reflect variations on the larger theme of the generic adaptation. While recognizing this point of view, in this book we have chosen to use only a single genus, *Australopithecus*.

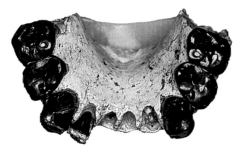

Australopithecus afarensis, KT12/H1. Described as a new species in 1996, this mandible from Bahr el Ghazal, Chad extends the range of australopithecines to 2,500 kilometers west of the Great Rift Valley. It is now considered to belong to *A. afarensis*. Actual size. *Courtesy of Michel Brunet, Université de Poitiers.*

***Australopithecus africanus,* STS 71 cranium and STS 36 mandible.** A child's skull from Taung served as the type specimen for this species in 1925, but several hundred specimens, including these two, have since been recovered from the site of Sterkfontein, South Africa. Actual sizes. *Photograph by David L. Brill; courtesy of Transvaal Museum.*

Efforts to dig even deeper into the past have begun to pay off with the recovery of purported pre-australopithecine fossils dating back to nearly 7 million years. The discoveries are, for the most part, fairly fragmentary, but they are providing us with a glimpse into creatures that may not be very distant from the common ancestor to both the extant African apes and hominids. Until these finds from Chad and Ethiopia our evidence for pre-4 million-year-old hominids was restricted to a few jaw fragments and teeth from the Kenyan sites of Tabarin and Lothagam and perhaps Lukeino (Kanapoi hominids are considered above).

When initially published, fossil hominid remains from 4.4 million-year-old deposits from the site of Aramis in the Middle Awash region of Ethiopia were assigned to *Australopithecus ramidus* (*ramid* is from the Afar word for root). The remarkably primitive aspects of the teeth and jaws, such as the thin dental enamel, relatively large canines, and several other features prompted the discoverers to transfer these finds to a new genus, *Ardipithecus ramidus* (*ardi* is from the Afar word for floor). In doing so, paleoanthropologists Tim White from Berkeley, Gen Suwa from Japan, and Berhane Asfaw from Ethiopia removed the Aramis hominids from designation as an australopithecine.

An even more ancient species of *Ardipithecus*, also known from the Middle Awash of Ethiopia, ranges in age from 5.2-5.8 million years and is assigned to *Ard. kadabba*. Recovery of a toe bone suggests this taxon was bipedal, but the evidence is inconclusive. An upper canine assigned to *Ard. kadabba* is taller than that of *Ard. ramidus* and slightly more primitive in showing traces of a canine shearing complex like that seen in African apes. Taken together the two species of *Ardipithecus* suggest that alterations in canine size and function as well as perhaps bipedalism were features already present in some of the earliest putative hominids.

Another example of a pre-australopithecine comes from Miocene deposits in Chad and has been dubbed *Sahelanthropus tchadensis*. The very crushed cranium, perhaps as old as 6-7 million years, with a remarkably small brain shows an amalgam of ape and humanlike features. The base of the cranium suggests that the skull was balanced on an upright vertebral column, but whether this is evidence for bipedalism is questionable. The upper canines appear to be worn on the tip, like humans, and lack the shearing function typical of apes.

In the Tugen Hills of Kenya 6 million-year-old fossil remains, thought to be hominid, have been assigned to *Orrorin tugenensis*. In this taxon, as opposed to the other pre-australopithecines, the canines are very pointed, resembling the ape condition. Anatomy of the femoral bones appears to resemble hominids and not apes, thus suggesting a gait pattern more similar to humans than apes.

The 4 million year barrier has been breached and hopefully additional finds will bring us a crisper picture of what the earliest hominids looked and acted like. Molecular evidence based on the relationships between modern humans and apes suggests a divergence date of near 7 million years ago between chimpanzees and hominids. It now appears that fossil finds are converging on that divergence date.

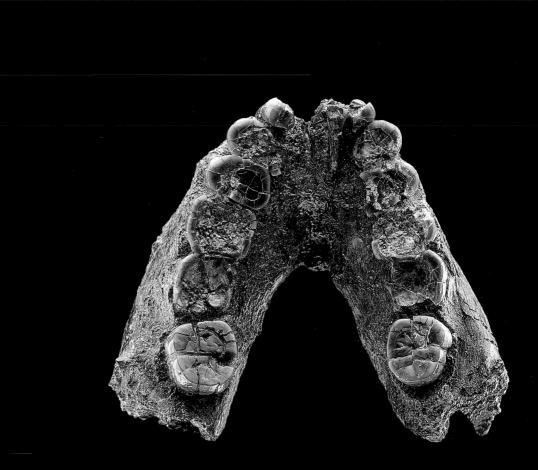

Australopithecus boisei, **L. 7a-125 from Omo.** An occlusal view of the mandible best illustrates the enormous premolars and molars found in this species. Actual size. *Photograph by David L. Brill; courtesy of National Museum of Ethiopia.*

SPECIMEN	LOCALITY	AGE	DISCOVERERS	DATE	PUBLICATION
Adult mandible	Kanapoi, Kenya	4.1 million years	Peter Nzube	September 10, 1994	Leakey, M.G., C.S. Feibel, I. McDougall and A.C. Walker. 1995. New four million year-old species from Kanapoi and Allia Bay, Kenya. *Nature* 376: 565-571

Australopithecus anamensis

KNM-KP 29281 TYPE SPECIMEN

The wind-swept desert southwest of Lake Turkana (formerly Lake Rudolf) was first explored in 1965 when an expedition from the Museum of Comparative Zoology at Harvard University collected 4 million-year-old fossils from this area at a site called Kanapoi. Bryan Patterson, expedition leader, was the son of Lieutenant-Colonel Patterson, an Indian army officer who played a critical role in the construction of the railway from Mombasa to Uganda at the end of the nineteenth century and beginning of the twentieth. Patterson's group found only a single hominid, a left distal humerus, KNM-KP 271.

In 1965, recovery of a 4 million-year-old hominid, even an isolated specimen, drew considerable attention, but its taxonomic assignment was ambiguous. Because of its great age, some scientists referred the specimen to *Australopithecus*, but others pointed out anatomical affinities with *Homo*. As more has become known about the anatomy of early hominid humeri, it has become apparent that due to extensive overlap in anatomy between *Homo* and *Australopithecus*, it was impossible to assign the Kanapoi humerus to a particular genus with certainty. More diagnostic specimens were necessary to resolve the evolutionary affinities of the Kanapoi hominid.

Because of the remoteness of Kanapoi, fieldwork did not begin again until zoologist Meave Leakey of the National Museums of Kenya organized a research team in 1994. Her extensive field experience, gained in part by working with her husband, Richard Leakey, in the Lake Turkana Basin, provided a vital background for the work at Kanapoi. As other members of the search team, Meave Leakey selected Peter Nzube and Kamoya Kimeu, veteran fossil hominid finders from the National Museums of Kenya.

Now 40 years after the initial hominid find at Kanapoi, the total number of specimens, the majority of which are isolated teeth, has grown to 47. The teeth and jaws exhibited anatomical features reminiscent of *A. afarensis*, but there were obvious differences. Although they were clearly best placed in *Australopithecus*, a host of primitive features demanded that a new species be erected for the hominids: *A. anamensis*. The species name derives from the Turkana word anam, which means lake, and was chosen because of the proximity of Kanapoi to Lake Turkana, and as a reference to the paleolake, Lake Lonyumun, which existed in the area millions of years ago.

The type specimen consists of a mandible (KNM-KP 29281), and a fragment of a left temporal bone of a skull (KNM-KP 29281B). The mandible, lacking the ascending rami but possessing a full complement of teeth, is rather small and narrow, reminiscent of the condition seen in Miocene apes. The tooth rows are even straighter and more parallel than in *A. afarensis*. In lateral view, the front of the mandible in *anamensis* is markedly receding, with the internal buttress, or torus, as it is called, heavily developed and elongated posteriorly. These and many other dental features substantiate claims of a more primitive status for the Kanapoi hominids than that seen in *afarensis* (see page 133).

A partial tibia (KNM-KP 29285), consisting of proximal and distal ends, has anatomical features indicating bipedalism. The proximal condyles of the tibia, which articulate with the distal femur, are concave, as in humans, and the distal tibia, where it articulates with the ankle, has a thickening of the bone that acts as a built-in shock absorber during bipedal locomotion (see page 39).

An additional 31 fossil hominid specimens from Allia Bay, on the eastern shores of Lake Turkana, have been placed in the same species as those from Kanapoi. The sedimentary sequences at Kanapoi and Allia Bay have been very adequately dated using the argon-argon method. The majority of the vertebrate fossils at Kanapoi come from two levels. The lower one, dated to approximately 4.2 million years, is where the type mandible for *anamensis* was found. The upper one, containing Patterson's humerus and the tibia, is older than 3.5 and younger than 4.1 million years. The Allia Bay hominids are roughly 3.9 million years old.

The enlarged, although still fragmentary, sample of fossil hominid specimens recovered from Kanapoi and Allia Bay is a testament to dedicated field work under extremely challenging conditions. Thankfully we now have a tantalizing snapshot of the most primitive (apelike) known species of *Australopithecus*. For example, in *A. anamensis* the upper canine/lower premolar region is more primitive than in *A. afarensis* in exhibiting a larger canine (relative to the cheek teeth) and a lower first premolar (P3) that has only a single cusp. In *A. afarensis* the upper canine shows some degree of reduction in size, and the sample of P3s in some specimens have only a single cusp and some have two cusps (a more advanced condition). The more primitive status of *A. anamensis* is also obvious in the straight, parallel tooth rows and the strongly receding nature of the anterior outline of the mandible. These and other anatomical features distinguish *A. anamensis* as a more primitive and presumably ancestral species to *A. afarensis*. Detailed study is underway assessing the veracity of accepting *A. anamensis* as a direct ancestor to *A. afarensis*, suggesting perhaps recognition of an evolving lineage in eastern Africa.

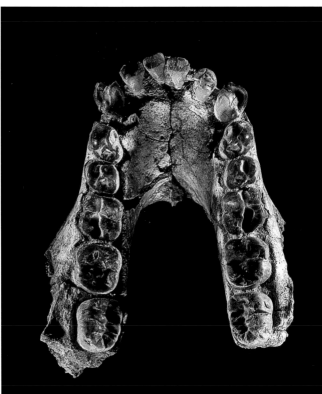

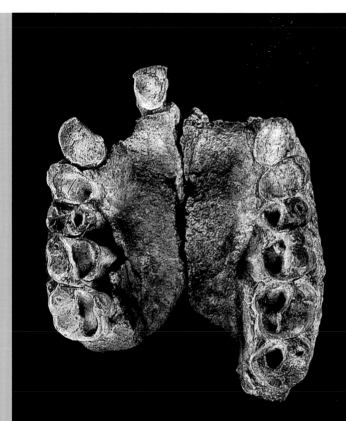

Australopithecus anamensis, **KNM-KP 29281** (see previous page). Straight, parallel tooth rows, seen in occlusal view, and the receding symphysis beneath the front teeth in lateral view distinguish this mandible from that of other hominid species. Actual sizes. *Photographs by Robert I.M. Campbell; courtesy of National Museums of Kenya.*

Australopithecus anamensis, **KNM-KP 29283.** In this species, the maxilla has a very shallow palate, seen in occlusal view, and the canine tooth with its robust root and crown, seen in lateral view, exceeds in size that of *A. afarensis.* Actual sizes. *Photographs by Robert I.M. Campbell; courtesy of National Museums of Kenya.*

SPECIMEN	LOCALITY	AGE	DISCOVERERS	DATE	PUBLICATION
Cranium	Bouri, Ethiopia	2.5 million years	Y. Haile-Selassie	November 20, 1997	Asfaw, B., T. White, O. Lovejoy, B. Latimer, S. Simpson, and G. Suwa. 1999. *Australopithecus garhi*: A new species of early hominid from Ethiopia. *Science* 284: 629-635

Australopithecus garhi
BOU-VP-12/130

The study of paleoanthropology is known not only for being highly controversial, but also because so many discoveries seem to come as a total surprise. Dart's Taung discovery (see page 154) startled everyone in Europe because the Eurocentric view stressed the notion that the earliest human ancestors would be found in Europe or Asia, but certainly not Africa. The recovery of diminutive humans with ridiculously small brains on the Island of Flores (see page 248) was also a stunning revelation to anthropologists, particularly since these people lived only 18,000 years ago. The list goes on and on, but the 1999 publication of a 2.5 million year old cranium from Bouri, Ethiopia was such a surprise to the discoverers they actually named it "the southern ape surprise," *Australopithecus garhi*—"garhi" means surprise in the Afar language.

The fossil hominid record between 2 and 3 million years ago is pivotal for comprehending the origins of *Homo* as well as specialized vegetarians like the robust australopithecines. While this period of time is well-known from fossil discoveries assigned to *A. africanus* in South Africa, the East African hominid record is lamentably patchy for this segment of time. Anthropologists have long focused on ascertaining the beginnings of *Homo*, and the Bouri specimen is found virtually half way between *A. afarensis* and early *Homo*, offering a glimpse into the this murky, yet critical time interval.

While some anthropologists consider *A. africanus* in South Africa a plausible ancestor to *Homo*, others point out that with a series of proto-robust features, such as expanded teeth and jaws, it is more acceptable to see *A. africanus* as an exclusive ancestor to *A. robustus*, also confined to South Africa. With *A. garhi* it is now possible to test the hypothesis that non-robust hominids from this time period would not resemble those assigned to *A. africanus*. This is indeed the case. *A. garhi* is distinguished from *A. africanus* on the basis of primitive features in the palate, the subnasal region, the face, and frontal bone, reminiscent of the conditions typical of *A. afarensis*.

Interestingly, the Bouri specimen does not share the derived and specialized anatomy typical of the robust australopithecines such as a flat, dish-shaped face. In contrast it possesses the prognathic face of the more ancient *A. afarensis*. It is also noteworthy that at 2.5 million years ago in East Africa, the Black Skull, *A. aethiopicus* (see page 164), signals the presence of an independent robust lineage. With enormous cheek teeth and a massive sagittal crest *A. aethiopicus* is allied with the robusts, but retains many primitive features such as a very prognathic face that must have been inherited from its presumed *A. afarensis* ancestor. In fact, it seems inescapable that *A. afarensis* was the last common ancestor to the robusts and *A. garhi*.

The Bouri specimen featured on the cover of the August 23, 1999 issue of *Time* magazine was considered a surprise precisely because of a unique and unexpected combination of primitive and derived anatomical features seen in the cranium. Like *A. afarensis*, the cranium has a rather small capacity, 450 cc, a U-shaped dental arcade with divergent dental rows, strong (subnasal) prognathism, a convex area below the nasal aperture (clivus), and even a small, posteriorly placed sagittal crest. But presence of a very large canine—larger than any *Australopithecus* or early *Homo* species—and the huge molars and premolars makes for a strange combination of anatomical features unique to the Bouri cranium, distinguishing it from all other early hominids.

Postcranial elements, presumably from a single individual, recovered in geological horizons contemporaneous with the *A. garhi* cranium, offer insight into the evolution of limb proportions. The femur is relatively long compared with the upper arm bone (humerus), similar to modern humans and unlike the more primitive, shorter femur seen in *A. afarensis*. The forearm, however, is elongated as it is in *A. afarensis* and apes. Although the postcranial bones cannot be assigned to a species, due to the lack of diagnostic cranial material, they reveal that the mode of evolutionary change in the postcranial skeleton was lengthening of the lower limb prior to the shortening of the forearm.

Only isolated stone tools in the shape of cores and flakes were recovered during survey at the Bouri site, a circumstance unlike the concentrated stone tool occurrences at Gona and Hadar, sites of similar age to Bouri. The discovery, however, of antelope bones bearing stone tool cut marks and even hammerstone impacts demonstrate that hominids were processing animal carcasses for meat and or marrow. It is currently impossible to ascertain if these early hominids were hunting or scavenging, but the lack of suitable raw materials at the open lake margin habitat of Bouri suggests that they were transporting raw stone and perhaps tools to specific places where carcasses were exploited. The appearance of this novel shift in behavior, that is meat eating and stone tool manufacture, suggests that it was precisely this behavioral change that stimulated and supported brain enlargement: tools proceeded big brains, big brains did not proceed stone tools!

While *A. garhi* is temporally positioned between *A. afarensis* and early *Homo*, not all scholars accept that it was ancestral to our own genus. *A. garhi* may simply be a side branch on the human family tree that evolved megadontia (large teeth) in parallel with the robust australopithecines. Certainly more reliable associations between the stone tools, and especially the postcranial remains, would go a long way towards clarifying the position of *A. garhi* on the tree, but so would the recovery of limb bones associated with the earliest *Homo* jaws, such as the 2.33 million year old palate from Hadar (see page 180). If *A. garhi* at 2.5 million years was the precursor to *Homo* at 2.33 million years, then very rapid dental reduction would have transpired over a short time. There seems to be little problem, however, with accepting an ancestral-descendent relationship between *A. garhi* and *A. afarensis*, because in many ways the former species looks like the latter one, only with enlarged teeth.

One thing is for certain *A. garhi* will not be the last surprise for paleoanthropologists.

***Australopithecus garhi,* ARA-VP-12/130**
(see previous page). This 2.5 million-year-old
cranium from the Middle Awash, Ethiopia
possesses features like a projecting face
that resemble the more ancient *A. afarensis*,
yet has an extremely enlarged postcanine
dentition, like robust *Australopithecus*. Actual
size. *Photograph by David L. Brill; Courtesy of
National Museum of Ethiopia.*

SPECIMEN	LOCALITY	AGE	DISCOVERERS	DATE	PUBLICATION
Partial adult skeleton	Hadar, Ethiopia	3.2 million years	Donald Johanson	November 24, 1974	Taieb, M., D.C. Johanson, and Y. Coppens. 1975. Éxpédition internationale de l'Afar, Ethiopie (3 ème campagne 1974), découverte d'Hominidés plio-pléistocènes à Hadar. *C.R. Acad. Sci. Paris* 28: 1297-1300

Australopithecus afarensis
A.L. 288-1 LUCY

Most fossil finds are known simply by a catalogue number or the place name where the remains were found. In the case, however, of Lucy, a partial skeleton, few outside the profession of paleoanthropology would recognize her from the catalogue entry Afar Locality (A.L.) 288-1.

A celebrity among fossils, Lucy is better known than her discoverer. She shows up in crossword puzzles, on "Jeopardy!," in cartoons, poems, rock music lyrics, feminist plays, and even as a tattoo. In Ethiopia, where she was found, the government issued a commemorative stamp with her Amharic name, Dinquinesh, which means "wonderful thing." In remote Ethiopian towns it is not unusual to find a "Lussy Bar" or "Lucy Cafe." The scientific name, *Australopithecus afarensis*, derives from the Afar region and the nomadic Muslim Afar people. The Afar take great pride in Lucy and, although Muslim, some Afars believe that the first human was Lucy, so that all humanity is descended from the Afars.

Worldwide, Lucy has become an ancestral ambassador of sorts, acting like a magnet, drawing people to the study of human origins. Even among those who know practically nothing about human evolution, Lucy is vaguely familiar. In fact, bringing up her name is like referring to a distant relative, which of course she is. The affectionate name given to the partial skeleton comes from the Beatles' song "Lucy in the Sky with Diamonds."

Although more complete skeletons and much older fossils than Lucy have now been found, she is still the touchstone, the reference point, to which other discoveries are compared. Fossil hominids are "older than" or "more complete than" Lucy, for example. And, most important, like the derivation of her name from the Latin *lux*, she has thrown much light on one of the earliest stages of human evolution.

Referred to a new species of *Australopithecus* in 1978, she made her debut at a Nobel Symposium on early hominids as *A. afarensis*. Her species appears to be the last common ancestor to the several branches of hominids that emerged

between 2 and 3 million years ago. A long-lived species, extant between 3 and 4 million years ago, *afarensis* remains are presently recognized in Tanzania, Kenya, and Ethiopia. Lucy's kind left the magnificent footprints in a 3.6 million-year-old volcanic ash at Laetoli, Tanzania.

Lucy's skeleton consists of some 47 out of 207 bones, including parts of upper and lower limbs, the backbone, ribs, and the pelvis. With the exception of the mandible, the skull is represented only by five vault fragments, and most of the hand and foot bones are missing. Because there is no duplication of any skeletal element (for instance, there are not two right humeri), the remains are from a single individual.

The erupted third molar, or wisdom tooth, as well as the closed epiphyseal lines (growth lines) attest that she was a mature adult, despite her diminutive stature of just over 1 meter (3.5 feet). Comparative observations of other fossils assigned to her species suggest that these ancestors were sexually dimorphic, and her small size is characteristic of her female status (see page 73).

In 1995 two Swiss researchers, Martin Häusler and Peter Schmid, challenged assignment of the A.L. 288 skeleton to female status and concluded, after a detailed obstetric analysis of the pelvis (and that of Sts 14; see page 148), that it was more likely the specimen was male, and furthermore that the nickname "Lucy" should be changed to "Lucifer." According to their analysis of a reconstructed pelvis, the very short anterior-posterior diameter of the birth canal was not sufficiently large to permit the head of a neonate to pass through. In addition they suggested that A.L. 288-1 and Sts 14 from Sterkfontein, South Africa were of the same species, and the latter was a female and the Hadar pelvis a male.

Owen Lovejoy and Robert Tague, of Louisiana State University, countered these conclusions in a point-by-point refutation of the Häusler-Schmid analysis. First of all, the Hadar hominids, of which A.L. 288-1 is an integral part, belong to *A. afarensis*, a species different from Sts 14,

which is *A. africanus*. If these two pelves are considered to be the same species, it would mean that multiple hominid species were living at Hadar; a conclusion which is totally unsupported. Secondly, it would ignore the accepted view that *A. afarensis* is a highly sexually dimorphic species with small-bodied females and large-bodied males.

The 2002 discovery of what is now the oldest, most complete skull of *Australopithecus* from Hadar has thrown some light on size variation within *A. afarensis*. The A.L. 822-1 skull, dated to 3.1 million years, possesses light muscle markings, a small canine, and a facial structure like other female *A. afarensis* specimens from Hadar. The mandible, the most complete ever found for *A. afarensis*, is significantly larger than Lucy's, indicating that by chance Lucy is among the smallest individuals of *A. afarensis* known, and therefore best interpreted as a female of the species.

More than 90 percent of *A. afarensis* specimens derive from Hadar, but this species is also known at other sites such as Maka and Belohdelie, Ethiopia, Koobi Fora, Kenya, Koro Toro, Chad, and Laetoli, Tanzania. *A. afarensis* represented by nearly 400 specimens is geographically widespread and spans perhaps nearly a million years making it arguably the best represented species of *Australopithecus* thus far discovered. Recent work under the direction of Zeresenay Alemseged in the Dikika Region, situated immediately south of Hadar and across the Awash River, is promising to further expand our knowledge base for this species. Fossil hominid specimens already found at Dikika are attributed to *A. afarensis*. Some of the Dikika specimens are older than those known from Hadar and offer the possibility to help span the temporal gap that exists between the oldest Hadar hominids at 3.4 million years and those from Laetoli dated to 3.6 million years.

The center of vigorous debate, Lucy has spawned over a quarter century of research in paleoanthropology. For example, some primitive features such as her relatively long arms and curved hand bones suggest that she was still

an agile arboreal climber. Others see her as a terrestrial biped and interpret these features as evolutionary baggage, left over from the time when her ancestors lived in trees. It was difficult at first to find a place for Lucy on the hominid family tree, but few today would deny her an important role in hominid evolution. To some she is the "mother of all humankind" and to others she is the "woman who shook up man's family tree."

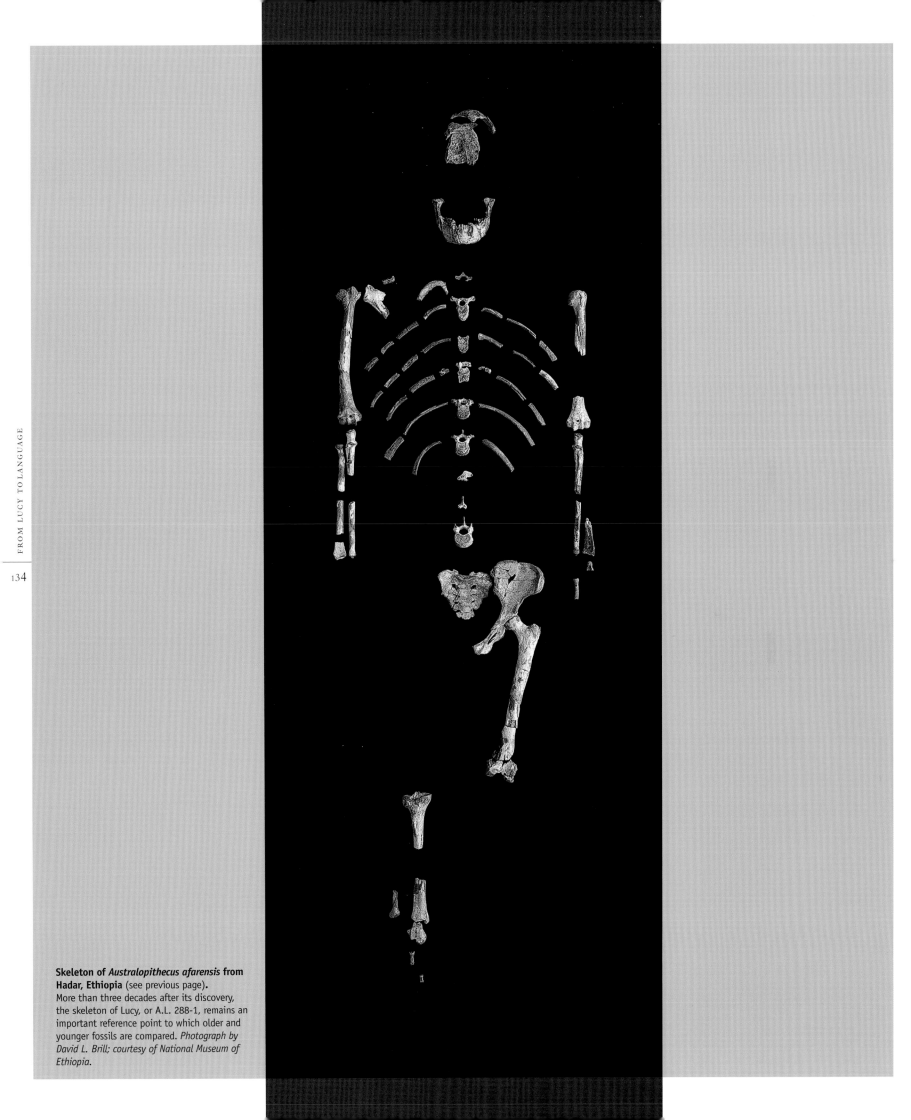

Skeleton of *Australopithecus afarensis* from Hadar, Ethiopia (see previous page).
More than three decades after its discovery, the skeleton of Lucy, or A.L. 288-1, remains an important reference point to which older and younger fossils are compared. *Photograph by David L. Brill; courtesy of National Museum of Ethiopia.*

SPECIMEN	LOCALITY	AGE	DISCOVERERS	DATE	PUBLICATION
Fragments of 13 individuals	Hadar, Ethiopia	3.2 million years	Michael E. Bush	November 2, 1975	Taieb, M., D.C. Johanson, Y. Coppens, and J.-J. Tiercelin. 1978. Expédition internationale de l'Afar, Éthiopie (4 ème et 5 ème Campagne 1975-1977): Chronostratigraphie des gisements a hominidés pliocène de l'Hadar et correlations avec les sites prehistorique de Kada Gona. *C.R. Acad. Sci. Paris* 287: 459-461

Australopithecus afarensis

A.L. 333 *THE FIRST FAMILY*

In the fall of 1975, about a year after Lucy was unearthed, a unique discovery was made at Hadar. At Afar Locality (A.L.) 333, on a steep hillside, fossil remains of a number of hominid fossils were found eroding from a single geological horizon. At the outset it was thought that perhaps this was another partial skeleton, but when two identical left distal fibulae were collected, it was clear that more than one individual was represented. Following two field seasons of work at locality 333, more than 200 hominid fragments were recovered. Most were found during screening operations on the hillside, but 19 specimens were excavated in situ.

Continued work at locality 333 has resulted in the recovery of an additional 23 postcranial fragments and three mandibular and dental hominid specimens. It is now thought that the minimum number of individuals represented by the more than 240 specimens at 333 is 17, representing nine adults, three adolescents, and five juveniles.

Where geological strata are rich in fossil specimens, the assemblage usually consists of a mixture of different animal groups. In the case of locality 333, however, with the exception of a few fish and rodent remains, all the bones were hominid. In paleontology the study of death assemblages is called taphonomy, and when it is biased in such a way as at 333, it is referred to as a catastrophic assemblage. In other words, some disaster overcame a group of *afarensis*, which ultimately became buried in a single geological horizon.

The precise cause of the death assemblage has not been ascertained, but various scenarios such as a flash flood have been suggested. Close inspection of the bone surface reveals a lack of widespread carnivore damage, such as tooth marks or crushing, which rules out systematic predation. Moreover, the articular ends of the bones are intact, not chewed off by scavengers. The bones have, however, been broken, perhaps by water transport before final deposition and fossilization. Lack of advanced weathering of the bone surfaces implies rapid burial.

Since the 333 collection is a catastrophic one, in all probability it samples a group of related individuals. This unique and important collection, therefore, allows some insights into the biology and structure of a single fossil hominid species not available from any other locality in Africa.

Aside from size, the 333 specimens are nearly identical in their anatomy. In fact, the entire size range of hominids found at Hadar and assigned to *A. afarensis* is represented in the 333 collection, adding strength to the notion that only a single species existed in the pre-3.0 million-year-old deposits at Hadar, and that that species, *afarensis*, was markedly sexually dimorphic.

A highly significant benefit gained from the 333 sample is the appreciation of growth and development in a fossil species. A number of fairly complete upper and lower jaws as well as a partial cranium (A.L. 333-105) of a child allow the tracing of development from youth to maturity. Some anatomical features, such as the shape of the nasal region and mandible, which distinguish adult *afarensis* specimens from other species of *Australopithecus*, are already indicated in the immature specimens.

Geologically, the locality is capped by a resistant sandstone, which indicates an ancient meandering river. The geological layer from which the hominids derive is fairly fine-grained, suggesting slow moving water. Ongoing field work at locality 333 has prompted a clearer picture of the paleodepositional history at this unique locality and provided insight into how the hominids came to be interred there. Decisive geological work at the site under the guidance of Kay Behrensmeyer at the Smithsonian Institution suggests that the hominids were buried in a shallow depression (swale) in the upper reaches of a river channel that flowed northwards. Her view is that this interment, a naturally occurring process, was rapid, perhaps within weeks after the death of the individuals. According to Behrensmeyer, however, the under representation of the axial skeleton as well as a few instances of postmortem carnivore damage is not inconsistent with the notion of feline predation.

One final distinction for the hominids at 333 is their geological age. Argon dating of the Hadar deposits brackets these hominids between two volcanic ashes. Precise laboratory analysis shows that the disaster that killed and interred the earliest evidence of our ancestors living in groups transpired between 3,180,000 and 3,220,000 years ago!

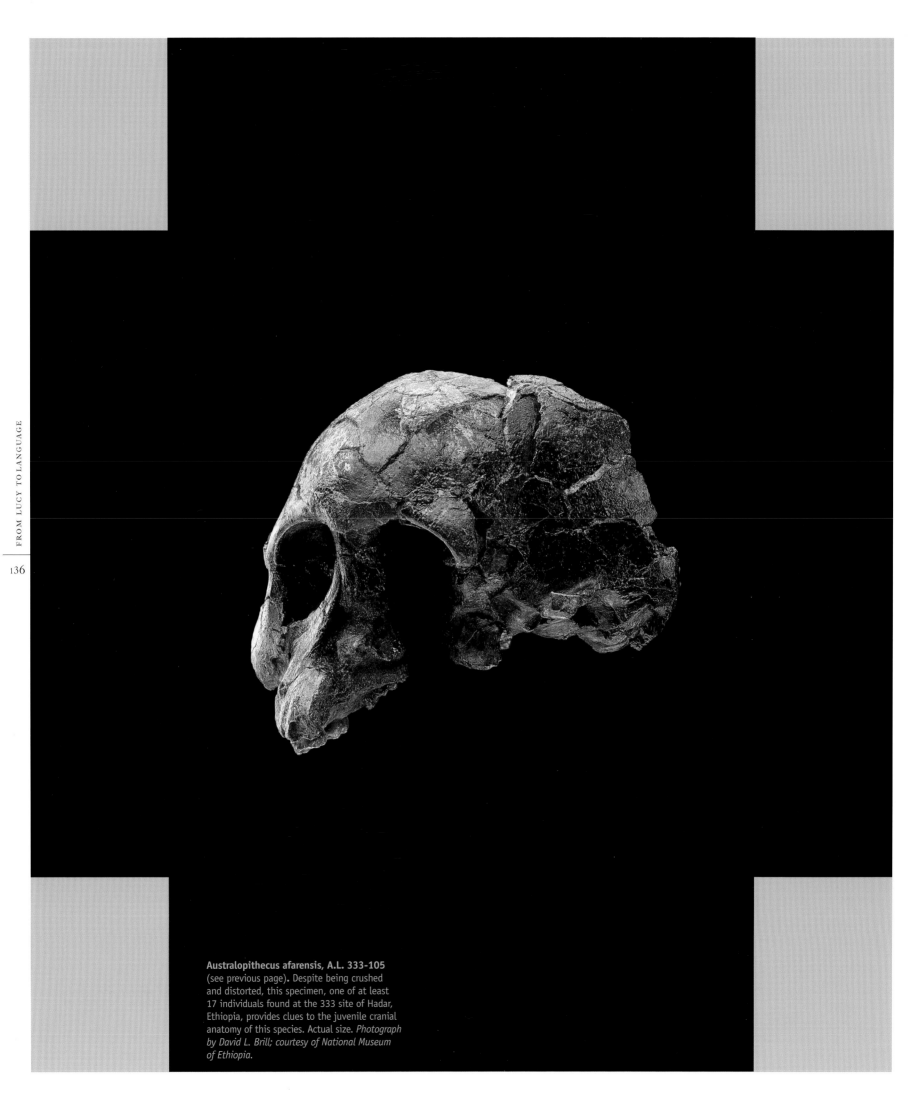

Australopithecus afarensis, A.L. 333-105
(see previous page). Despite being crushed
and distorted, this specimen, one of at least
17 individuals found at the 333 site of Hadar,
Ethiopia, provides clues to the juvenile cranial
anatomy of this species. Actual size. *Photograph
by David L. Brill; courtesy of National Museum
of Ethiopia.*

SPECIMEN	LOCALITY	AGE	DISCOVERERS	DATE	PUBLICATION
Adult skull	Hadar, Ethiopia	3.0 million years	Yoel Z. Rak	February 26, 1992	Kimbel, W.H., Y. Rak, and D.C. Johanson. 2004. *The Skull of* Australopithecus afarensis. New York: Oxford University Press

Australopithecus afarensis
A.L. 444-2

Humans recognize each other by the distinctive constellation of features seen in our skulls, especially our faces. Although other parts of our anatomy are also unique to each of us, it is our face that immediately identifies us. This is also the case for fossil hominids, in which species are most easily identified by a unique combination of anatomical features in the skull. Paleoanthropologists, therefore, generally place much confidence in species for which skulls are known.

Although the vast majority of anatomical traits marshaled to support the announcement, in 1978, of *Australopithecus afarensis* as a new species are seen in the skull, acceptance of this species was not universal, owing in part to the lack of a fairly complete skull specimen. Nearly every anatomical region of the skull was known, but from fragments of different individuals. In 1983 these fragments formed the basis of a composite reconstruction of a male *A. afarensis*. But the lack of association of the different anatomical parts in a complete skull led some paleoanthropologists to suggest that more than one species of hominid was combined into the reconstruction. The implication was that the face of one species and the braincase of another had been assembled into one skull.

An opportunity to test the veracity of the composite skull reconstruction came with the discovery in 1992 of a fairly complete *A. afarensis* skull by Yoel Rak of Tel Aviv University, Israel. Thirteen major pieces of skull and hundreds of small fragments were found in a gully at Hadar. After these fragments were reassembled, the Afar Locality (A.L.) 444-2 skull closely resembled the composite reconstruction, supporting the hypothesis that only a single species was incorporated into the reconstruction.

The skull is slightly older that 3.0 million years and measurements of its overall size distinguish it as the largest *Australopithecus* skull ever found, yet its brain size is estimated to be 530 cc. Because of its size, big canine teeth, massive mandible (not pictured here), and strong muscle markings, the skull is presumed to be a male. Other specimens from Hadar, such as A.L. 417-1, with small canines and a less projecting (prognathic) face, are presumed females. The degree of difference between male and female *A. afarensis* specimens is similar to that seen between the sexes of great ape species.

Based on the heavy dental wear, A.L. 444-2 was an old individual, for much of the enamel is worn away, exposing large areas of dentine. The anterior teeth exhibit heavy wear, suggesting the stripping of vegetation and perhaps manipulation of food with the lips and anterior teeth. Such a view gains support from the posteriorly positioned but low sagittal crest. Fibers of the temporalis muscle toward the back of the skull are oriented fairly horizontally and were enlarged to resist the pulling forces generated by food preparation at the front of the prognathic face. The condition in *A. afarensis* is more like that seen in chimps and especially gorillas. This contrasts with the typical condition in robust australopithecines, where the anterior fibers of the temporalis muscle, acting vertically to generate powerful grinding forces between the cheek teeth, are sometimes so enlarged that an anteriorly placed sagittal crest is formed.

Fortunately, the frontal region of the skull, a previously poorly known cranial region in *A. afarensis*, is well preserved in A.L. 444-2. Details of this anatomy, such as the low, sloping forehead and the bony torus above the eye orbits, add further to the distinctiveness of this species. In addition to affirming the presence of only a single species at Hadar, the skull extends the temporal range of *A. afarensis* from 3.4 to 3.0 million years at this site.

Recovery of the A.L. 444-2 skull permits positive identification of a partial frontal bone found in 1981 at Belohdelie, a site situated roughly 72 kilometers directly south of Hadar. The Belohdelie specimen shares a series of diagnostic features with the A.L. 444-2 specimen such as a thick torus just above the orbits and absence of a sulcus behind the torus, so typical of chimpanzees. Because the Belohdelie frontal is dated to roughly 3.9 million years, the temporal range of *A. afarensis* is extended to 900,000 years, suggesting a long period of anatomical stasis for this region of the skull.

The 2002 discovery at Hadar of a rewardingly well-preserved female skull of *A. afarensis* from A.L. 822-1 offers a long-anticipated view of female cranial anatomy. Lacking only two teeth, the mandible is the most complete thus far discovered for *A. afarensis*. Taxonomic assignment of the 3.1 million-year-old skull to Lucy's species is supported by the diagnostic anatomy of the both the mandible and the cranium that is so characteristic of *A. afarensis*. The small canine size and the lack of strongly marked muscle insertions attest to the female nature of A.L. 822-1. This specimen is especially important since comparisons with male *A. afarensis* skull specimens, especially A.L 444-2, promise to contribute vital knowledge of the pattern of sexual dimorphism in this taxon. Some obvious differences are apparent in the occipital and subnasal regions that undoubtedly reflect subtle differences between males and females. Size wise it is interesting to note that A.L. 822-1 is intermediate between Lucy's mandible and that of A.L. 444-2, highlighting the fact that Lucy is among the smallest of all *A. afarensis* specimens thus far recovered. Added insight into the tempo and mode of evolutionary change, as well as the pattern of sexual dimorphism, in *A. afarensis* skulls is now made available by the A.L. 822-1 specimen, which is temporally intermediate between the 3.2 million-year-old First Family collection and the 3.0 million year old A.L. 444-2 skull.

Using a fairly complete skull like A.L. 444-2, John Gurche, an artist trained in anthropology, was able to reconstruct, in three dimensions, a male *A. afarensis* skull. Beginning with a cast of the skull, and drawing on knowledge gained from numerous dissections of humans and apes, he sculpted layers of clay into the shape of muscles. After adding fatty tissue and salivary glands, he then positioned glass eyes and the nasal cartilage. Finally, he covered the entire reconstruction with pliable urethane for skin. Although such details as wrinkles, skin tone, and the placement of hair will always be open to debate, the resulting image is undoubtedly a close facsimile of a male of Lucy's species. At long last, *A. afarensis* received a face, something all of us can look at and appreciate how different that species was from our own. Yet, in spite of the overall apeness of the skull, there is a sense of concern in the eyes that reaches out to us across millions of years.

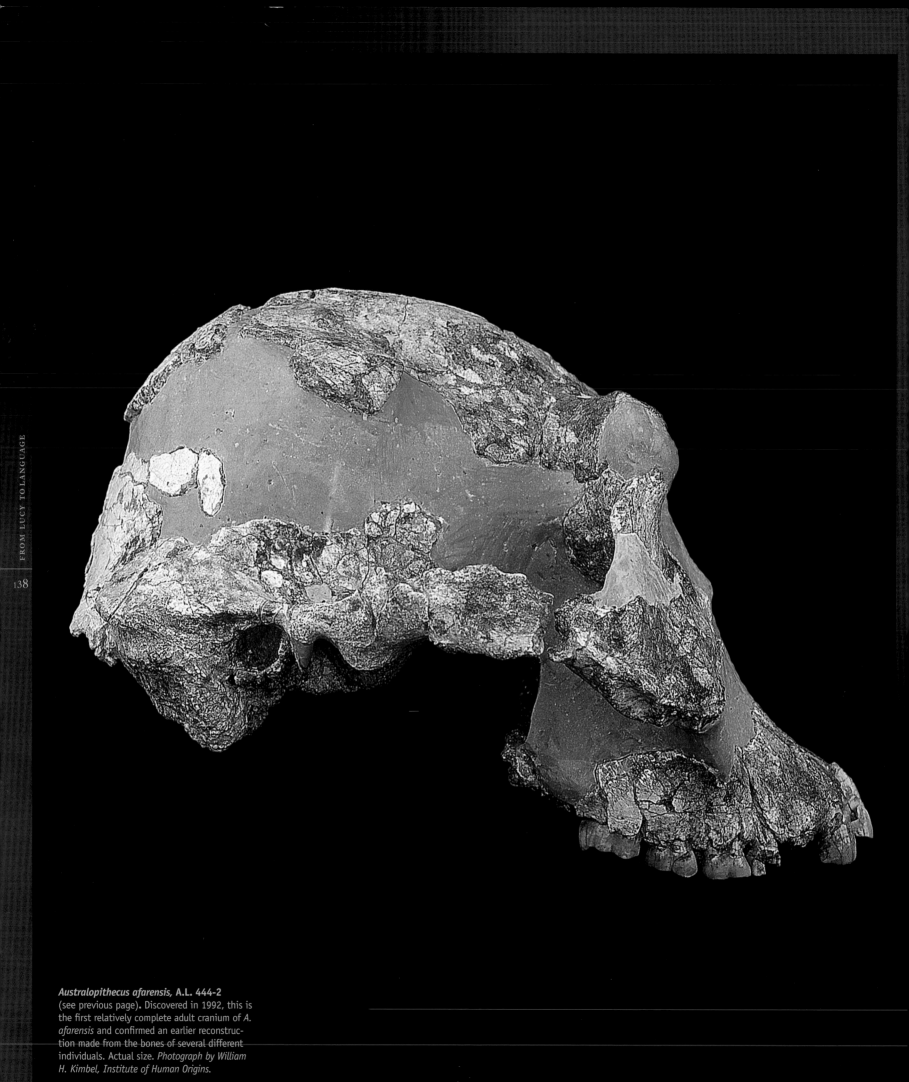

Australopithecus afarensis, **A.L. 444-2**
(see previous page). Discovered in 1992, this is
the first relatively complete adult cranium of *A.
afarensis* and confirmed an earlier reconstruction made from the bones of several different
individuals. Actual size. *Photograph by William
H. Kimbel, Institute of Human Origins.*

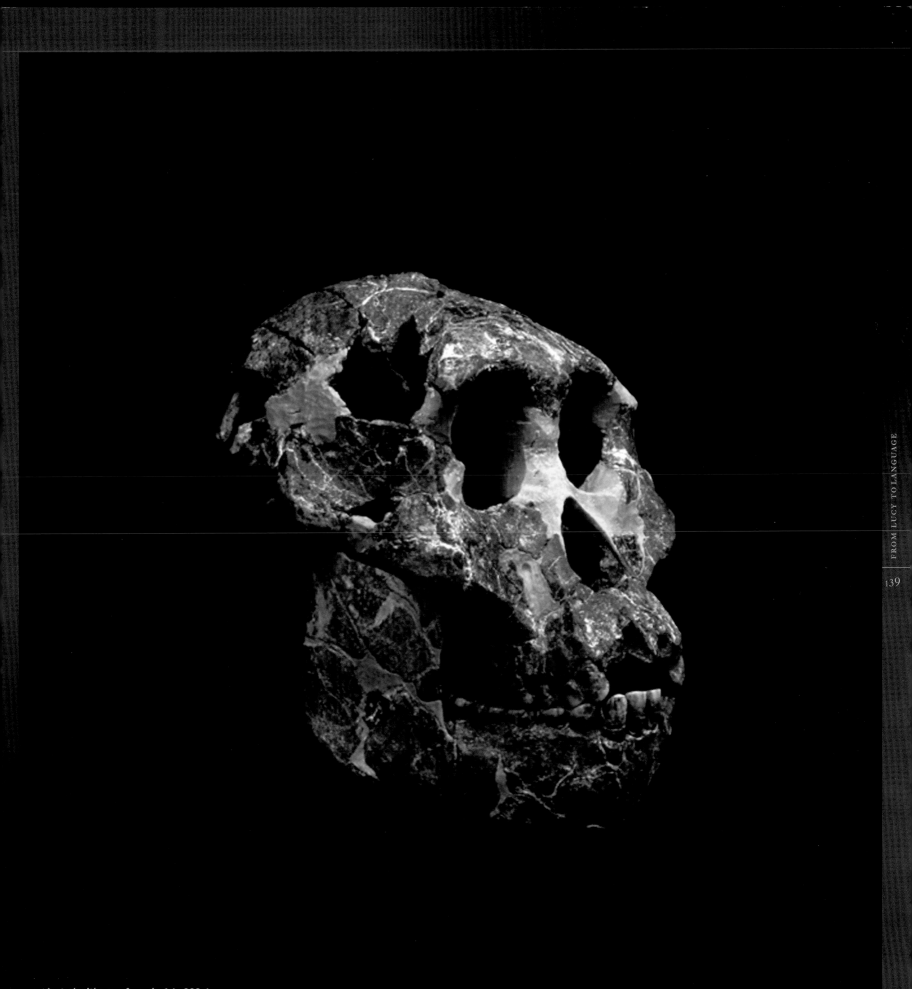

Australopithecus afarensis, **A.L. 822-1**
(see page 137)**.** This 3.1 million-year-old
female skull from Hadar, Ethiopia is wonderfully
complete and adds significant knowledge to the
pattern of sexual dimorphism in *A. afarensis.*
Actual size. *Photograph by William H. Kimbel;
courtesy National Museum of Ethiopia.*

SPECIMEN	LOCALITY	AGE	DISCOVERERS	DATE	PUBLICATION
Adult female knee joint	Hadar, Ethiopia	3.4 million years	Donald Johanson	October 30, 1992	Johanson, D.C., and Y. Coppens. 1976. A preliminary anatomical diagnosis of the first Plio/Pleistocene hominid discoveries in the central Afar, Ethiopia. *Am. J. Phys. Anthropol.* 45: 217-234

Australopithecus afarensis
A.L. 129-1a+1b

During a brief visit of only a few days to the site of Hadar, Ethiopia, in 1972, an international team of scientists was stunned by the incredible number of well-preserved fossils eroding from the geological strata. Evaluation of some of the fossil fauna, particularly the pigs and elephants, suggested an age well in excess of 3 million years for the site. Only a handful of hominids dated to earlier than 3 million years had been found in Africa, and the team was hopeful that fossil hominids would eventually be found at Hadar. With great optimism, plans were made to launch the inaugural expedition to Hadar the following year.

Members of the International Afar Research Expedition, as it was then called, began a systematic survey of the Hadar badlands, with the paramount objective being the discovery of fossil hominids, for any significant finds could add immeasurably to a poorly known segment of time in hominid evolution. As fortune would have it, the first discovery of a hominid at the site was made in some of the oldest geological levels at Hadar.

A small fossil fragment was spotted poking out from soft sediments, and on closer inspection it turned out to be broken, exposing a cross section of bone. The diagnostic portion, the articular end, was hidden in the loose sediment, but after careful removal it was identified as the proximal (top) end of a tibia (A.L. 129-1b). Quite small in size, it could easily be mistaken for a monkey tibia. The tibia was left in the exact place where it had been found and the searchers began a slow survey on hands and knees, which led to the finding of a distal (bottom) end of a femur (A.L. 129-1a). It matched the color of fossilization in the tibia exactly, and when the two bones were fitted together, it was obvious that they were from the same individual. The articulating surfaces were perfectly congruent, having been molded to each other during the life of the individual roughly 3.4 million years ago.

The anatomy of the knee joint is a reliable indicator of locomotion, and from a number of diagnostic features it was apparent that this fossil knee was from a biped.

Three obvious characters verified that it was a hominid knee: first, the femoral shaft rose at an angle that allowed the head of the femur to fit into the pelvis; second, the lateral condyle (one of two knobs on the femur, which sits on the tibia) was flattened and elongated from front to back, which enhances the transfer of body weight to the tibia during walking; and third, the lateral border of the patellar notch on the femur was raised, which prevented the patella (kneecap) from dislocating.

The immediate importance of the knee joint was that it confirmed the presence of hominids in the Hadar geological sequence and, at the time of discovery, provided the oldest evidence for a cardinal feature in human evolution—bipedality. The distal femur fits comfortably into one's palm, suggesting that the individual represented by the knee joint was fairly small, similar in size to Lucy. When first discovered, it was not possible to assign the specimen to a genus or species, but we know now that it belongs to *Australopithecus afarensis*.

The proximal (top) ends of the right and left femora were also found and are assumed to be from the same individual because they are identical in color and fossilization to the knee joint. Both proximal fragments are broken through the neck—a typical fracture pattern in modern humans who break their hip. Furthermore, the anterior and posterior surfaces of the bones show areas of crushing. The splintered nature of the damage suggests that it was inflicted when the bone was still fresh, most likely by a carnivore. Even more interesting is the fact that the left proximal femur fragment was found about 18 meters away, suggesting that perhaps these bones were scattered by carnivore activity on the prehistoric landscape.

Although the Hadar knee joint is no longer the oldest evidence for hominid bipedality, its discovery at Hadar in 1973 provided the impetus for continuing research not only at this important fossil site, but in other regions of the Afar Triangle as well.

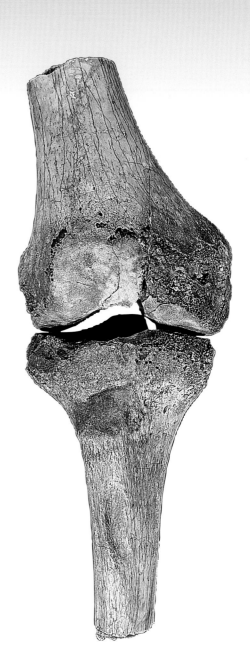

Australopithecus afarensis, **A.L. 129-1a+1b.** The first hominid fossil found at Hadar, Ethiopia, this femur and tibia preserve a complete knee joint, which indicates that this species walked on two legs. Actual sizes. *Courtesy of Donald Johanson, Institute of Human Origins.*

SPECIMEN	LOCALITY	AGE	DISCOVERERS	DATE	PUBLICATION
Adult mandible	Laetoli, Tanzania	3.6 million years	Maundu Muluila	1974	Leakey, M.D., R.L. Hay, G.H. Curtis, R.E. Drake, M.K. Jackes, and T.D. White. 1976. Fossil hominids from the Laetolil Beds, Tanzania. *Nature* 262: 460-465; and Leakey, M.D., and R.L. Hay. 1979. Pliocene footprints in the Laetolil Beds at Laetoli, northern Tanzania. *Nature* 278: 317-328

Australopithecus afarensis

L.H. 4   TYPE SPECIMEN
FOSSIL HOMINID FOOTPRINTS

The preconceptions we carry with us, no matter how scientifically objective we try to be, can strongly influence the way we conduct research. This was probably the case when Louis Leakey visited the fossil site of Laetoli, Tanzania, in 1935, in search of possible hominid ancestors. Although the site is rich in fossil mammalian remains, Leakey spent little time there, for the absence of stone artifacts signaled to him that there could not possibly be hominid remains at the site. Humans are distinct because we make tools, and Olduvai Gorge, some 30 kilometers to the north, where stone tools were oozing out of the geological layers, held much more promise for finding human ancestor remains.

After studying with Sir Arthur Keith, then England's most accomplished and respected anatomist, Leakey was convinced that "true man" did indeed have a very ancient ancestry. Keith was fanatically dedicated to the veracity of the Piltdown skull, which had been found in 1912 in a gravel pit in Sussex, England. There, a hominid skull with a brain of modern size had been recovered in association with the teeth of extinct animals like the extinct pygmy elephant stegodon, with an age of 5 million years. The jaw, however, was like that of a primitive ape. The combination of a modern-appearing cranium and a primitive apelike jaw, now known to be a hoax, justified Keith's view that it was the brain that led the way in human evolution.

As for the teacher, so for the student: Leakey took the view of a very ancient occurrence for "true man" back with him to his native Kenya, and when he learned of the discovery at Olduvai Gorge, in neighboring Tanzania, of a human skeleton, which was found by the noted German paleontologist Hans Reck, from Berlin, Leakey was certain that the find would support Keith's views. In 1931, Leakey and Reck mounted an expedition to Olduvai, where Leakey immediately found stone tools in abundance. He also found something else he was looking for—a very ancient but modern-looking hominid, Olduvai Hominid (OH) 1, in association with stone tools. Back in England, Keith was elated and congratulated Louis Leakey on his discovery. The die was now

cast that would continue to color Leakey's belief in very ancient "true *Homo*," right up to the time of his death.

Even though it was soon discovered that the Olduvai skeleton was the burial of a modern Maasai into older geological horizons, Leakey was not deterred. At the sites of Kanam and Kanjera, in Kenya, he recovered remains of anatomically modern humans in deposits of presumed great antiquity. Geological investigation showed, however, that the finds were not ancient but were in fact intrusive burials into older sediments, and the specimens were those of modern humans. The Piltdown hoax was unveiled in 1953, by anatomists Joseph Weiner and Wilfred Le Gros Clark of Oxford University and Kenneth Oakley, an archaeologist at the British Museum of Natural History. The Piltdown specimen was a practical joke: the cranium of a modern human and the mandible of a modern orangutan had been buried together, along with the bones of extinct animals.

Perhaps it was Louis Leakey's *a priori* assumption about the nature of early "true man" that explains why he did not recognize a left lower hominid canine collected at Laetoli in 1935 along with numerous other Pliocene vertebrate fossils. The tooth was too primitive, not modern in form, and there were no associated stone tools. The tooth, misidentified as that of a monkey, rested in obscurity until it was recognized at the Natural History Museum in London as a hominid, and published in *Science* in 1981.

Except for a brief period in 1939, when the German explorer Ludwig Kohl-Larsen recovered two hominid specimens at Laetoli (the Maasai word for a red lily), little attention was paid to the potential of the site for elucidating our past. In 1974, however, Mary Leakey's friend George Dove, who owned the Lake Ndutu Lodge, some 40 kilometers from Olduvai, brought to Mary Leakey's attention his discovery of fossil teeth in a load of sand he had collected for building. He and Mary Leakey traced the source of the sand back to Laetoli. In 1974 she returned to the site, which led to the discovery of some 30 hominid specimens.

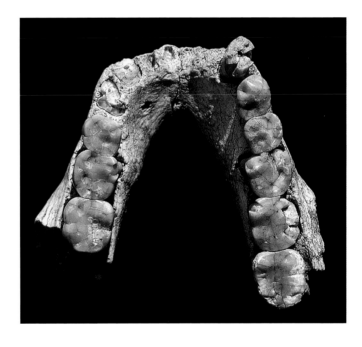

Australopithecus afarensis, L.H. 4.
This mandible was chosen as the type specimen for this species to affirm the connection between it and other fossils from Laetoli, Tanzania with hominid fossils from Hadar, Ethiopia, 1,500 kilometers away. Actual size. *Courtesy of Donald Johanson, Institute of Human Origins.*

Geological studies at Laetoli confirmed a date of 3.5 million years for the hominids, roughly equivalent to the date for the hominids at Hadar. This dictated a much closer examination and comparison of the Laetoli hominids with those from Hadar by Donald Johanson and Tim White. It was soon apparent that the hominids from the two sites were nearly identical in their anatomy and should properly be placed in the same species. A whole suite of anatomical features in the teeth, jaws, and cranial material highlighted the australopithecine affinities of the Hadar and Laetoli hominids. The more primitive character states (see page 133) in the specimens, however, resulted in the recognition, in 1978, of a novel species, A. afarensis. Even though the sites were separated by roughly 1,500 kilometers, the virtual anatomical identity of the individuals dictated that they be placed in the same species. Although the Hadar specimens were more complete, the L.H.-4 mandible, lacking the ascending rami but containing nine teeth, was selected to be the flagship fossil for the species in 1978 because of its distinctive and diagnostic anatomy, as well as the fact that detailed descriptions and photographs of it had already been published. Furthermore, choosing a Laetoli specimen for the holotype of the larger Hadar collection helped solidify the connection between Laetoli and Hadar hominids.

While the fossil hominids collected at Hadar and Laetoli were attracting headlines, a totally unanticipated discovery in 1976 at Laetoli later would distinguish the site as one of the true wonders of the prehistoric world. In a volcanic ash, dated to roughly 3.6 million years, thousands of exquisitely preserved animal footprints were found. While the layer of volcanic ash, Tuff 7, was still wet, monkeys, antelopes, elephants, two species of rhinos, three-toed horses, a small cat, giraffes, guinea fowl, francolins, and even perhaps a passing dung beetle left their imprints in the ash.

It was, however, the 1978 discovery by Paul Abell, a geochemist, of undeniable hominid footprints at Laetoli that ignited the imagination of scientists and lay people alike. At first the discovery was met with disbelief, but after the hominid nature of the prints was verified, excavations in 1978 and 1979 uncovered a 24-meter-long trackway with more than seventy hominid footprints.

After detailed inspection, there was no doubt that the imprints were made by a hominid. The impression in the ash revealed a strong heel strike, and the deep indentation made by the great toe was identical to one modern humans would leave in beach sand during toe-off. But most important about the great toe was that it was in line with the lateral four toes, and not divergent as in a quadrupedal primate like a monkey. In some of the clearer, better-defined prints, even the longitudinal arch of the foot and the ball of the foot were visible.

Which hominid made these footprints? The answer seems obvious, because the only hominid present in the Laetoli deposits was A. afarensis, and by inference the footprints must have been made by members of this species. One scientist, Russell Tuttle, anatomist at the University of Chicago, looked at the foot fossils from Hadar and said they were too large to fit into the Laetoli impressions. The Laetoli prints, according to Tuttle, were so modern that they could not have been made by a large-footed A. afarensis with somewhat curved toes. The implication was that another hominid, more Homo-like, made the prints, but strangely, not a single bone of such a hominid has been found at Laetoli. One would also have to explain why A. afarensis left bones at Laetoli but no footprints.

To address Tuttle's assertions, anthropologists Gen Suwa at the University of Tokyo, Japan, and Tim White at the University of California, Berkeley, took a closer look at the Hadar foot bones in 1983. They scaled the anatomy of the bones down to Lucy's size, and the reconstructed foot fit perfectly into the Laetoli prints. Until the phantom hominid at Laetoli, who left footprints and no fossil remains behind, is found, the best bet is to infer that A. afarensis was responsible for the Laetoli footprint trail. To find afarensis footprints was nothing short of a miracle!

In-depth geological study of the Laetoli strata by geologist Richard Hay revealed that the volcanic eruption that produced the ash in which the footprints were preserved occurred at the end of the dry season, when the grass had been grazed down to a stubble. The pocked surface, a result of the impact of falling raindrops, indicated that the rainy season had started. Higher up in the geological sequence, apparently deeper into the rainy season, heavier rainfall washed away many of the smaller prints, but game trails left by large migrating herds of antelopes are evident.

It took a unique series of events to create this marvelous portrait of the past. The volcano Sadiman spat out a cloud of ash with the consistency of fine beach sand. The ash blanketed the ground like new-fallen snow. Then, it rained. The ash became mushy, and animals left their impressions in the mud. The sun came out and quickly dried the mud. Because the ash was rich in carbonates, it hardened into an almost cement-like layer. Another puff of ash, and the footprints were sealed for posterity. Sadiman was not quiet until some 16 to 20 centimeters of layered ash had accumulated. The final event occurred 3.6 million years later, when scientists recognized the Footprint Tuff that erosion had exposed.

We will never know what Lucy's relatives thought when Sadiman began to erupt. The footprint trail, however, shows a steady progression from south to north of two hominids, perhaps walking side by side. One set of prints is large, the other small. Were they male and female? Were the smaller prints made by a child? Partway along the trail, the hominids appear to have paused, turned, and looked westward. What caught their attention? The answer is lost in time, but after the pause they resumed walking in the original direction—as if they knew exactly where they were going.

Australopithecus afarensis, **Laetoli footprint.**
The site of Laetoli, Tanzania attracted worldwide attention after the discovery in 1978 of a hominid footprint trail that demonstrated a humanlike gait for this early hominid.
Actual size. *Photograph by John Reader, Science Source/Photo Researchers.*

SPECIMEN	LOCALITY	AGE	DISCOVERERS	DATE	PUBLICATION
Skeleton	Sterkfontein, South Africa	3.0 million years	Ron Clarke	September 6, 1994	Clarke, R.J. 1998. First ever discovery of a well-preserved skull and associated skeleton of *Australopithecus*. *S. Afr. J. Sci.* 94: 460-463

Australopithecus cf. africanus
STW 573

Very often the science of human origins in books, TV specials, and popular articles is likened to a detective story with the paleoanthropologist playing Sherlock Holmes searching for clues to the origins of humankind and solving the puzzle of "who dunnit," that is to say, who really was our ancestor and how did it evolve into us? Well, the story of the truly remarkable discovery of a fossil *Australopithecus* from South Africa by Ron Clarke reads like a page-turning, investigative thriller with "who dunnit" being the oldest, most complete skeleton of a human ancestor ever found.

The story begins when excavations at the famous cave of Sterkfontein targeted an area called the Silberg Grotto, where some of the oldest section of cave breccia is exposed in what is called Member 2. Attention turned to the grotto because it was hoped that hominids significantly older than the near 500 specimens of *Australopithecus africanus*, which had been recovered from less ancient levels, might be found in this area of Sterkfontein. In October 1992 a large block of breccia was blasted out, but curiously it contained an over abundance of carnivore and monkey fossils, unusual for a cave that has produced an untold number of antelope fossils. Ron Clarke thought this strange and decided to investigate. He began to scour the fossil assemblage that was collected from the grotto in 1978-80 from breccia blasted away by lime workers in the 1920s and 1930s.

On September 6, 1994 Ron went to the storage shed at the site of Sterkfontein and opened a small cardboard box labeled D 20, the number of the lime miners' dump from Silberg Grotto. He selected a plastic bag marked "toe and ankle" bones. Much to his astonishment he spotted a small white bone, an ankle bone called the talus, upon which rests the shin bone (tibia) and exclaimed to his wife, working nearby, "This is a hominid. It's not a carnivore!"

A meticulous worker Ron identified other parts of a foot including a navicular and cuneiform from the ankle and a first metarsal that forms the base of the great toe. He was convinced they all came from the same left foot, the foot of a hominid now cataloged as StW 573. After carefully examining the foot bones in collaboration with his colleague Phillip V. Tobias, it was decided that the hominid foot showed humanlike features near the heel, but apelike features in the great toe. Following a careful examination of the articular surfaces of the bones, they concluded that the hominid, presumably an australopithecine, had a slightly divergent great toe, implying that it was a grasping foot useful, as in apes, for climbing.

The story of "Little Foot," as the specimen came to be known, to distinguish it presumably from "Big Foot," continued and in May 1997, two years after the initial publication in the journal *Science*, new evidence came to light. Ron entered the large walk-in safe at the University of Witwatersrand in Johannesburg where many valuable fossils reside in close company with the famed Taung specimen. In a box stored in a cupboard purporting to contain fossil monkey bones from D 20, Ron was surprised to see white hominid ankle bones—the second metatarsal and the bottom end of a fibula, that little, long bone that rides adjacent the larger tibia.

All of these bones fit together nicely since they belonged to Little Foot's left foot. When examining another plastic bag a couple of days later, Ron's trained eyes fell upon a vital clue, a small fragment of the bottom end of a tibia, the shin bone, but it did not really fit the articular surface of the talus. Inspired by the presence of a fresh break in the shaft of the bone he prophesized that somewhere in the bone collections there must be more of the tibia. Unrelentingly Ron, systematically began to search through bag after bag of D 20 material from the Silberg Grotto. A week later he found the lower part of a left tibia that snuggly fitted onto Little Foot's ankle. He realized that the other tibia was from the right leg.

By June 1997 Ron had assembled 12 foot and lower leg bones of an early hominid. Based on this evidence he made the bold prediction that the rest of the hominid skeleton must still be in the Silberg Grotto, since his search through other boxes and bags came up empty.

He briefed his assistants Stephen Motsumi and Nkwane Molefe about his suspicions, armed them with small miner's lamps, a cast of the right tibia, and sent them into the cavern to search for a bone section to which the tibia would fit. This was no mean task since the Silberg Grotto is the size of a medium sized church in the English countryside.

Against all odds, just two days later, on July 3, 1997 Ron's assistants struck pay dirt. They found a perfect join to the broken shaft of the tibia and just next to it was the broken cross section of the fibula. Amazing, all of the meticulous detective work was beginning to pay off.

Ron moved on to the next stage of investigation at the findspot. He and his coworkers began to chisel away, ever so carefully, the dense, rock-hard cave breccia, in which the bones were interred. Gradually they uncovered the shin bones and the lower halves of the thigh bones. Months later after a demanding period of excavation they uncovered the lower end of a left upper arm bone and next to it, the glint of a tooth. Ever so slowly more of the specimen was revealed, which turned out to be a mandible, attached to a cranium still in occlusion with the upper teeth. An investigation that began with a few ankle bones in a laboratory vault had lead Ron and his team to a beautifully preserved skull, resting on its right side, undisturbed since the creature fell into the cave more than 3 million years ago.

The team continued to chip away at the stubborn rock face using a mini-jackhammer called an airscribe and by the end of August 1999 a left forearm, connected to a set of wrist and hand bones, was revealed. The palm of the hand faces upwards and the fingers are tightly clenched almost giving you the impression that this was exactly how the creature died.

Still today the uncovering of StW 573, deep in the bowels of the Sterkfontein cave complex continues with pelvis, vertebrae, and the right arm and hand slowing appearing after millions of years. It is mind-boggling to think of the long

journey spanning the time when an *Australopithecus* tumbled through a cave opening and the arduous day-in day-out task that is the on-going investigation of Ron Clarke.

It will still be sometime before the StW 573 skeleton is completely excavated and cleaned for study in the laboratory and debate will continue about whether it really had a divergent big toe and might have been somewhat arboreal as Ron suggests, but it is worth the wait. When the forensic team thoroughly evaluates all the clues from the death scene, we may be rewarded with a more complete picture of not only how this hominid came to die in the Silberg Grotto but invaluable details of how our ancestors lived.

Australopithecus cf. africanus, **StW 573.** Still undergoing excavation, this specimen from Sterkfontein, South Africa promises to be the most complete skeleton ever found of *Australopithecus*. Actual size. *Photograph by Ron Clarke, University of the Witwatersrand.*

Australopithecus africanus
STS 5 *MRS. PLES*

Raymond Dart's 1925 publication of the Taung Child (see page 154) as a new kind of human ancestor was not well received. Instead of eliciting praise, the announcement drew strong criticism from nearly everyone except paleontologist Robert Broom. Broom, one of the more eccentric characters ever to have worked in the field of paleoanthropology, burst into Dart's lab just after publication of *A. africanus* and fell to his knees "in adoration of our ancestor," as he put it.

Robert Broom, a Scotsman, arrived in South Africa in 1897 via Australia, drawn by his fascination with the mammal-like reptiles of the Karoo Desert. Supporting himself with a medical position, he collected, cleaned, drew, and published extensively. In 1925, the year Dart announced the Taung Child, Broom was 58 and had 250 publications to his name. The level of his output was almost incomprehensible: he named some 70 new genera and close to 200 new species of reptiles.

Broom caught the hominid fever, but his medical work and the Great Depression prevented him from doing much about it. Then, in 1934, he was given a post at the Transvaal Museum in Pretoria in paleontology and physical anthropology. He continued to focus on reptiles, publishing 16 papers in 18 months, but in 1936 he vowed to find an adult version of the Taung Child.

Broom was led by two of Dart's students to a cave being mined for limestone, some 64 kilometers southwest of Johannesburg. G. W. Barlow, who was in charge of the quarrying operation, on the side sold to tourists fossils that were found at the cave, known as Sterkfontein. Broom told Barlow to keep a keen eye open for anything even vaguely resembling a hominid.

On August 17, 1936, only eight days after Broom's appeal, he was presented with two thirds of an endocast, not unlike that of the Taung Child. Broom searched through the broken breccia and found the base of a skull, parts of the vault, some teeth, and a badly crushed maxilla. Here was the hominid he wanted. With a proclivity for naming new taxa, he dubbed the find, TM 1511, *Australopithecus*

transvaalensis (interestingly, the name is mentioned only in a figure caption), but in 1938 he placed it in *Plesianthropus transvaalensis*, "near man from the Transvaal."

During World War II, fossil hunting essentially ceased, but in 1947 Broom, now over 80 years old, and his assistant John T. Robinson, later renowned for his work on the dentition of the australopithecines, reinitiated work at Sterkfontein. The bone breccia at the site is so consolidated that they used dynamite to blast open the deposits. On April 18, after the dust had cleared, they found an adult skull of *Australopithecus* in the breccia. The blast had split the skull into two major fragments across the braincase, exposing a crystallined brain cavity.

After weeks of laboratory preparation and cleaning, the specimen, Sts 5, was found to be a nearly complete cranium, lacking teeth. Robinson and Broom concluded that it was the skull of a middle-aged female, and since then it has been known as "Mrs. Ples." From examination of anatomical details on the inside base of the skull and the bones in the orbit, they concluded in June that "these small-brained man-like beings were very nearly human."

In some ways the recovery of "Mrs. Ples" was a vital turning point in the broader acceptance of the South African australopithecines as hominids. With an adult cranium like Sts 5, the case could no longer be made that the Taung Child was simply a juvenile ape. Sts 5 demonstrated without doubt that had the Taung Child grown up, it would not have developed into an ape.

Some of Broom's more cavalier approaches to excavation, such as his lack of attention to stratigraphic details, drew criticism from the Historical Monuments Commission. In fact, he had been forbidden to work at Sterkfontein, but he defied the ruling. Again the Commission banned him, but scientific and public outcry quickly forced a reversal of the ban, allowing him to further his excavations at Sterkfontein.

Australopithecus africanus, **Sts 5** (see also page 3). Although it has long carried the nickname "Mrs. Ples," this cranium actually belongs to a male according to recent anatomical studies. Note its very projecting lower face in lateral view. Actual size. *Photograph by David L. Brill; courtesy of Transvaal Museum.*

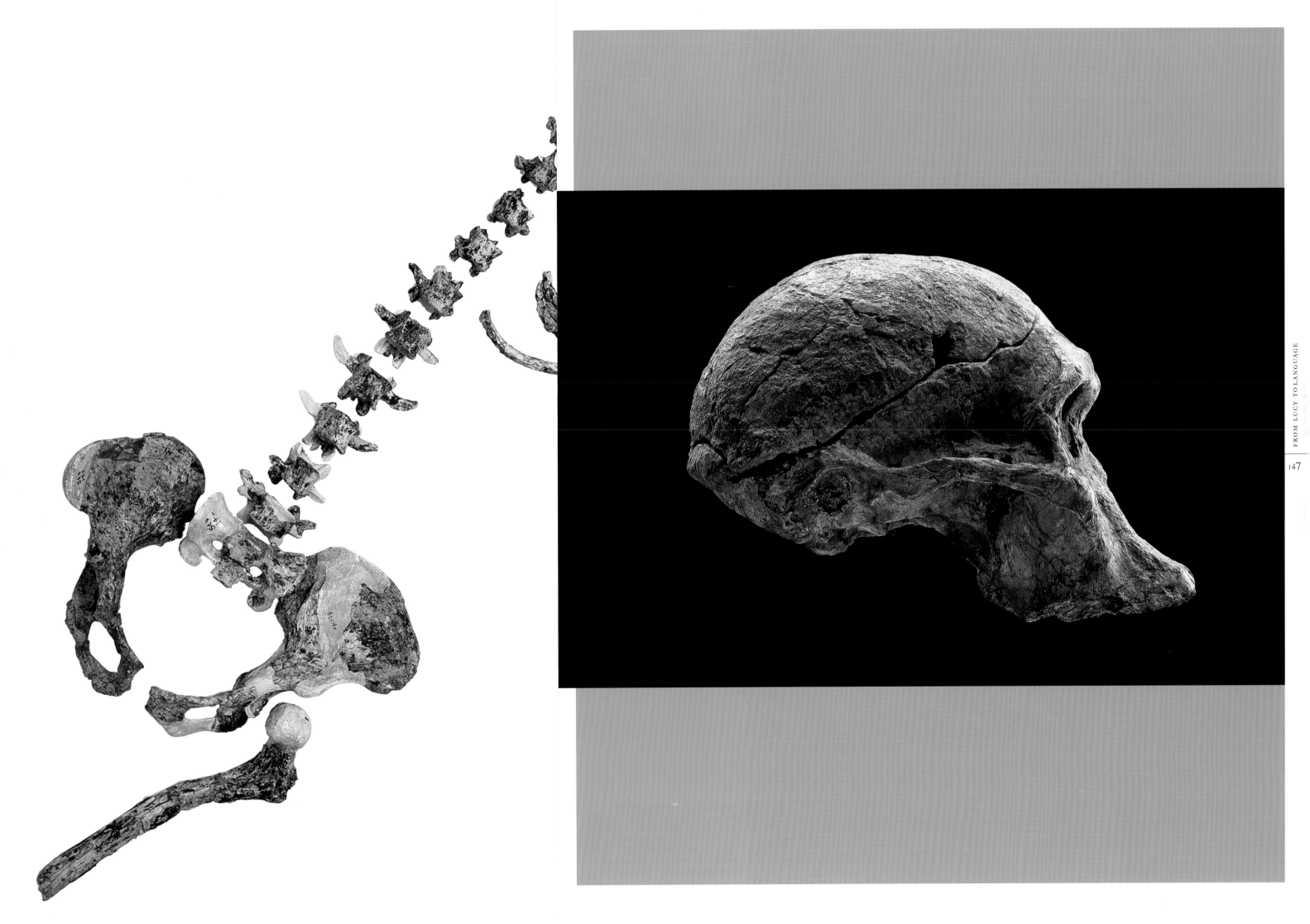

SPECIMEN	LOCALITY	AGE	DISCOVERERS	DATE
Partial adult skeleton	Sterkfontein, South Africa	2.5 million years	Robert Broom and John T. Robinson	August 1, 1947

Australopithecus africanus
STS 14

Robert Broom, a Scottish paleontologist at the Transvaal Museum in Pretoria, South Africa, was one of the more determined and colorful characters ever to have participated in the hunt for human origins. In 1947, following his short period of banishment from the important fossil cave site of Sterkfontein by the Historical Monuments Commission (see page 146), he was more determined than ever to find significant hominid fossils at the site. On August 1, he and his assistant at the Museum, John Robinson, later distinguished by his all-encompassing study of the dentition of the australopithecines, blasted out a slab of cave breccia containing a portion of a thigh bone, several vertebrae, and, most important, both innominates (pelvic blades) of an australopithecine. This was a remarkable find when judged by any criterion. It was also the vital evidence Broom needed to convince those who doubted that the australopithecines were indeed hominids. This partial skeleton attested to the bipedal nature of these early creatures, which assured them a place in the family Hominidae.

Teeth and jaws are always abundant at early hominid sites, teeth because they contain the hardest substance in the body, enamel, and jaws because the bone surrounding the teeth is so compact and dense. This is fortunate, because teeth and jaws are extremely useful for understanding dietary adaptations and taxonomic affinities of hominid remains. The more delicate parts of the skeleton, however, such as the vertebrae, ribs, and pelvis, tend to be a rarity. The paper-thin bone, especially of the pelvic blades, is easily broken and destroyed before being buried and ultimately fossilized. Vertebrae have the unique hydrodynamic property of easily floating away from a decaying carcass and seldom become fossilized. Ribs, containing nourishing marrow, are readily crunched up as hors d'oeuvres by hungry scavengers.

The August 1947 find received the catalogue number Sts 14. After extensive laboratory work, a partial sacrum, 15 consecutive vertebrae, and four ribs were extracted by meticulously chipping away the hard matrix. Comparisons of the Sts 14 pelvis with a modern human pelvis revealed slight differences, as was to be expected, but the overall anatomy was distinctly different from that of any ape. The iliac blade is short and wide, with a well-developed sciatic notch and a strong anterior inferior iliac spine. Broom was quick to point out that this was undeniable proof of the hominid affinities of the australopithecines and confirmed the bipedal nature of these early creatures.

Because the pelvis was noticeably small, Broom and Robinson considered the specimen to be female. The pelvis differs from a modern human pelvis in having a forward projecting anterior superior spine, a very small articular surface for the sacrum, and a marked outward flare of the iliac blades. Robinson concluded that this individual probably had wide hips and a bulging abdomen.

Unlike modern humans, who normally have five lumbar vertebrae, Sts 14 had six. The lumbar region in the modern African apes tends to be reduced, with the number of lumbar vertebrae varying from three to five but averaging between three and four. In Sts 14 the first lumbar vertebra is transitional with a rib facet, like a thoracic vertebra, but with superior articular facets typical of a lumbar vertebra. The six lumbar vertebrae show increased dorsal wedging from the higher to the lower vertebrae, consistent with a well-developed lumbar curve (lordosis).

In 1950 Broom and Robinson provided only preliminary descriptions of Sts 14, and more detailed study was not published by Robinson until 1972. There is always pressure to publish new finds. It is a delicate balance between satisfying those awaiting publication and not publishing too hastily, leaving oneself open to criticism that the information is incomplete. Broom was acutely aware of this problem, and in 1950 he wrote, in an obviously defensive mode, "We think it much preferable to issue even inadequate descriptions and let other workers know something of our finds than to keep them secret for 10 years or more." Broom was reacting to scientists who held off publishing their finds for many years but who were swift to criticize those, like himself, who did publish quickly.

Reading Broom's publications, one gets a sense of urgency, no doubt in response to his advancing years. In the end he was fanatically driven to finish his monograph on the Swartkrans hominids. On April 6, 1951, after completing the final corrections, he allegedly said, "Now that's finished . . . and so am I." He died that evening.

SPECIMEN	LOCALITY	AGE	DISCOVERERS	DATE	PUBLICATION
Adult cranium and mandible	Sterkfontein, South Africa	2.5 million years	Robert Broom and John T. Robinson	November 13, 1947; August 10, 1948	Broom, R., Robinson J.T., and G.W.H. Schepers.1950. Sterkfontein ape-man *Pleisanthropus. Trans. Mus. Mem.* no. 4; and Broom, R., and J.T. Robinson. 1949. A new mandible of the ape-man *Plesianthropus transvaalensis. Am. J. Phys. Anthropol.* 7: 123-127

Australopithecus africanus
STS 71 AND STS 36

The 1947-49 excavations at Sterkfontein, under the direction of Robert Broom, distinguished paleontologist at the Transvaal Museum, and his student John Robinson, resulted in the discovery of 70 early hominid specimens. Work at the cave proceeded at a rapid pace, and in 1950 Broom and Robinson, along with G.W.H. Schepers, an anatomist from the University of the Witwatersrand in Johannesburg, published detailed descriptions of these finds in a Transvaal monograph. Originally catalogued as Skull No. 7, one of the better preserved crania at Sterkfontein, and now known as Sts 71, this specimen was an important addition to the hominid species at the site. Sts 71 consists of more than half of the face and the entire right side of the braincase, permitting a cranial capacity estimate of 428 cc. In lateral view, and probably as a result of slow, plastic deformation during fossilization, the cranium appears more globular, less elongated, than "Mrs. Ples" (Sts 5), which has a cranial capacity of 485 cc.

In a figure caption Broom referred to Sts 71 as a female. It was a strange assignment because the facial robusticity, especially the forwardly placed root of the zygomatic process (where the cheekbone leaves the side of the face) and the large size of the postcanine dentition, is more indicative of a male status. An interesting resolution of the correct identification of the sex of Sts 71 came in 1972 when John Wallace, the late student of Phillip Tobias, professor emeritus of anatomy at the University of the Witwatersrand, completed his doctoral thesis on the form and function of the South African australopithecines.

Wallace knew that some of the upper and lower dentitions from the South African caves, such as Sts 52a (maxilla) + 52b (mandible), were from the same individuals, since details of their occlusion and of their occlusal wear patterns matched. He identified ten individuals for which he was able to match upper and lower dentitions. Nine were from the cave of Swartkrans, with the sole specimen from Sterkfontein being the association of the cranium, Sts 71, with the mandible, Sts 36. As in Sts 71, the teeth in Sts 36 were heavily worn, exposing large islands of dentine. In an evaluation of the Sts 36 mandible, Broom and Robinson concluded that the jaw was male because of its massive size and very large teeth. It is this important association of the Sts 71 cranium and the Sts 36 mandible that offers a striking picture of a male skull from Sterkfontein.

Recognition of Sts 71 as male solved one problem but created another. Compared with female faces at Sterkfontein, Sts 5 and Sts 52, the facial profile of Sts 71 is less prognathic. This is unusual, because among the modern apes the males are regularly more prognathic than the females. If sexual dimorphism cannot explain the variation in facial morphology seen at Sterkfontein, then, according to paleoanthropologists William Kimbel of the Institute of Human Origins and Tim White of the University of California, it is possible that more than one hominid species may be present in the collection.

Substantial cranial variation in the Sterkfontein sample is further indicated by the cranium Stw 505 (Stw designates Sterkfontein, University of the Witwatersrand excavations), found in 1989. Due to significant distortion of the Stw 505 cranium, an accurate cranial capacity cannot be offered. A published estimate of 515 cc is certainly an underestimate of the actual value for this large specimen, which would make the cranial capacity of Stw 505 significantly larger than the mean of 441 cc based on six adult *A. africanus* crania. Morphologically the Stw 505 cranium is very similar in shape to other crania from Sterkfontein (especially Sts 71) and the tooth roots suggest massive cheek teeth and very reduced anterior teeth. The root of the zygomatic (cheekbone) is very thick, and there is even a hint of a low sagittal crest toward the back of the cranium.

The continued reevaluation of specimens such as Sts 71, which have been known for quite sometime, is an excellent example of how interpretations may change as a result of more detailed study. If Sts 71 and the associated mandible Sts 36 represent a species distinct from *A. africanus*, it may be extremely challenging to sort the Sterkfontein fossil hominid specimens into the two different species, because they are so similar in so many respects.

Australopithecus africanus, Sts 71. A lateral view shows the braincase of this male cranium, with a cranial capacity of 428 cc, and a less projecting face than that of Sts 5. Actual size. *Photograph by David L. Brill; courtesy of Transvaal Museum.*

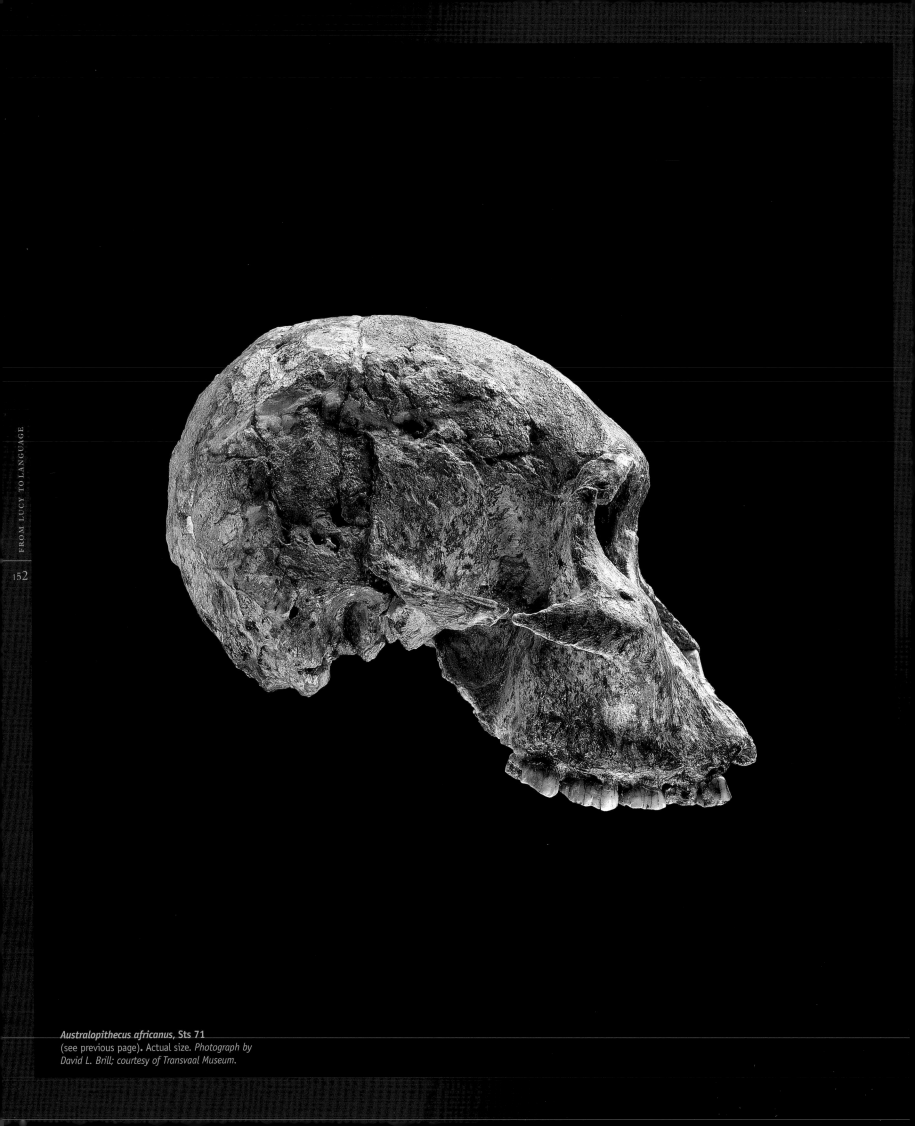

***Australopithecus africanus*, Sts 71**
(see previous page). Actual size. *Photograph by*
David L. Brill; courtesy of Transvaal Museum.

Australopithecus africanus, **STW 505**
(see page 150). This specimen has a significantly larger cranial capacity than other *A.
africanus* crania at Sterkfontein. Furthermore,
the presence of a weak sagittal crest, and
prominent glabella and brow ridges suggests it
is male. *Photograph by Ron Clarke, University of
the Witwatersrand.*

Australopithecus africanus
TAUNG CHILD   TYPE SPECIMEN

At a town called Taung, then located in the South African protectorate of Bechuanaland, the Northern Lime Company was actively mining limestone. M. de Bruyn, a quarryman who was overseeing the mining activities, knew of anatomy professor Raymond Dart's keen interest in the fossil animals that occasionally turned up in the dolomitic limestone. He saved these fossils and from time to time sent a box or two to Dart at the University of the Witwatersrand in Johannesburg. Little did de Bruyn know that his shipment of specimens collected in the fall of 1924 contained something even more valuable than the diamonds being found at Kimberley, some 130 kilometers to the north. In that shipment Dart recognized an endocranial cast that differed from those of the usual baboons found at the site. Most of the face and an attached lower jaw were preserved, and an endocast fit perfectly into the front part of the skull. There was no mistake: the anatomy confirmed his initial impulse that the Taung specimen was clearly not a monkey. After more than two months of carefully cleaning away the limestone, Dart wrested from the block of breccia the skull of an immature primate that could only be an early hominid. On February 7, 1925, in the pages of *Nature*, he announced evidence that harkened back to Darwin's prediction that Africa would eventually yield remains of the earliest stages of human evolution.

In a bold stroke, Dart named the first new genus in paleoanthropology *Australopithecus*. He chose the Taung specimen as the holotype for the species *Australopithecus africanus*. Despite mixing Latin, *australo*, with latinized Greek, *pithecus*, his name for this "southern ape of Africa" has stood the test of time.

This "man-ape of South Africa", as Dart referred to it, contained quite a number of humanlike features. These included a rounded, high forehead lacking browridges, delicate cheekbones, a relatively flat profile in lateral view, and a lightly built mandible. The child's estimated cranial capacity is 405 cc, with a projected adult size of 440 cc. The teeth were also rather modern looking, lacking the space between the canine and the first lower premolar that is seen in the apes, and

the canine was small and did not project much above the adjacent teeth. The foramen magnum (the hole in the base of the skull from which the spinal cord emerges) was forwardly placed on the base of the skull and not positioned toward the back, as in quadrupeds. This anatomical arrangement convinced Dart that the Taung Child was bipedal.

Once published, grave doubts and outright dismissal of Dart's claims issued forth from the established British anthropological community. For one thing, the find was all wrong: it had the jaw of a human and the brain of an ape, just the reverse of the situation predicted by those who saw the brain as leading the way in human evolution. After all, the English already had Piltdown Man, with an enlarged brain and apelike jaw, and how could an English fossil be cast aside in favor of one from darkest Africa?

One of the most widespread criticisms of Dart's Taung Child was merely that it was a child. The first permanent molar was just beginning to erupt, attesting to the juvenile nature of the fossil. Because most of the distinguishing features such as large canines, a sloping forehead, a large jaw, and a projecting face appear later in the growth of apes, it was thought unacceptable to use an immature specimen to diagnose a new species of ancient hominid. Given a chance, the Taung Child might have grown up to be an ape.

Additional circumstances were stacked against acceptance of the Taung Child as a human ancestor. Dart was considered young and inexperienced. Few scientists actually traveled the long distance to Johannesburg to see the original, and casts were not easily available. Even when Dart traveled with the Taung specimen to England, he was overshadowed by the more eloquent expositions on the Peking finds, which were beginning to take center stage.

Unfortunately, no other hominid specimens have ever been found at Taung, despite recently renewed excavations. Why was only one specimen recovered from the site? Puncture marks and depression fractures on the Taung specimen are similar

to those seen in animal bones found in the nests of eagles. It is possible that the Taung Child represents the leftovers from the meal of a large bird of prey.

For whatever reason, Dart's monograph on the Taung Child, except for the passage on the teeth, was rejected by the Royal Society. Adding insult to injury, the influential English anatomist Sir Arthur Keith and the Austrian paleontologist Othenio Abel each published approximately 100 pages on the Taung specimen, with Keith concluding it was a chimpanzee and Abel that it was a gorilla.

Perhaps for these as well as other reasons, Dart chose not to pursue discovery of other fossil evidence to bolster his ideas about Taung, but instead dedicated himself to developing the anatomy department at the University of the Witwatersrand. It was not until adult specimens of *A. africanus* were found at sites like Sterkfontein (see pages 146 148, and 150) twenty years later that Dart's views about the Taung Child were confirmed. Oddly, the type specimen for *Australopithecus africanus* still languishes undescribed in a university vault, more than 80 years after its discovery.

155

Australopithecus africanus, **Taung child**
(see also page 79). The first early hominid
found in Africa, this skull provided the basis
for a new genus and species. Note that the first
molar has only partially erupted, indicating
that this individual died as a juvenile. Actual
sizes. *Photographs by David L. Brill; courtesy of
University of the Witwatersrand.*

Australopithecus robustus

TM 1517 TYPE SPECIMEN

After the initial discovery of an australo-pithecine at Sterkfontein, in 1936, Robert Broom made repeated visits to the lime quarry to see if Barlow had located any more hominid specimens. Broom purchased some 18 additional fossil hominids from Barlow, but most were much more fragmentary than the original endocast and partial cranium. One specimen, TM 1513, a distal femur, however, was quite complete, and from its anatomy, especially the carrying angle of the femoral shaft and the elongated lateral condyle, Broom became convinced that *Plesianthropus transvaalensis* was bipedal.

On a visit to Sterkfontein on June 8, 1938, Broom bought, for two pounds, a maxillary fragment containing the first molar. This specimen was of particular interest to him for it was quite large with round, puffy cusps that he surmised might represent a species different from the *transvaalensis* fossils recovered earlier from the site. Broom further noticed that the matrix adhering to the maxilla was different from that at Sterkfontein, and he asked Barlow where the specimen had been collected. A week later Barlow finally admitted that the specimen had been given to him by a schoolboy who worked as a guide at the cave on Sundays.

Broom immediately went off to visit the boy's home and found that Gert Ter-blanche was at school. He was eager to find the boy, since fresh breaks on the maxilla suggested that additional frag-ments might be found at the place of dis-covery. Tracking the boy down at school, Broom purchased four additional teeth from him and, after lecturing the school on the cave sites, he was led by Gert to a place called Kromdraai, located only 1.5 kilometers east-northeast of Sterkfontein.

Just as Broom had suspected, more fragments were found on the hillside where the initial find had been made. These fragments included most of a palate with teeth, the left side of a cranium with the zygomatic arch, and the right half of a mandible with the premolars and molars. Broom's original hunch was correct: the hominid from Kromdraai was quite different from the one at Sterkfontein.

The Kromdraai specimen, known as Trans-vaal Museum (TM) 1517, had a larger face and more powerful jaws housing much larger premolars and molars. Broom was quick to publish, and in August 1938, just slightly more than three months after the initial find at Kromdraai, TM 1517 became the holotype for *Paranthropus robustus*. For the genus name Broom combined Greek terms meaning "beside man," and to emphasize the more massive nature of the fossils he chose the Latin *robustus*, meaning "strongly and stoutly built," for the species designation. (In this book we use *Australopithecus* in place of *Paran-thropus*, but some paleoanthropologists prefer Broom's genus name for the robust australopithecines.)

When World War II brought an end to the mining operation at Sterkfontein, Broom turned his attention to laboratory work. He chipped away at blocks of breccia from Kromdraai and found hand and foot bones, a right elbow joint, and an ankle bone, the talus. Broom became preoccupied with completing a monograph on the austra-lopithecines, and he even made the bold assertion that they belonged in their own zoological subfamily, the Australopith-ecinae. His insightful descriptions of the fossil hominids continue to be invaluable reference for anyone working in paleoan-thropology. Descriptions, thought by some to be fairly fanciful, of the australopith-ecine endocasts were presented in Part II of the monograph by G.W.H. Schepers, an anatomist, and one of Dart's students, the same boy who had originally led Broom to Sterkfontein in 1936.

Broom's monograph, published on January 31, 1946, received the U.S. National Academy of Sciences award as the most important book of the year in biology and went a long way toward altering the view that the South African ape-men were not human ancestors. The year 1946 was also a pivotal year for paleoanthropology, because Sir Wilfred E. Le Gros Clark, renowned professor of anatomy at Oxford University, made a careful study of the original australo-pithecine fossils and became convinced that they were indeed human ancestors. Le Gros Clark's stamp of approval carried substantial weight, and at the First Pan-

African Congress, convened by Louis Leakey in January 1947 in Nairobi, he championed referral of the australopith-ecines to the Hominidae.

***Australopithecus robustus*, TM 1517.**
The large face, jaws and teeth of this specimen
relative to *A. africanus* convinced Robert Broom
that it belonged to a different species, the
first of the robust australopithecines to be
discovered. Actual size. *Photograph by John
Reader, Science Source/Photo Researchers.*

SPECIMEN	LOCALITY	AGE	DISCOVERERS	DATE	PUBLICATION
Adult cranium	Sterkfontein, South Africa	ca. 2.5 million years	A. Hughes	June 6, 1984	Clarke, R.J. 1988. A new *Australopithecus* cranium from Sterkfontein and its bearing on the ancestry of *Paranthropus*. In *Evolutionary History of the "Robust" Australopithecines*, F.E. Grine, ed., pp. 285-292

Australopithecus sp.
STW 252

Most paleoanthropologists consider the large sample of fossil hominids from the Transvaal cave known as Sterkfontein as belonging to a single species, *Australopithecus africanus*. But the recovery of a fragmented partial cranium in 1984, catalogued as Stw 252, from the site raises the possibility of multiple species at Sterkfontein as some investigators had already suggested.

Although the geological age of Stw 252 cannot be determined with any degree of accuracy since it extracted from an undated cave deposit, University of the Witwatersrand paleoanthropologist Ron Clarke reported that it is from deposits deep in the cave, raising the possibility that it is older than many other Sterkfontein specimens. Clarke has studied and reconstructed Stw 252 which consists of nearly a complete upper dentition, most of the palate, small fragments of the face, and portions of the frontal, parietal and occipital bones. He noted aspects of the anatomy of the specimen which differ significantly from typical *A. africanus* specimens such as Sts 5 (see page 146) and more closely resemble the condition in *Paranthropus*. (Clarke strongly advocates retention of paleontologist Robert Broom's generic designation for the robust australopithecines which in this book are designated *Australopithecus*.) These features are a thin supraorbital margin, temporal lines which converge toward the midline atop the cranium, a hint of a frontal trigone (a flat hollowed area on the frontal bone, on the midline, just above the nasal bones), and a relatively flat face.

Clarke also noted the unusual combination of very large anterior teeth and very large posterior teeth in Stw 252 which distinguishes this specimen from other australopithecines. The maxillary anterior teeth are much larger than any other Sterkfontein hominid, and in fact larger than all other australopithecines from South Africa. The canine of Stw 252 is remarkably large, as well as very pointed and projecting. The cheek teeth, premolars and molars, are very large with "puffy cusps" and by and large their dimensions fall into the size range of robust australopithecines, exceeding those of the

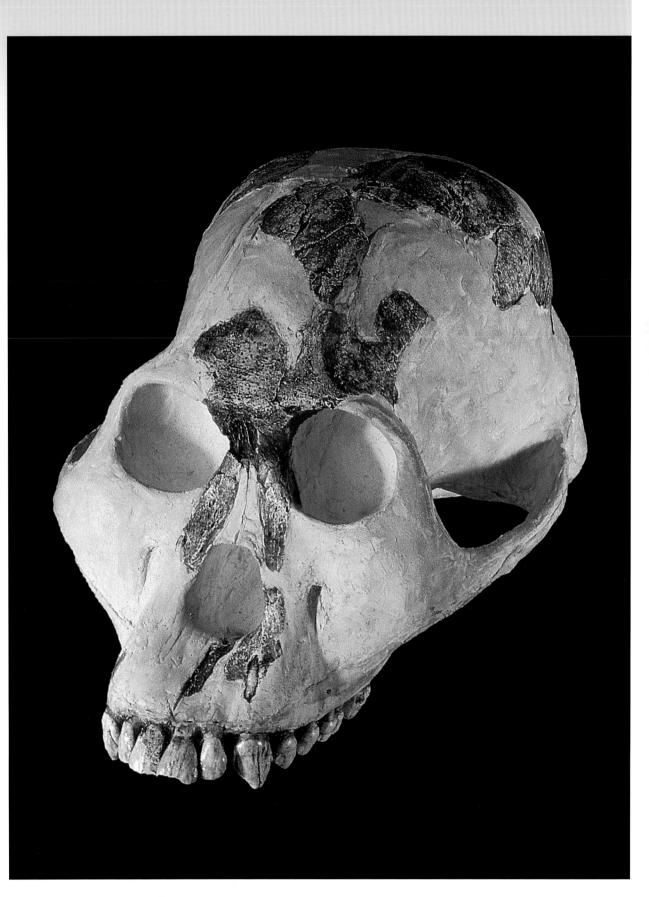

more gracile *A. africanus*. Clarke contends that Stw 252 is a precursor to the robust australopithecines, and although he does not put it into a new species, he leans strongly in favor of this designation.

Typical robust australopithecines are characterized by having greatly expanded back teeth, associated with heavy mastication, and reduced front teeth. Clarke postulates that Stw 252 samples an early part of the robust lineage when natural selection had increased cheek tooth size, but had not yet reduced the anterior dentition. This condition is similar to that seen in the Black Skull (see page 164) which, judging from the tooth roots, had large anterior and posterior teeth like Stw 252. Unlike the Black Skull, Stw 252 lacks the blade-like sagittal crest, reflecting the marked enlargement of the muscles of mastication called the temporalis muscles.

The large anterior teeth are in many ways reminiscent of those see in the east African species *A. afarensis*, notably the large canine similar to a specimen from the site of Laetoli, in Tanzania, dated to 3.5 million years. And like *afarensis*, Stw 252 has a diastema, a gap in the tooth row, between the canine and the first upper premolar. The cheek teeth of *afarensis* are, however, dwarfed in comparison with the large molars and premolars of Stw 252.

Stw 252, along with the Black Skull, contributes to the notion that in the evolution of robust australopithecines the sequence of morphological change was first enlargement of the crushing and grinding cheek teeth, and only later the reduction of the anterior teeth. Both the Black Skull and Stw 252 have prognathic (projecting) mid-faces with enlarged teeth, not unlike the more ancestral condition seen in *A. afarensis*. One interpretation, therefore, is that an *afarensis*-like hominid may have been ancestral to both Stw 252 at Sterkfontein and the Black Skull (see pages 158 and 160) in eastern Africa.

Australopithecus sp., **StW 252 cast** *(left)* **and fossil fragments** *(right)*. The case for a second species at Sterkfontein is strengthened by this discovery, reconstructed from the original teeth and cranial fragments shown here. It has a flat face and other features similar to robust australopithecines, but the combination of large front and back teeth is unlike any other members of this genus. Actual sizes. *Photographs by Ron Clarke, University of the Witwatersrand.*

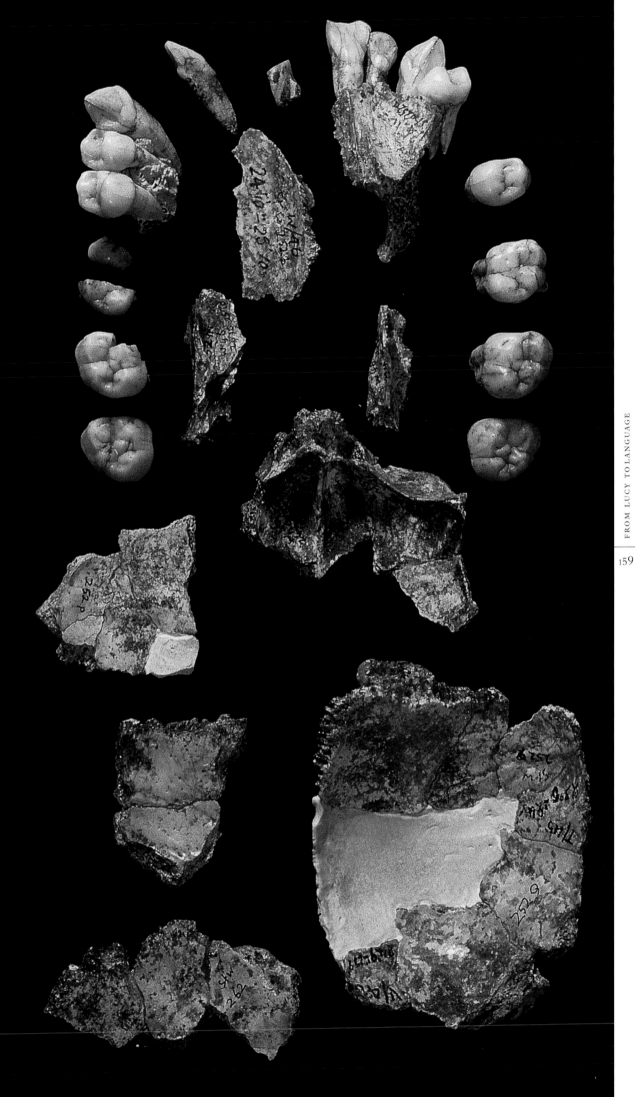

159

SPECIMEN	LOCALITY	AGE	DISCOVERERS	DATE	PUBLICATION
Adolescent mandible	Swartkrans, South Africa	1.5-2.0 million years	Robert Broom and John T. Robinson	November 16-26, 1948	Broom, R. 1949. Another new type of fossil ape-man. *Nature* 163: 57
Adult cranium	Swartkrans, South Africa	1.5-2.0 million years	Fourie	June 30, 1950	Broom, R., and J.T. Robinson. 1952. Swartkrans Ape-man, *Paranthropus crassidens. Trans. Mus. Mem.* no. 6
Adult cranium	Swartkrans, South Africa	1.5-2.0 million years	Robert Broom and John T. Robinson	1949	Robinson, J.T. 1956. The Dentition of the Australopithecinae. *Trans. Mus. Mem.* no. 9

Australopithecus crassidens [*A. robustus*]
SK 6 TYPE SPECIMEN / SK 48 / SK 79

Funding for paleoanthropological research has never been abundant, and whereas institutional and governmental research grants do cover a portion of the cost, most scientists sooner or later must turn to private sources for financial support. This was especially true in the early days of the science, and it was definitely the case for the Scottish paleontologist Robert Broom, when he worked at the Transvaal Museum in South Africa. In 1948, sadly in need of funding, he began working with a wealthy young American by the name of Wendell Phillips, who, according to Broom, had "more money and egotism than scientific ability." Phillips was leading and financing the University of California African Expedition and was eager to find ape-man fossils for the university.

Broom, having worked for many years at Sterkfontein, and highly knowledgeable about the entire valley, knew of another cave site, situated only 1,170 meters directly west of Sterkfontein, called Swartkrans. Hoping to find hominids there he initiated work in November 1948. Almost immediately most of the left half of an adolescent hominid mandible was found, containing five cheek teeth, as well as the isolated right second premolar and three molars.

It was immediately obvious to Broom that the massive jaw and large, thick-enameled teeth were quite different from those found across the valley at Sterkfontein. Broom succumbed to his usual temptation to give an entirely new genus name to the jaw, and less than two months later he published the mandible, SK 6, as the holotype of *Paranthropus crassidens*. The species name combines the Latin *crassus*, meaning "thick" or "solid," and *dens*, meaning "tooth." Broom was struck by the puffy cusps and the thick-enameled caps on the dental crowns.

Although most paleoanthropologists now place the Swartkrans remains in *A. robustus*, there is strong sentiment for restoration of *Paranthropus* as a distinct genus, and based on detailed study by Fred Grine, a paleoanthropologist at the State University of New York at Stony Brook, who considered especially the deciduous teeth, the Swartkrans remains may be a different species from the *P. robustus* that Broom had named eleven years earlier from a hominid fossil found at the cave of Kromdraai (see page 156).

Broom's excavations at Swartkrans were handsomely rewarded, for the hominid collection was dominated by teeth and jaws. His work at Swartkrans is an excellent example of the interruption of serious scientific research by economic interests. Although priority was given to commercial mining of the cave for calcite, Broom, intrepid as always, scoured the breccia piles, which were growing rapidly, owing to the use of heavy charges of explosives at the site. On June 30, 1950, a "fine skull" was blasted out of the deposits. Only slightly crushed, but somewhat damaged in the explosion, the SK 48 specimen contained the right canine, first premolar, and left three molars.

The importance of this cranium, SK 48, was that it added significant knowledge about the distinctive morphology of the robust australopithecines. Considered a female by Broom, SK 48 sports a slight sagittal crest, and in lateral view the zygomatic arches are massive and project far forwards, hiding the sunken nasal area. The canine is quite small, as is the entire region for the incisors, while the posterior teeth are relatively large—a typical robust australopithecine condition.

The Transvaal cave sites were not places of occupation but simply traps that collected bones that fell into them and ultimately fossilized them in very hard cave breccia. The slow process of burial and fossilization often put tremendous pressure on the bones, cracking and distorting them. An excellent example of this can be seen in SK 79, where only portions of the face are preserved.

Even in death Broom would not be silenced, for in March 1952, nearly a year after his death, in a 123-page monograph entitled *Swartkrans Ape-Man, Paranthropus crassidens*, with numerous drawings and plates, was published as Transvaal Museum Memoir No. 6, authored by Broom and Robinson. Robinson continued to work at Swartkrans for only a year after Broom's death. Fortunately, however, under the able leadership of C. K. Brain, former director of the Transvaal Museum, work began again in 1966. Brain replaced the destructive dynamiting of the cave site. He developed an excavation technique employing metal wedges that were placed in holes drilled into the breccia and driven in with a 16-pound hammer. The resulting breccia blocks were carefully broken using hammers and chisels. When a promising bit of fossil was detected, smaller and more controllable drills were used to remove the breccia.

After 26 years, Brain ceased his excavations at Swartkrans, and in September 1992 the site was declared a National Monument, ensuring its protection.

Australopithecus crassidens, **SK 6.**
Robert Broom designated this mandible as the type specimen of *Paranthropus crassidens*. The majority opinion today is that SK 6 should be placed in *A. robustus*. Actual size. *Photograph by John Reader, Science Source/Photo Researchers.*

***Australopithecus robustus*, SK 48** (see previous page and see also page 108)**.** The frontal view *(above)* shows the slight sagittal crest and broad, flat face of this specimen, and the basal view *(opposite)* displays the characteristically small canine and enlarged rear teeth.
Photograph by David L. Brill; courtesy of Transvaal Museum.

SPECIMEN	LOCALITY	AGE	DISCOVERERS	DATE	PUBLICATION
Adult cranium	Lake Turkana, Kenya	2.5 million years	Alan C. Walker	August 29, 1985	Walker, A.C., R.E. Leakey, J.M. Harris, and F.H. Brown. 1986. 2.5-Myr. *Australopithecus boisei* from west of Lake Turkana, Kenya. *Nature* 322: 517-522

Australopithecus aethiopicus

KNM-WT 17000 *BLACK SKULL*

The Black Skull, so called because manganese-rich minerals had stained the fossil blue-black, is a dramatic example of species diversity in hominid evolution. Here is a specimen, KNM-WT 17000, recovered from west of Lake Turkana and dated to 2.5 million years, which belongs to a distinct species, *Australopithecus aethiopicus*, and appears to have absolutely nothing to do with our genus, *Homo*. Even though it is not on a direct evolutionary line to modern humans, the Black Skull has much to tell us about the speed of morphological change, the character of that change, and the possibility of multiple lineages in hominid evolution.

The Black Skull is most likely an early example of a very successful group of australopithecines that were geographically widespread and temporally successful but ultimately went extinct. Known as robust australopithecines, these fascinating bipeds developed a host of masticatory specializations, such as massive jaws, huge crushing and grinding cheek teeth, and sometimes sagittal crests, useful in processing vast amounts of low-quality food. The best-known examples are Robert Broom's *Paranthropus robustus*, from South Africa (see page 160), and Louis Leakey's *Zinjanthropus boisei*, from East Africa (see pages 5 and 168), now both placed in the genus *Australopithecus*.

When the Black Skull was announced, the geometry of the human family tree resembled a Y, with *A. afarensis* as the stem species and ancestral to two divergent lineages: one leading to robust australopithecines and extinction and the other to our own genus, *Homo*. Because of the general lack of fossil hominids in the interval between 2 and 3 million years ago, the discovery of a fairly complete cranium from this time range promised to throw light on this dark period in hominid evolution.

A. aethiopicus possesses a host of primitive features reminiscent of its presumed ancestor, *A. afarensis*. For example, the face is quite prognathic (projecting), the cranial capacity is very small (410 cc), and cresting features, especially on the back of the skull, are typical of *A. afarensis*. On the other hand, there are a number of derived features, such as a prominent sagittal crest, a flat and concave face, and markedly expanded cheek teeth (indicated by the preserved molar and premolar roots), which are characteristic of geologically younger robust australopithecine species.

Interpretations of the amalgam of primitive and advanced features in the Black Skull are divided. On one hand, the discoverers placed KNM-WT 17000 in *Australopithecus boisei*, choosing to emphasize the few derived anatomical features shared between them. They argued that the lineage of robusts in East Africa was continuous and that to avoid the difficult choice of where to slice the lineage into different species, the Black Skull should simply be considered a more ancient representative of *A. boisei*.

On the other hand, paleoanthropologists William Kimbel, Tim White, and Donald Johanson compared some 32 cranial features observed in KNM-WT 17000 with other species of *Australopithecus* and came to a different conclusion. They noted that whereas only two features were shared with *A. boisei*, twelve were shared with *A. afarensis* (of the remaining, six were shared with *A. africanus, robustus*, and *boisei* and twelve with *A. robustus* and *boisei*).

Based on the strength of the derived robust australopithecine features in KNM-WT 17000, the option of calling it *A. afarensis* was simply not defensible. Furthermore, the argon date of 2.5 million years for the skull placed it midway between *afarensis* and *boisei*. It was therefore concluded that 17000 was deserving of the species name *aethiopicus*. So, how was the name *aethiopicus* chosen for a fossil find from Kenya?

Almost forgotten, a mandible lacking teeth from the Shungura Formation in southern Ethiopia, west of the Omo River, which had been found in 1967 by a French expedition, suddenly took on new meaning. Initially this 2.5 million-year-old mandible was assigned to *Paraustralopithecus aethiopicus* by French paleontologists Camille Arambourg (now deceased) and Yves Coppens at the Collège de France. They thought the V-shaped jaw,

among other features, distinguished it from other robust forms, but the majority of paleoanthropologists ignored the new name. Now resurrected, the Omo 18-1967-18 specimen, although a mandible and therefore not directly comparable with a cranium (17000), provided a species name (the genus designation was dropped) for the spectacular Lake Turkana specimen: *A. aethiopicus*.

The geographical range for *A. aethiopicus* is now extended to the Laetoli area, some 700 kilometers south of Lake Turkana where an maxilla lacking teeth was found in the Ndolanya Beds (c. 2.6 million years in age). The specimen exhibits resemblances with the Black Skull such as strong prognathism, a large canine root, and a shallow palate.

Aside from the nomenclatural debates surrounding the find, the evolutionary implications of 17000 are most intriguing. Because of its antiquity it might be an ancestor to *boisei, robustus*, or both species. Even more interesting is the observation that since *A. africanus* does not have the primitive traits seen in *afarensis* and *aethiopicus*, it can no longer be a precursor to all later robusts. This dictates addition of a third branch to the family tree.

It is now highly possible that *A. africanus* gave rise to *A. robustus* in southern Africa and that *A. aethiopicus* evolved into *A. boisei* in eastern Africa. If this is borne out by further finds, it implies a case of parallel evolution. In other words, robust australopithecines whose ancestors diverged from a common ancestor, *A. afarensis*, subsequently evolved similar adaptations independently, presumably in response to environmental change and subsequent dietary specialization.

There is a thought-provoking ramification of parallel evolution. The evolutionary specialization of the robusts was uncommonly successful in that they survived, over a very large geographic area, for as much as 1.5 million years, yet they met a dead end. Even though their adaptation was highly successful and apparently even arose twice in different lineages, this did not assure survival for the robust australopithecines.

Australopithecus aethiopicus, **KNM-WT 17000** (see also page 125). Compared to other robust australopithecines, the striking so-called Black Skull has a very projecting face, and seen in lateral view *(see next page)*, a tall sagittal crest along the middle of the cranium and compound bony crests at the rear. Actual size. *Photograph by Robert I.M. Campbell; courtesy of National Museums of Kenya.*

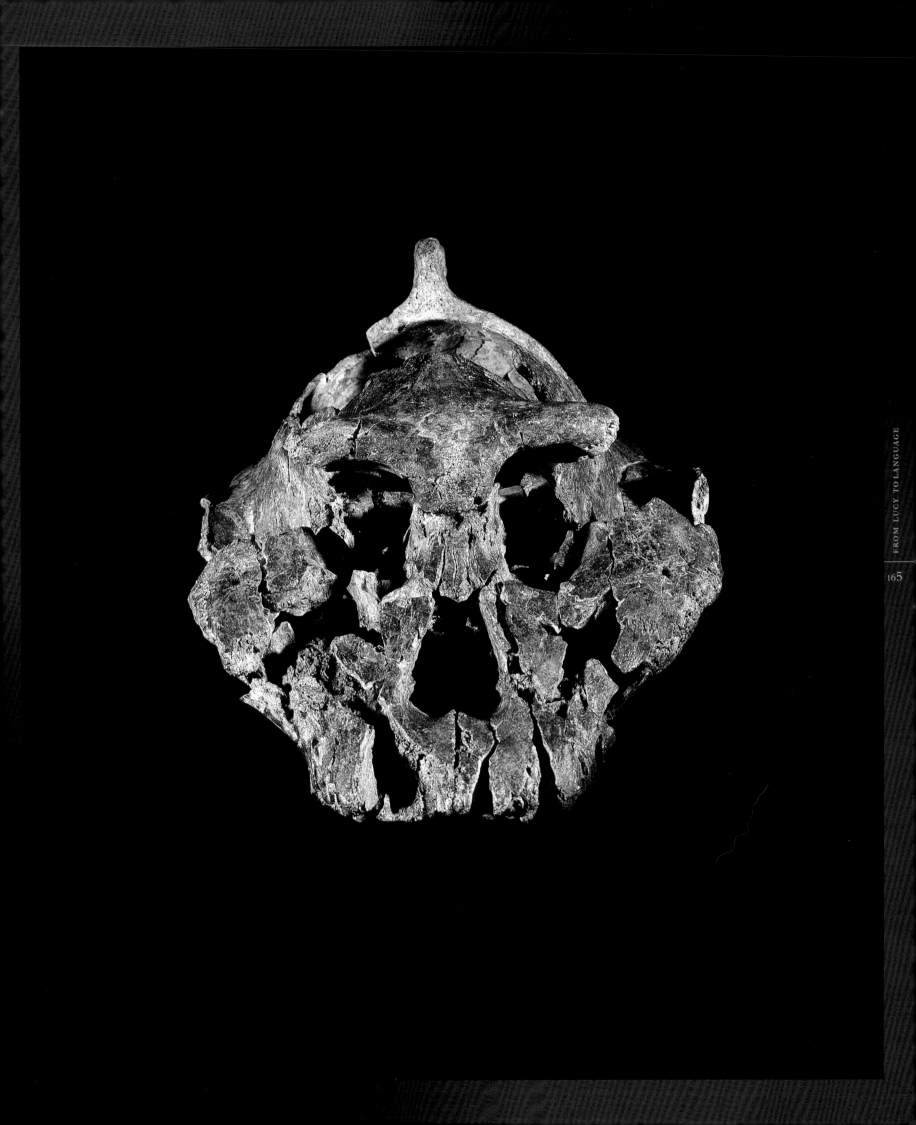

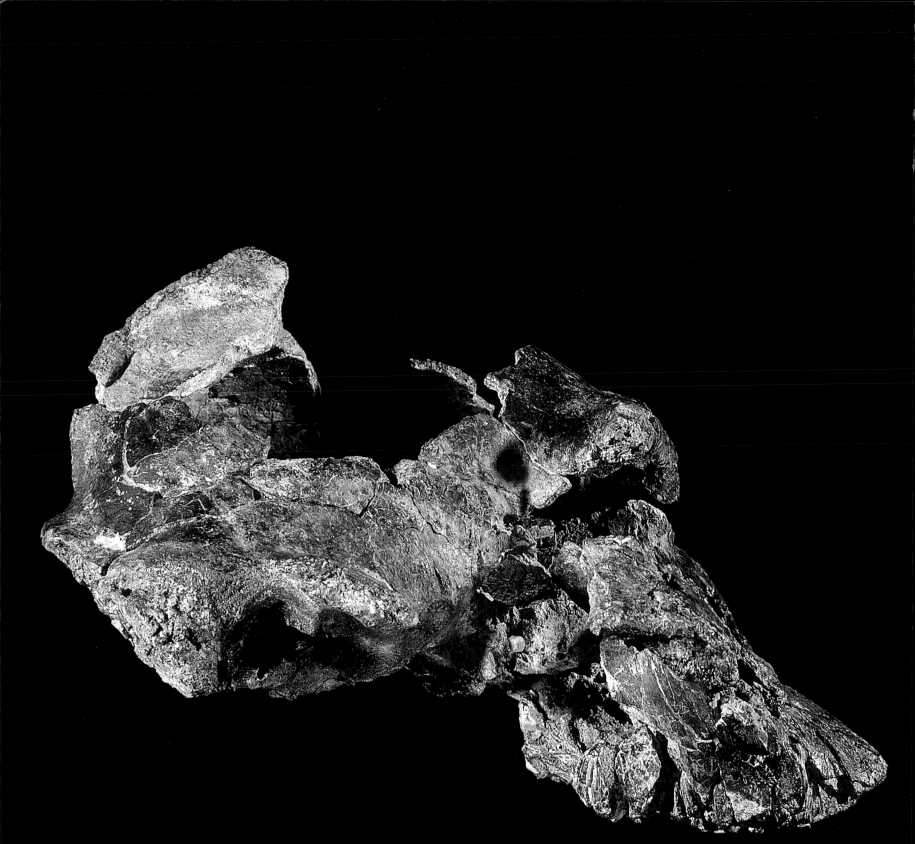

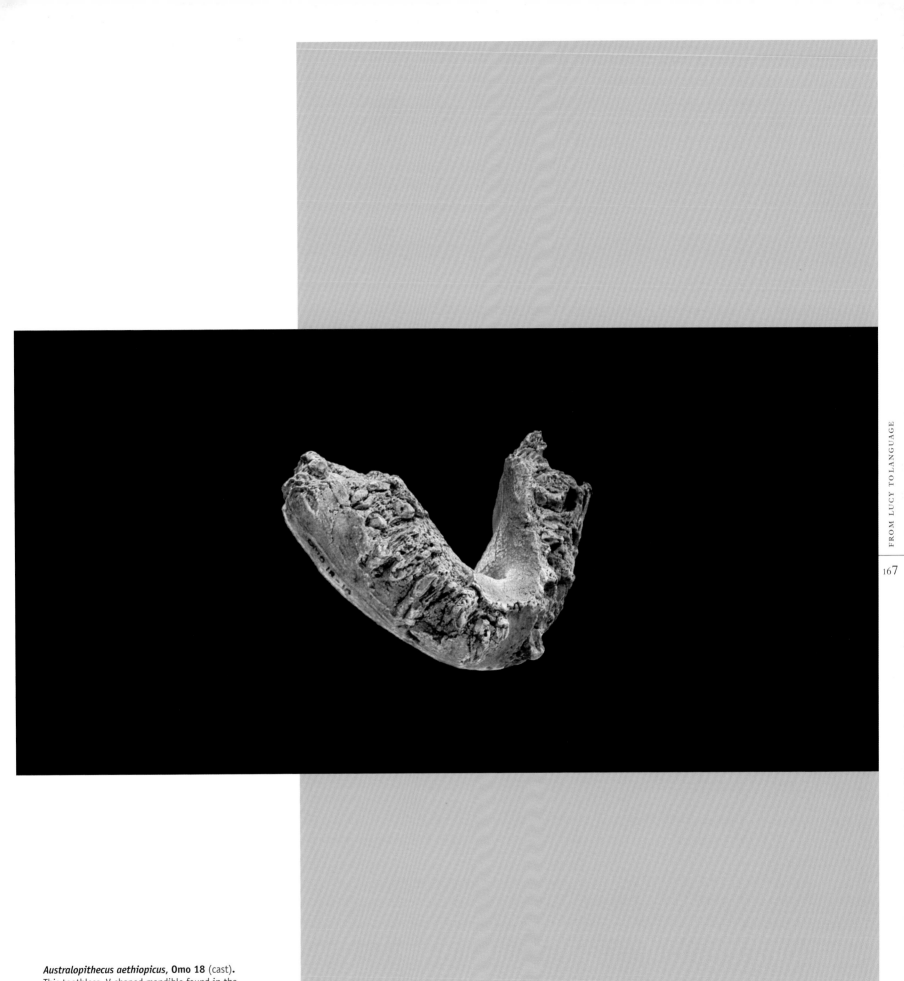

167

***Australopithecus aethiopicus,* Omo 18** (cast).
This toothless, V-shaped mandible found in the
Omo region of southern Ethiopia in 1967 is the
type specimen for *Australopithecus aethiopicus*,
which includes the KNM-WT 17000 cranium
as well as some isolated teeth. Actual size.
*Photograph by David L. Brill; courtesy of Donald
Johanson, Institute of Human Origins.*

Australopithecus boisei

OH 5 TYPE SPECIMEN NUTCRACKER MAN · ZINJ

Too often, important hominid fossil finds lie securely locked away with nothing more than a brief initial announcement in an obscure scientific journal. This was certainly not the case for the type specimen of the species *Australopithecus boisei*, discovered by Mary Leakey in 1959, almost exactly a century after Darwin published *On the Origin of Species*. Soon after its discovery, on the edge of the Serengeti Plain at Olduvai Gorge in northern Tanzania, Mary and her husband Louis Leakey placed the wonderfully complete cranium in the hands of a remarkable anatomist, Phillip V. Tobias. After years of dedicated descriptive, comparative, and interpretive work, in 1967 Tobias published a remarkable 264-page monograph on the find, a landmark in paleoanthropology.

Discovery of the celebrated specimen from Olduvai Gorge was a critical event in paleoanthropology. Not only did it launch modern paleoanthropology as a multi-interdisciplinary endeavor (see page 21), but it was the pivotal find that focused attention on East Africa, resulting in a seemingly endless series of hominid discoveries in the Great Rift Valley.

Louis Leakey, virtually the embodiment of human paleontology, believed that humans, large brained and toolmaking, originated millions of years ago. In fact, it was his initial discovery of tools at Olduvai Gorge in 1931 that convinced him that hominid fossils would eventually be found at the site. For Louis, tools were the defining feature of being human. How ironic, then, that the landmark find at Olduvai would be an *Australopithecus*—too small brained and primitive to be the toolmaker.

Most revealing about Louis Leakey's convictions are his first impressions of the skull, which he recorded on July 17, 1959, in a field notebook. Despite the obvious australopithecine features, such as small brain size and a well-developed masticatory system, he called the find *Titanohomo mirabilis* ("miraculous giant man"). Leakey considered the skull "near man," the fabricator of stone tools and therefore our direct ancestor. Some scientists might have accepted the find as the toolmaker, but to Leakey the Olduvai toolmaker could not be an *Australopithecus*.

Less than one month after the discovery, Leakey, as sole author, published in the pages of *Nature* a new genus and species for the Olduvai skull: *Zinjanthropus boisei*. The species name derives from one of Leakey's benefactors, Charles Boise, who helped finance the Olduvai work. Zinj is an old Arabic name for East Africa, and *anthropus* comes from the Greek for "man." Leakey's claim that Zinj was a new genus and a direct ancestor of modern humans generated little support among paleoanthropologists. The great geological antiquity of the cranium—at first only thought to be in excess of 600,000 years old, it is now dated to 1.8 million years—its completeness, and the simple fact that stone artifacts derived from the same geological horizons were, however, all reasons to rejoice when Leakey and Zinj made their appearance at the University of Chicago's Darwin Centennial, celebrating the anniversary of *On the Origin of Species*.

More than 100 specimens of teeth, mandibles, and some fairly complete crania of *Australopithecus boisei*, as it is now known, have been found in Ethiopia, Tanzania, and Kenya. The oldest of these was unearthed at Omo, Ethiopia, and is dated to about 2.3 million years ago (L. 74a-21, a fragmentary mandible) with the youngest, from Olduvai Gorge, dated to about 1.2 million years (OH 3 and OH 38, isolated teeth).

A. boisei was a truly impressive creature, dominated by a massive skull with a broad, concave face—a face like no other in human evolution. Zinj resembled its cousin in South Africa, *A. robustus*, but was "hyper-robust" in cranial adaptations. Although males and females showed marked sexual dimorphism (for example, OH 5 was a male and KNM-ER 732 was a female), the identical masticatory anatomy is diagnostic of both sexes. Males had prominent crests on the skull, especially a longitudinal crest along the skull midline, for anchoring huge temporalis muscles. These are the ones you feel on the side of your skull when you move your jaw up and down. Strong facial buttressing, seen in prominent, flaring cheekbones, projected anteriorly to hide the nasal area in lateral view. The gum-chewing muscles, the masseters, on the side of the face were bulging and massive. Enormous molars, four times the size of

our own, with low cusp relief and fairly flat surfaces were deeply set into the jaws by massive roots. The greatly reduced incisors and canines suggest that slicing and piercing were not significant in eating. To afford enhanced power to the chewing muscles, the mandible, in some instances ten times as thick as our own, was retracted and placed under the braincase, allowing for the production of more powerful vertical forces by the chewing muscles. The repositioning of the mandible and the palate produce a flat, slightly dished face.

The deep, thick mandible (see page 128) is a response to large masticatory forces that were necessary for processing low-quality food. In short, here was a creature in which a craniofacial architecture had been honed over generations to produce the most efficient chewing machine ever—the Cuisinart of human evolution. From the powerful vertical forces evident in the skull and jaw comes this individual's affectionate nickname, "Nutcracker Man."

Louis Leakey's suggestion that Zinj was a distinctive sort of early human ancestor is now universally accepted, even though the genus name, *Zinjanthropus*, has been dropped in favor of *Australopithecus*. Detailed examination of other finds, especially from Koobi Fora, substantiate the unique facial skeleton of *A. boisei*. In a notable study of comparative and functional anatomy, The Australopithecine Face, Yoel Rak defined a host of morphological features found in the face that make boisei a unique species. The palate is strongly retracted; the zygomatic (cheek) bones have migrated forward, extending into a visor-like support for the masseter muscles; and the supraorbital tori slope away from the midline, giving the face a "sad" expression.

Recovery of a partial skull from Konso in southern Ethiopia is the first association of a cranium and mandible of *A. boisei*. Dated to 1.4 million years ago, it was recovered from strata rich in Acheulean artifacts and a partial mandible of *H. erectus*. The skull, KGA-525, with a cranial capacity of 545 cc is slighter larger than OH 5 with 530 cc. Furthermore, the Konso specimen, showing advanced dental wear, differs slightly from the typical *A. boisei* condition in having

a posteriorly placed sagittal crest (like *A. aethiopicus*), a shallow palate, and a face lacking the visor-like support for the masseter muscles. Discoverers of the Konso specimen suggest that rather than giving a new species name to the find, they prefer to see the differences as indicative of intraspecific variation.

A. boisei was the end point of an evolutionary trajectory that can now be traced back to 2.5 million years ago (see *Australopithecus aethiopicus*, page 164). Extinction of the robust australopithecines may have been due to changes in the environment that made it more difficult for them to compete with other animals for diminishing food resources. Perhaps the diversification and population increases of baboons were important factors. Or perhaps the robusts were easy prey for the different members of the carnivore guild, which was developing its modern character. C. K. Brain showed that the robusts at Swartkrans in South Africa appear to have often been taken by leopards, as is evidenced in the puncture marks in one cranium from the site. Most likely the robusts made the evolutionary mistake of becoming too specialized, superbly adapted at what they did but not adaptable enough to keep up with the changes around them.

Australopithecus boisei, OH 5 (see also page 5). The most famous fossil from Olduvai Gorge, Tanzania, this cranium, nicknamed "Zinj," belongs to a specialized East African hyper-robust australopithecine. Actual size. *Photograph by David L. Brill; courtesy of National Museum of Tanzania.*

SPECIMEN	LOCALITY	AGE	DISCOVERERS	DATE	PUBLICATION
Adult male and female crania	Koobi Fora, Kenya	1.7 million years	Richard Leakey and H. Mutua	1969/1970	Leakey, R. 1970. New hominid remains and early artifacts from North Kenya. *Nature* 226: 223-4; and Leakey, R. 1971. Further evidence of lower Pleistocene hominids from East Rudolf, north Kenya. *Nature* 231: 241-245

Australopithecus boisei

KNM-ER 406
KNM-ER 732

It took the discovery of two important crania, KNM-ER 406 and KNM-ER 732, from Koobi Fora, Kenya, to resolve a controversial hypothesis proposed in the 1960s by University of Michigan anthropologists C. Loring Brace and Milford H. Wolpoff. Wolpoff championed the proposition that all South African australopithecines belonged to a single species. His reasoning was that the differences in tooth size, robusticity, cresting patterns, and facial shape were merely a reflection of sexual dimorphism (size and shape differences seen in the two sexes of one species). For Wolpoff, the males had been assigned to *A. robustus* and the females to *A. africanus*.

There were a number of obvious criticisms of what became known as the single species hypothesis. First, why did all the females (*A. africanus*) live and die at Sterkfontein and all the males (*A. robustus*) at Swartkrans? Second, why did the males wait half a million years to assemble at Swartkrans for a communal death? Despite these obvious problems in interpretation, Wolpoff stressed the high degree of overlap in morphology between *A. africanus* and *A. robustus* and continued to promulgate the single species hypothesis.

A solution to the single species versus multiple species argument required the recovery of an enlarged fossil hominid record from a well-dated geological sequence. The highly fossiliferous deposits east of Lake Turkana in northern Kenya, known as Koobi Fora, fit the bill perfectly, and in 1968 Richard Leakey, director of the National Museums of Kenya, initiated a series of highly successful expeditions to the region. In late August of the next year, Leakey found a virtually complete but edentulous (lacking teeth) cranium. The specimen, KNM-ER 406, although partially encased in sandstone, exposed sufficient anatomy to warrant placement in *A. boisei*, initially named after Olduvai Hominid 5 (see page 168).

KNM-ER 406 shares with OH 5 the following anatomical features: (1) sagittal and nuchal crests; (2) a deep palate with massive tooth roots and presumably very large cheek teeth; (3) a dish-shaped face that was astonishingly wide (due to the

flaring cheekbones) but short in height; (4) just behind the thin supraorbital tori, a frontal bone that is severely pinched (postorbital constriction), and immediately above the nasal bones, a projecting area known as the glabella, behind which is a flattened area called the frontal trigone; (5) a heart-shaped foramen magnum (as is also seen in OH 5); and (6) the root of the zygomatic (cheekbone) placed far forward over the second upper premolar.

Evidence for bone disease is relatively rare in early hominids, but KNM-ER 406 has a small, round hole on the frontal bone, just behind the left orbit. Apparently this was a pathology of some sort and may have been a metastatic abscess.

In 1970 a partial cranium catalogued as KNM-ER 732, sometimes referred to as a demicranium since a large portion of the left half of the cranium is missing, was found at Koobi Fora in geological deposits identical in age to the 1.7 million-year-old KNM-ER 406 specimen. Although clearly smaller in size and lacking the sagittal and nuchal crests of KNM-ER 406, the 732 cranium otherwise displayed the suite of anatomical traits typical of *A. boisei*. These features include: (1) a prominent glabella, (2) a frontal trigone, (3) a dish-shaped face, (4) a forwardly placed zygomatic root, (5) a very wide face, owing to the flaring zygomatics, and (6) strong postorbital constriction. Both crania have thin vault bones, typical of the australopithecines, with reconstructed cranial capacities of 510 cc for KNM-ER 406 and 500 cc for KNM-ER 732.

Employing the great ape model of sexual dimorphism, in which male chimps and, especially, male gorillas have larger, more heavily built crania than females, it is possible to assess the differences between KNM-ER 406 and 732. These two crania share an overwhelming number of anatomical characteristics. Those absent in KNM-ER 732, such as sagittal and nuchal crests, are precisely those missing in female great ape skulls. Therefore, the most parsimonious explanation of the difference in size between these two specimens is that KNM-ER 406 is a male and KNM-ER 732 is a female *A. boisei*.

In large part it was the recovery of these two crania from Koobi Fora that demonstrated without doubt the presence of sexual dimorphism in the australopithecines. Employing this model of sexual dimorphism seen in *A. boisei*, it was possible to reexamine the South African australopithecine sample. This resulted in two important conclusions: first, the anatomical differences between the Sterkfontein and Swartkrans samples were more marked than those seen in a single species such as *A. boisei*, and second, at each of these sites in South Africa, male and female specimens could be identified using as a guide the differences in size and rugosity seen in KNM-ER 406 and 732. The single species hypothesis was now effectively dead.

Australopithecus boisei, **KNM-ER 406.** Note the flat face, enormous arching cheek bones, and prominent crest atop this male robust australopithecine cranium. *Photograph by David L. Brill; courtesy of National Museums of Kenya.*

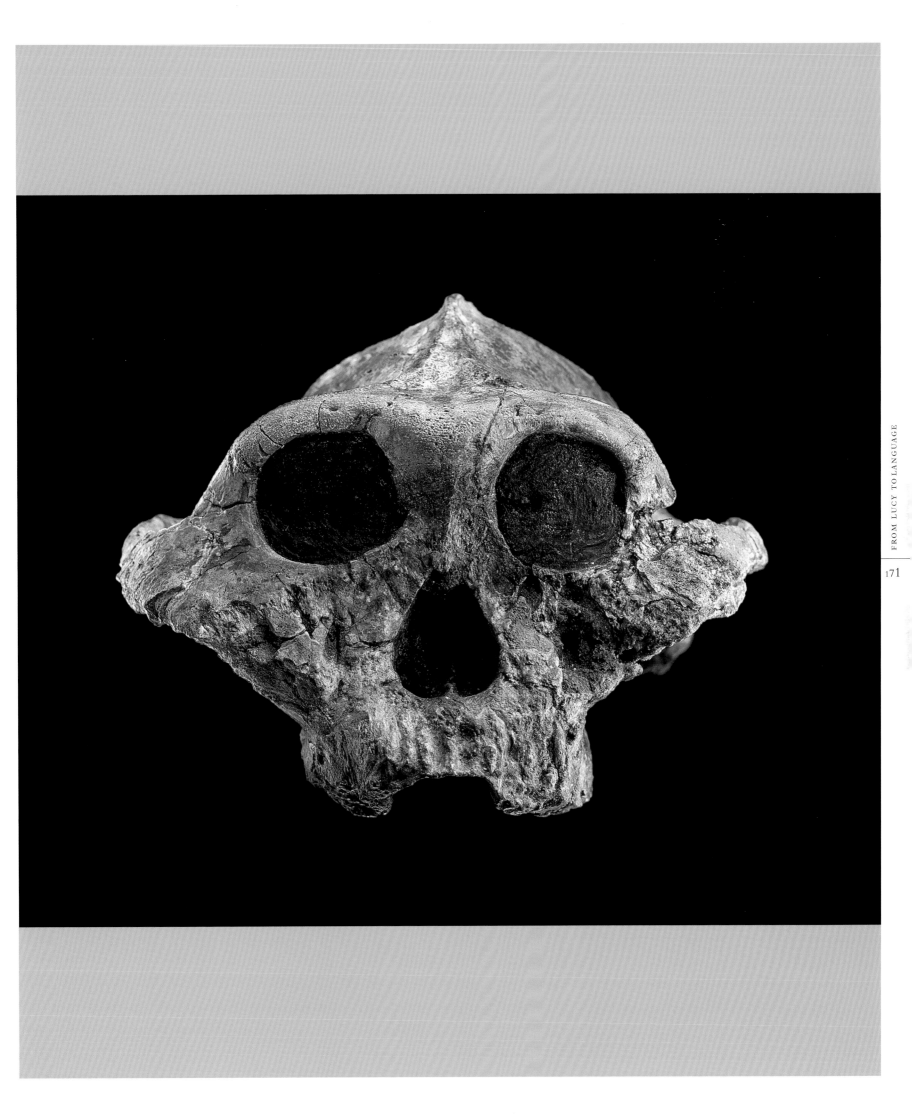

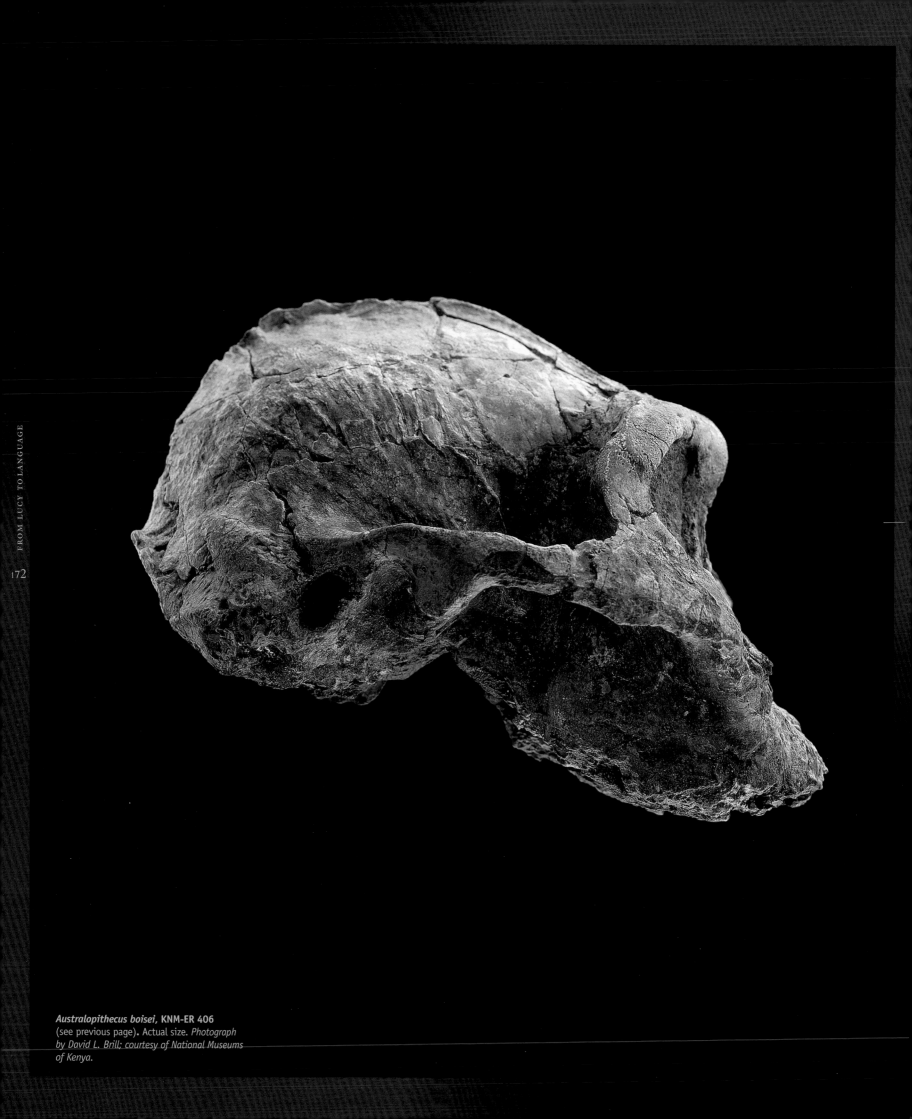

Australopithecus boisei, **KNM-ER 406**
(see previous page). Actual size. *Photograph
by David L. Brill; courtesy of National Museums
of Kenya.*

Australopithecus boisei, **KNM-ER 732**.
In a case of strong sexual dimorphism, an apparent female cranium of the same species has a similar brain size but much smaller facial features and no crests. Actual size. *Photograph by David L. Brill; courtesy of National Museums of Kenya.*

HOMO

For decades since its initial discovery in 1924, our understanding of *Australopithecus* greatly overshadowed knowledge of our own genus, *Homo*. Until the early 1960s, the oldest known species of *Homo* was *Homo erectus*, named by Dutch anatomist Eugène Dubois following his 1891 find of a skullcap in Java. What is emerging now as a result of discoveries of *Homo* fossils, particularly in East Africa, is a more comprehensive look at the diversity of early *Homo* during Pliocene and Pleistocene times (roughly 1.5 to 2.5 million years ago). At long last the significant temporal and morphological gap separating *H. erectus* and the australopithecines is beginning to be bridged. Yet at this stage of our understanding, because of the paucity of specimens, paleoanthropologists are far from agreeing on the roots and diversity of early *Homo*.

Australopithecines are characterized as having relatively small brains, large cheek teeth, a postcranial skeleton with some apelike features, and an absence of culture. In contrast, species of the genus Homo have relatively and absolutely large brains, a more modern postcranial skeleton, a significant reduction in tooth and jaw size, and, most important, culture.

The first mention of an early *Homo* in Africa that was distinct from *Homo erectus* came in 1964 with the announcement of *Homo habilis* from Olduvai Gorge, Tanzania. For the type specimen, Louis Leakey, Phillip Tobias, and John Napier chose Olduvai Hominid (OH) 7, a mandible associated with parts of the cranial vault and some postcranial bones. Although their landmark paper drew attention to features different from *Australopithecus* such as the narrow cheek teeth and a humanlike foot, their main reason for putting the Olduvai hominids into *Homo* was brain expansion. Until their article appeared in *Nature*, a cerebral Rubicon of anywhere between 700 and 800 cc had existed as the lower limit for the genus *Homo*. But in order to include the Olduvai specimen in *Homo*, which had an estimated cranial capacity of 680 cc, Leakey, Tobias, and Napier lowered the cerebral Rubicon to 600 cc. They also stressed a nonanatomical rationale for placing OH 7 in *Homo*—the presumed ability of these hominids to manufacture the stone tools found at the site. With this in mind it was Raymond Dart, author of *Australopithecus*, who suggested the name *habilis*, which implies "able, handy, mentally skilful [sic], vigorous," resulting in the popular moniker for the species, "handy man."

The overwhelming emphasis placed on the supposed cultural abilities of handy man in defining a species did not sit well with many paleoanthropologists. A major objection was that another hominid, *A. boisei*, a robust australopithecine, was already known from the same 1.8 million-year time stratum that had yielded the *habilis* material, and many critics thought it impossible to determine which hominid was responsible for the tools. According to most paleoanthropologists, the naming of a species should rest on its anatomical distinctiveness and not on presumed cultural abilities.

Louis Leakey led the charge on this issue, for not only did he believe that "man makes tools" but, even more important, that "tools maketh man." *Habilis* was his man, and by the time of his death, in 1972, Louis Leakey had embraced the idea that *H. habilis* led directly to *H. sapiens*, putting *H. erectus* onto a side branch.

Acceptance of *H. habilis* was far from universal, with some researchers more willing to expand the definition of *Australopithecus* to include the Olduvai specimens and others thinking they were simply earlier representatives of *Homo erectus*. Additional discoveries from Olduvai, the Lake Turkana basin, and South Africa bolstered the notion of early *Homo*. Especially strong vindication came in 1972, when a fairly complete cranium, KNM-ER 1470, was recovered east of Lake Turkana (formerly Lake Rudolf) at Koobi Fora in northern Kenya. Unfortunately, the teeth were not preserved, but with a cranial capacity of just over 750 cc, nearly everyone accepted 1470 as evidence of *Homo* at approximately 1.9 million years, roughly equivalent to the age of *H. habilis* at Olduvai Gorge.

Other important Pliocene-Pleistocene *Homo* specimens include a lightly built mandible (KNM-ER 992) and a small toothed cranium (KNM-ER 1813), also from Koobi Fora, which were assigned to *H. habilis*. Some specimens, like the wonderfully complete adolescent male skeleton (KNM-WT 15000) and two crania (KNM-ER 3733 and 3883), were judged to have affinities with *Homo erectus*.

Not everyone thought that these specimens fell comfortably into *habilis* or *erectus*. Australian anthropologist Colin Groves and his Hungarian colleague Vratislav Mazák concluded in 1975 that

the KNM-ER 992 mandible, because of its very small cheek teeth, belonged to a new species, *Homo ergaster*. An obvious reference to the stone tools found in the same geological horizon as the mandible, *ergaster* is from the Greek for "workman". Recognizing this species designation, University of Liverpool anatomist Bernard Wood, an expert on early *Homo*, now includes KNM-WT 15000 and KNM-ER 3383 and 3733 in *H. ergaster*.

By announcing 1470 as *Homo*, but not placing it in a species, Richard Leakey, then director of the Kenya National Museums, left the door open for someone to assign it to a new species. This happened in 1986, when a Russian anthropologist, Valerii Alexeev, selected 1470 as the type specimen for a new species, *Homo rudolfensis* (he called it *Pithecanthropus rudolfensis*, but the genus name, no longer in use, was replaced by *Homo*). In spite of the incompleteness of Alexeev's description and his unfamiliarity with the original specimen, the name has stuck and is now in common usage.

An interesting new wrinkle in the understanding of *H. habilis* came in 1986 with the recovery of a partial skeleton at Olduvai Gorge, OH 62. Although very fragmentary, OH 62 presents the first definitive association of upper and lower limbs of *H. habilis*, which, surprisingly enough, indicate that the body build was rather apelike. The arms are relatively long compared to the legs, reminiscent of the condition in *A. afarensis* and very unlike the modern body build paleoanthropologists had envisioned.

The KNM-WT 15000 *H. ergaster* skeleton has, in contrast a postcranial skeleton that is modern-like in size and proportions. Unfortunately, the postcranial skeleton of *H. rudolfensis* is largely unknown, with the possible exception of two large femurs (KNM-ER 1472 and 1481). The size variation in early *Homo* has also been interpreted as representing sexual dimorphism, with the smaller specimens (*habilis*) being female and the larger ones (*rudolfensis*) male. Other detailed studies of facial and cranial anatomy suggest that KNM-ER 1813 and KNM-ER 1470 are too distinct to belong in a single species.

The current appreciation of early *Homo* in Africa suggests three plausible species. *Homo habilis* shows brain expansion relative to its small body size, but its

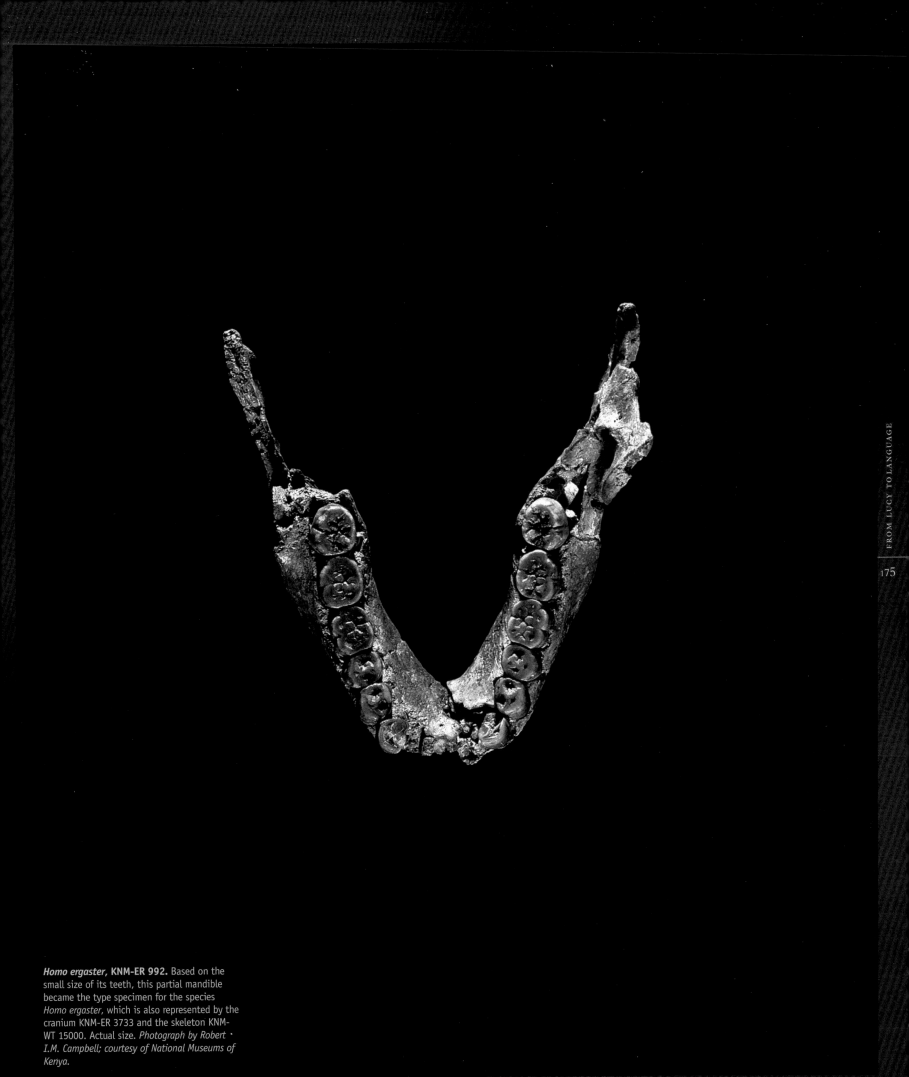

Homo ergaster, **KNM-ER 992.** Based on the
small size of its teeth, this partial mandible
became the type specimen for the species
Homo ergaster, which is also represented by the
cranium KNM-ER 3733 and the skeleton KNM-
WT 15000. Actual size. *Photograph by Robert*
I.M. Campbell; courtesy of National Museums of
Kenya.

postcranial skeleton is quite *Australopithecus*-like. A second species, *Homo rudolfensis*, with substantial brain expansion, has cranial features like pneumatization (large air cells in certain areas like the mastoid), a broad midface, large cheek teeth (as judged from the tooth roots), and relatively thick dental enamel—characteristics that are also *Australopithecus*-like. The third species, *Homo ergaster*, now acknowledged to be distinct from *H. erectus*, possesses a combination of a modern body build, large absolute and relative brain size, reduced teeth and jaws, and an apparent posture and locomotor abilities more like those of later *Homo*. Bernard Wood has concluded that *H. ergaster* is the most likely choice of an ancestor to later species of *Homo*.

For the moment, the evolutionary roots of *Homo* are still poorly understood, but they will ultimately be found in pre-2 million-year-old deposits. Despite the widely held view that *A. africanus* makes a good candidate for ancestor to *Homo*, equally convincing arguments can be mounted to support a unique link between *africanus* and *A. robustus* (see page 160). Should this be the case, then the three species of Pliocene-Pleistocene Homo are without an identifiable predecessor.

From the Pliocene (before 2 million years ago), there are precious few specimens attributable to *Homo*. Claims for early *Homo*, dated to roughly 2.5 million years, have been made for specimens from Lake Baringo, Kenya (a temporal bone), Uraha, Malawi (a mandible), and Sterkfontein, South Africa (a partial cranium). But there is a general lack of agreement as to the dating and evolutionary affinities of these specimens. Dating to 2.3 million years, the oldest well-dated specimen of *Homo* is A.L. 666-1 (see page 180), from Hadar, Ethiopia. This maxilla is not identical to any named species of *Homo*, but it does have affinities with *H. habilis*. Perhaps the A.L. 666-1 maxilla, 400,000 years older than other previously well-dated *Homo* specimens (for example, KNM-ER 1470), will lead the way toward clarifying the origins of our own genus.

Should further discoveries substantiate the presence of more than one species of early *Homo*, it may then be the case that *Homo* underwent an adaptive radiation. From the temporal distribution of the three proposed species, it would appear that *H. rudolfensis* is the most ancient, followed by *H. habilis* and finally *H. ergaster*. The evolutionary relationships between these three species and the Hadar *Homo* are extremely difficult to ascertain because of the relatively few specimens known for early *Homo*.

If *H. ergaster* gave rise to the later *Homo* species *heidelbergensis*, *neanderthalensis*, and *sapiens*, there is a considerable temporal gap between the most recent *ergaster*, at about 1.5 million years, and the oldest *heidelbergensis*, at about 0.6 million years. This gap is occupied by what is customarily referred to as *H. erectus*, but some paleoanthropologists, such as Ian Tattersall, of the American Museum of Natural History, New York, suggest that *erectus* as it is classically known from Asia is too specialized to be an ancestor to modern peoples. Hominids dating to 1.8 million years ago, presently only found in the Republic of Georgia (sometimes assigned to *H. georgicus*), document an early African exodus of a form very similar to *H. ergaster*, which may be the direct ancestor to *H. erectus* in Asia.

Homo heidelbergensis was long relegated to the dustbin, after having been named by Otto Schoetensack in 1908, based on the mandible from Mauer, near Heidelberg, Germany. The specimen lacks an exact geological age but it is a strong candidate for being among the oldest hominids in Europe, at more than 0.6 million years. Other specimens that might reasonably be assigned to *H. heidelbergensis* include Arago, in France, Atapuerca, in Spain, Petralona, in Greece, Kabwe and Bodo, in Africa, and perhaps Steinheim, in Germany, and Ndutu, in Tanzania. Discoveries at Gran Dolina in the Sierra de Atapuerca, Spain document the occurrence of humans in Europe close to 800,000 years ago. The specimens from Gran Dolina now assigned to a *H. antecessor* may represent the common ancestor to modern humans and Neandertals. This is in marked contrast to the widely held idea that *H. heidelbergensis* was the ancestor to Neandertals and modern humans. Neandertals, with their distinctive anatomy, are considered here to be a separate species, *Homo neanderthalensis*, which died out around 30,000 years ago.

Modern humans, *Homo sapiens*, probably arose in Africa between 200,000 and 100,000 years ago. This species is the only surviving hominid species of the genus, in fact the last surviving member of the zoological family the Hominidae. (Ironically, *Homo sapiens* is the only hominid species for which there is no type specimen.) Acceptance of a relatively recent origin for *sapiens*, the so-called "out of Africa" model (see page 46), implies that the Asian *H. erectus* is not an ancestor to modern-day *Homo*. A diminutive, small brained, but apparently culturally sophisticated hominid was found in Flores, Indonesia that survived apparently as recently as 12,000 years ago. This possible case of island dwarfism may document a population of *H. erectus* that became geographically and reproductively isolated in southeast Asia a few hundred thousand years ago and evolved into a unique species *H. floresiensis*.

This short review of the genus *Homo* serves to stress the complexity in the evolution of this genus. This is particularly true for the very earliest stages of the genus, when its evolution has customarily been thought to be uncomplicated and straightforward. Species diversity within *Homo* again draws attention to the view that the evolution of modern humans is not a directional or unilineal process but, as for other mammalian groups, one that exhibits branching and diversification. It is now necessary to dedicate as much attention to understanding the diversity in the genus *Homo* as has been afforded *Australopithecus*.

Homo erectus, **Peking Man reconstruction** (see also page 7). A third of the known fossils from this species come from the site of Zhoukoudian, China, but most of the original specimens, including those used in this reconstruction, have been lost. Actual size. *Photograph by David L. Brill; courtesy of American Museum of Natural History.*

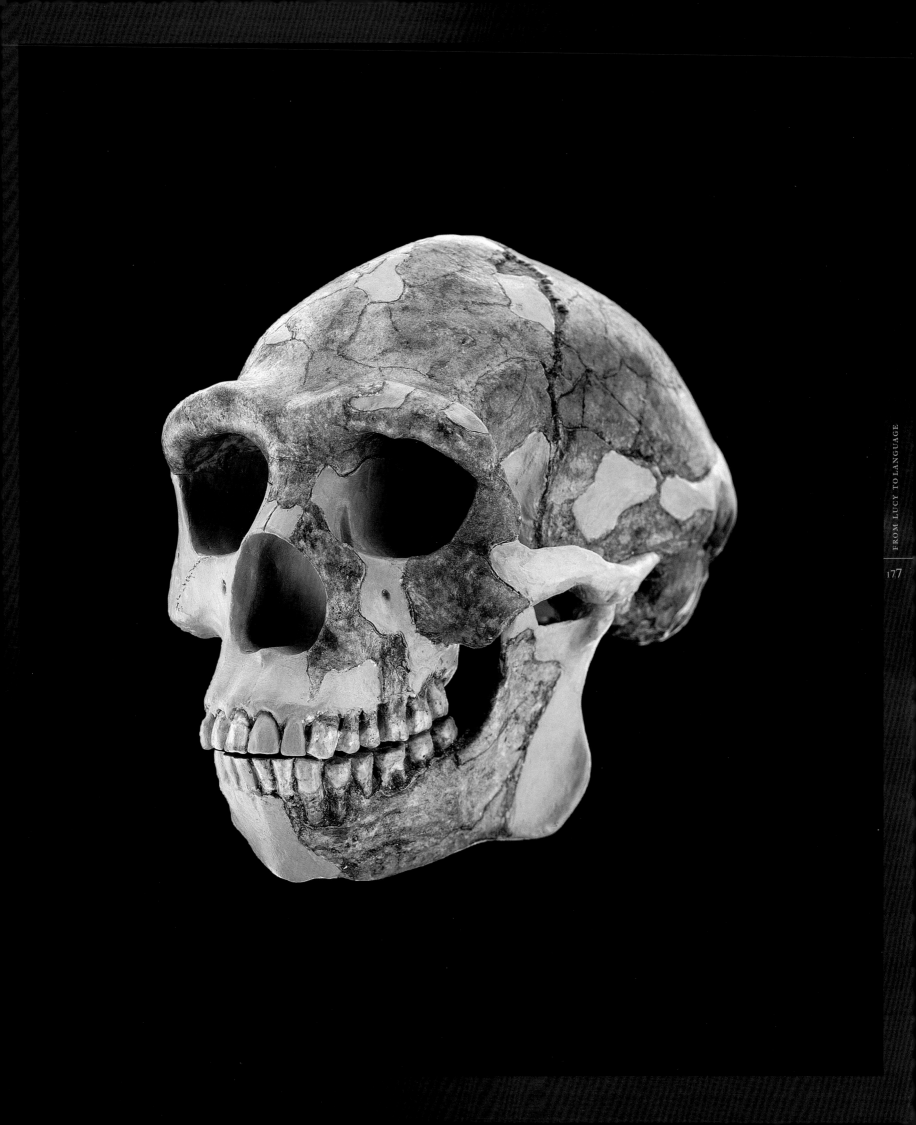

Homo neanderthalensis, Aı
page 235). Neandertals tenaeu to have a
stocky build, but the Amud 1 skeleton is the
tallest known individual—1.8 meters—and
has the largest brain capacity of any fossil
hominid specimen. Actual size. *Photograph
by David L. Brill; courtesy of Israel Antiquities
Authority, Rockefeller Museum.*

Homo sapiens, **Cro-Magnon I** (see also page 261)**.** Skeletons excavated from the French rock-shelter of Cro-Magnon have come to characterize, at least in popular literature, the earliest modern humans in Europe. Actual size. *Photograph by David L. Brill; courtesy of Musée de l'Homme.*

SPECIMEN	LOCALITY	AGE	DISCOVERERS	DATE	PUBLICATION
Adult maxilla	Hadar, Ethiopia	2.3 million years	Ali Yesuf and Maumin Allahendu	November 2, 1994	Kimbel, W.H., R. Walter, D. Johanson, K. Reed, J. Aronson, Z. Assefa, C. Marean, G. Eck, R. Bobe, E. Hovers, Y. Rak, C. Vondra, T. Yemane, D. York, Y. Chen, N. Evensen, and P. Smith. Late Pliocene *Homo* and Oldowan tools from the Hadar Formation (Kada Hadar Member), Ethiopia. 1996 *J. Hum. Evol.*, 31: 549-561

Homo sp.
A.L. 666-1

A surprising discovery was made at Hadar, Ethiopia, on November 2, 1994, when an unprepossessing hominid maxilla (upper jaw) was collected in an area called the Makaamitalu. Unlike the 362 other hominid specimens from Hadar, all of which have been attributed to *Australopithecus afarensis*, this specimen, catalogued as A.L. 666-1, had a distinctive parabola-shaped dental arch that immediately identified it as belonging to *Homo*.

The maxilla, found in two halves, was cleanly broken along the midline (intermaxillary suture), allowing the two parts to be easily fitted back together. The initial find was made on the slope of a steep knoll by Ali Yesuf and Maumin Allahendu, two Afar members of the Hadar Research Project from the Institute of Human Origins. Some 30 pieces of tooth crowns and roots as well as bits of the maxilla itself were recovered during careful surface collection and sieving operations at the locality. These pieces were all fitted back into the maxilla and included the right canine to first molar and left lateral incisor to second molar.

The anatomy of the 666 specimen is clearly distinct from other maxillas from Hadar assigned to *A. Afarensis*. In *afarensis* the maxilla is shallow and narrow with a rectangular dental arch, and there is marked prognathism (projection) below the nasal opening. In contrast, the 666 maxilla is relatively deep and wide with a bowl-shaped dental arch and exhibits only slight prognathism. Notably these features in the 666 specimen are common in species recognized as *Homo* rather than *Australopithecus*.

Also scattered on the side of the knoll where the maxilla was found were numerous manufactured stone flakes and cores (see page 89). These artifacts were very fresh with sharp edges, and, like the maxilla, presumably had not been on the surface long. Made predominantly of basalt and chert, the tools are most similar to those classified in the Oldowan industry, named after initial finds made at Olduvai Gorge, Tanzania.

A two-square-meter excavation was made into the knoll in an attempt to locate the precise horizon from which the maxilla and artifacts came. In a layer of silt identical to the silt matrix adhering to the maxilla, a total of 14 additional artifacts, similar to those on the surface, and three bone fragments were recovered. In patina and color these bones match the maxilla. In addition, a fossilized root cast in the sinus of the maxilla resembles the numerous root casts found in the excavation. It is therefore highly likely that the surface material came from the same horizon that yielded the *in situ* artifacts and bone fragments.

Geological studies of the locality in which 666 was found confirm that the artifact-bearing level is situated roughly 80 centimeters below a well-dated volcanic ash known as BKT-3. A very precise date of 2.27 ± .06 million years for BKT-3 was determined utilizing the single-crystal laser-fusion argon-argon technique (see page 26).

A date of 2.3 million years for the 666 maxilla is significant because no other occurrence of *Homo* in Africa is definitely dated to older than 1.9 million years. (An older date of 2.5 million years for a mandible from Malawi is imprecise because it is based on associated vertebrate species that range over a considerable time period.) While it is impossible to make a positive taxonomic assignment of the Hadar specimen with any of the other species of early *Homo*— *habilis*, *rudolfensis*, or *ergaster*—the overall anatomical nature of A.L. 666-1 has its strongest affinities with *H. habilis*. Due to the larger size of A.L. 666-1 compared with KNM-ER 1813, OH 13, OH 24, and OH 62, all considered to be female *H. habilis*, the Hadar specimen may represent a male individual of that taxon. Further support for the affinity of A.L. 666-1 with *H. habilis* comes from the discovery of a 1.8 million year old maxilla, OH 65, preserving all 16 teeth, from Olduvai Gorge. Assigned to *H. habilis*, OH 65, shares particularly strong similarities with A.L. 666-1 in details of the palate shape and depth as well as in details of molar morphology.

The importance of the A.L. 666 discoveries at Hadar is that this is the oldest known definitive association of hominid remains with stone tools and the oldest well-dated evidence for the genus *Homo*. Further finds of *Homo* will add substantially to our understanding of the roots of our own genus.

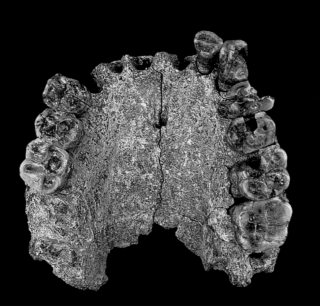

Homo sp., A.L. 666-1 (top) and *Australopithecus afarensis*, A.L. 200-1 (bottom).
Two maxillae from Hadar, Ethiopia show very different features. The more narrow, rectangular specimen with a shallow palate belongs to *A. afarensis,* while the deeper, shorter, and broader palate is *Homo.* Actual sizes. *Courtesy of Donald Johanson, Institute of Human Origins, and National Museum of Ethiopia.*

Homo sp., A.L. 666-1 (top) and *Australopithecus afarensis*, A.L. 200-1 (bottom).
A midline section view reveals further differences between these specimens, particularly the less projecting alveolar region of the *Homo* maxilla, which is the oldest well-dated fossil of our genus. Actual sizes. *Photograph by William Kimbel, Institute of Human Origins, courtesy of National Museum of Ethiopia.*

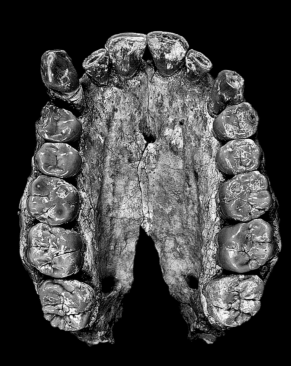

SPECIMEN	LOCALITY	AGE	DISCOVERERS	DATE	PUBLICATION
Juvenile male mandible	Olduvai Gorge, Tanzania	1.75 million years	Jonathan Leakey	November 4, 1960	Leakey, L.S.B., P.V. Tobias, and J.R. Napier. 1964. Recent discoveries of fossil hominids in Tanganyika; at Olduvai and near Lake Natron. *Nature* 202(4927): 7-9

Homo habilis

OH 7 TYPE SPECIMEN

OH 7 serves as the holotype for *Homo habilis*: the definitive single specimen used to describe the distinguishing attributes of this species, which has traditionally been placed at the root of our genus, *Homo*. The story of its discovery and analysis illustrates the process of naming a new hominid species and the obstacles that often block scientific acceptance of the new name.

The specimen consists of two dozen bones and 14 teeth that, judging from their proximity, similar size, and lack of duplication, come from the same individual: a nearly complete left parietal bone, a fragmented right parietal, most of the mandibular body, and all of the lower teeth, from the incisors back to the right first molar and left second molar. Twenty-one finger, hand, and wrist bones and an upper molar, all found nearby, probably come from the same skeleton, a juvenile that died at the age of ten to twelve years. OH 7 was nicknamed "Jonny's Child" in honor of the discoverer, Louis and Mary Leakey's oldest son.

An interesting aside about OH 7 is that a number of the bones show evidence for significant carnivore damage. The distal ends of the foot bones are gnawed away as is the lower margin of the mandible. The parietal bones show broad furrows that were perhaps produced by the canines or carnassials of a carnivore. The bone damage closely resembles the kind of damage inflicted by hyenas. It is very possible that OH 7 was killed or scavenged by a hyena and, as is often the case with hyenas, stashed under water to keep other carnivores from eating the prey.

Although OH 7 was selected to represent its species, it was not the first *Homo habilis* fossil found at Olduvai Gorge. In 1959 Heslon Mukiri, one of the Leakeys' fossil hunters, came across a molar and premolar at another site in Bed I, the oldest of Olduvai's major geological deposits. Weeks after Mukiri's discovery, however, Mary Leakey uncovered the spectacular OH 5 cranium (see page 168), a find that temporarily overshadowed the significance of the two teeth. With the discovery of OH 7 in 1960, attention again turned to the apparent presence of a more gracile hominid

that lived contemporaneously with the robust australopithecine represented by OH 5 and that may have made the site's many Oldowan stone tools instead of OH 5.

Louis Leakey informally referred to the OH 7 bones as "pre-*Zinjanthropus*," but he quickly became convinced that the fossils belonged to an early form of *Homo*. To help describe the bones, he enlisted paleoanthropologist Phillip Tobias and anatomist John Napier, who remained more cautious in their conclusions until detailed study could be made. Tobias noted the parietals' large size and determined that OH 7 had possessed a significantly larger brain than had *Australopithecus africanus* or *A. boisei*, yet much smaller than the brain of *Homo erectus* from Java and China. From a partial endocast, Tobias obtained a cranial capacity of 363 cc and estimated that the total cranial capacity would have been around 674 cc, or 50 percent larger than the average for six *A. africanus* crania.

Cranial capacity estimates on six crania attributed to *H. habilis* provide a mean value of 610 cc, some 30 percent larger than in *A. africanus*. Unfortunately the true significance of this increase is not fully understood because of a lack of associated postcranial material. If these individuals had small body size, then the relatively larger brains compared to body size suggests significant encephalization (ratio of brain to body size), but if the body sizes were large, then they do not document a high degree of encephalization.

Establishing such an obvious difference in brain size was a first step toward designating a new species. Further clues came from the teeth, the shape of which fell outside the known range for *A. africanus* and came much closer to later members of *Homo*. The incisors of OH 7 were relatively large compared to both *Australopithecus* and *H. erectus*. The premolars and molars appeared too narrow and elongated to belong to any australopithecine. The hand bones also more closely resembled other species of *Homo*, as did a nearly complete set of foot bones recovered nearby but from a different skeleton.

The combined evidence from three distinct parts of the skeleton, then, pointed toward this specimen being something new and different. Additional fossils found at Olduvai late in 1963 the specimens —OH 13 (Cinderella), 14, 15, and 16 (George)—clinched the argument that a second hominid species had lived beside *A. boisei*. In January 1964, more than three years after the discovery of OH 7, Leakey, Tobias, and Napier announced their new species of *Homo* with a name suggested by Raymond Dart, who had named the genus *Australopithecus* almost forty years earlier. *Homo habilis* means "handy man," a reference to this species's presumed aptitude for toolmaking.

After the name was published, several prominent anthropologists challenged its validity. The new fossils, they argued, were not distinct enough from *Australopithecus africanus* (or, alternatively, from *Homo erectus*) to be a separate species, but rather were some sort of subspecific variety. Other names and interpretations were introduced and debated, but habilis eventually became accepted in the late 1970s by a majority of paleoanthropologists. In large measure, acceptance came after discovery of the 1470 cranium (see page 189), which, ironically, has since been convincingly proposed as belonging to a second species of early *Homo*.

In contrast to the initial announcement of *H. habilis* that stressed the "humanlike" nature of the species, recent revaluation of the taxon by some observers has emphasized similarities with *Australopithecus*. These features include relatively large teeth and reduced encephalization as well as limb proportions (see O.H. 62; page 188) that are similar to *A. afarensis*. This opinion has not received wide support, but if it were substantiated, *H. habilis* would be transferred to *Australopithecus habilis*.

Homo habilis, OH 7. This mandible is one of two dozen bones from a juvenile skeleton found at Olduvai Gorge, Tanzania, that comprise the type specimen of this species. The relatively large incisors and narrow premolars and molars provided dental clues that this was a new kind of hominid. Actual size. *Photograph by John Reader, Science Source/Photo Researchers.*

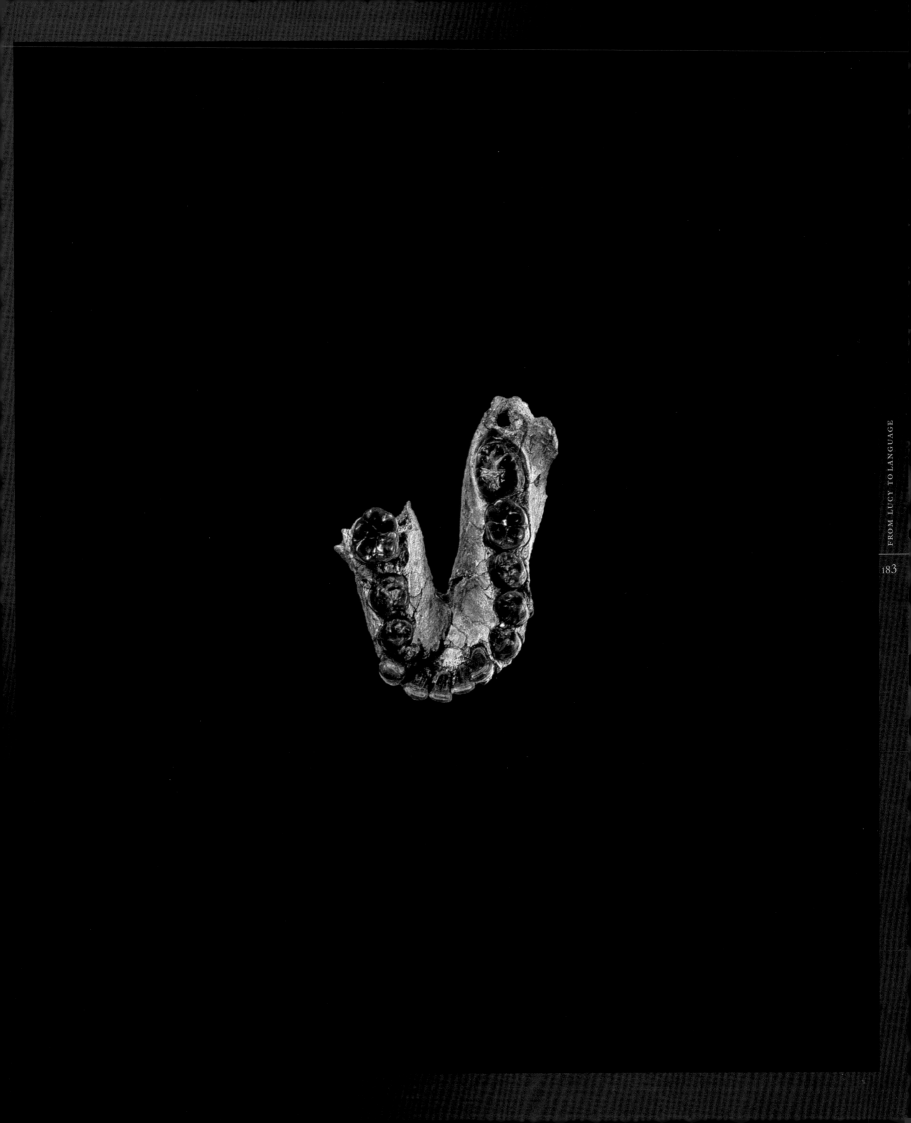

SPECIMEN	LOCALITY	AGE	DISCOVERERS	DATE	PUBLICATION
Adult female cranium	Olduvai Gorge, Tanzania	ca. 1.8 million years	Peter Nzube	October 1968	Leakey, M.D., R.J. Clarke, and L.S.B. Leakey. 1971. New hominid skull from Bed I, Olduvai Gorge, Tanzania. *Nature* 232: 308-312

Homo habilis
OH 24

Having enough distinctive fossil evidence is an important first step toward designating a new hominid species. For those scientists who considered it hasty to name the species *Homo habilis* from more fragmentary cranial fossils such as OH 7 (see page 182), this specimen provided important corroborating evidence for the species's validity. We include it here for that reason, and because it constitutes the oldest hominid from Olduvai Gorge and, with the exception of OH 5 (see page 168), the most complete cranium ever found there.

When it was found, however, the fractured and collapsed braincase, cemented together with a coating of limestone, give little hint of the specimen's value. Paleoanthropologist Phillip Tobias, co-author of the paper announcing *habilis*, remarked, on seeing the fossil, "Only Twiggy has been that flat!" The British model's name stuck as a nickname for OH 24.

Paleoanthropologist Ron Clarke's prolonged and painstaking reconstruction with dental pick, drill, and hammer restored as much of the original anatomical form as possible to OH 24. More than 100 tiny fragments could not be put back into place, and the cranium remains somewhat distorted. The top is still flattened, and the base is depressed around the foramen magnum. Behind the browridges, the vault should rise more vertically than it does. Despite these shortcomings the specimen can still tell us a lot about this hominid.

All the cranial bones are thin and lack the robusticity that tends to characterize australopithecine crania. Relative to australopithecines, the braincase has expanded from side to side, so that OH 24 approaches the striking parietal breadth of OH 7 (see page 182), but these bones of the braincase are shorter vertically than those of the habilis type specimen. The cranial capacity of OH 24 falls just under 600 cc, the minimum value for *Homo*, but the low estimate is probably due to the cranium's distortion.

The increased size of the braincase was countered by a reduction of the face, a pattern that continued with our species. From a short and relatively straight upper face, the maxilla of OH 24 slopes forward, imparting a hollowed profile like the face of *Australopithecus africanus* (see Sts 5, see page 146), although it is less projecting than in that species. The lightly built, somewhat projecting face of OH 24 stands in strong contrast to the massive, more vertical, dish-shaped face of OH 5.

Tooth crowns and roots in this specimen are small. The missing incisor and canine teeth probably eroded down the slope where the fossil was found long before its recovery. Two premolars and five molars have been preserved. Judging from the state of tooth eruption—the third molars had recently erupted and show no trace of wear—this individual was an adolescent or young adult at death.

In features of both the cranium and the teeth, OH 24 resembles more closely the specimen OH 13 than the presumed male specimen OH 7, even though OH 24 is closer in geological age to the latter. The sides of the cranium in OH 24 are clearly smaller than the parietal bones in OH 7. This suggests that the anatomy of early *Homo* incorporated significant individual and sexual variation.

Homo habilis, **OH 24.** Although it is heavily reconstructed, this cranium constituted important additional evidence in support of *Homo habilis* being a valid species. Actual size. *Photograph by John Reader, Science Source/ Photo Researchers.*

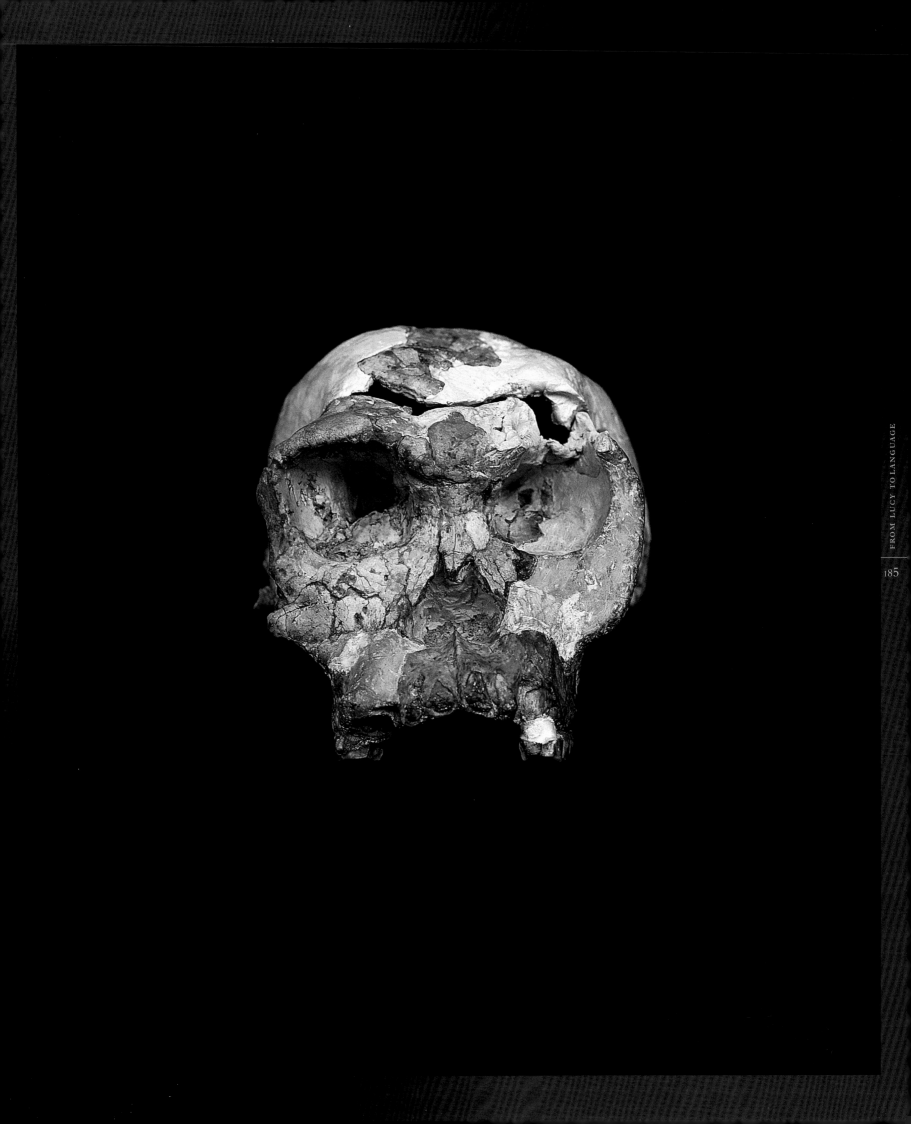

Homo habilis

KNM-ER 1813 *LUCY'S CHILD*

The discovery of a diminutive cranium at Koobi Fora posed a dilemma disproportionate to the specimen's size. A year earlier, KNM-ER 1470 (see page 189) had been found, revealing that a large-brained, large-faced, big-toothed species of *Homo* had inhabited Koobi Fora close to 2 million years ago. Then came this specimen of essentially equal antiquity and with a similar bell-shaped braincase but with a small face, small teeth, and a much smaller brain. Its small size was not due to immaturity, for the third molar had fully erupted and shows wear.

Could 1813 be a female of the same species as the male 1470, as some have suggested, or is the story more complicated? Although their obvious size differences make it likely that 1470 and 1813 represent respectively a male and female, it is increasingly unlikely that they belong to the same species. The differences in braincase size, facial size and shape, and overall robusticity are too great to ignore or to be explained by sexual variation. Instead, there were at least two species of early *Homo*, only one of which survived and led to *Homo erectus* and later to modern humans.

The smallest of the several hominid crania found at Lake Turkana, 1813 also has a slightly smaller brain capacity than any of the Olduvai *Homo* fossils. Its brain volume was about 510 cc (just above the australopithecine average and below the generally accepted cutoff point of 600 cc for *Homo*), compared to 1470's cranial capacity of 775 cc.

Although the nasal bones of this specimen have been crushed and the entire face is skewed somewhat to the left from its true anatomical position, the right side of the face is mostly complete. The upper jaw is particularly well preserved, with an intact set of left teeth from the canine to the third molar. Both the short, lightly built face and the small, arched, and rounded supraorbital torus are quite distinct from 1470's anatomy. A peculiar rounded protuberance on the back of the 1813 cranium may be the beginning of a transverse torus, as occurs in *Homo erectus*, so this feature could link 1813 with later species of *Homo*.

Although it can be readily contrasted with 1470, 1813 can be more easily compared to the less complete Olduvai crania generally included within *Homo habilis*. In size and shape, the teeth of 1813, as well as the palate and parts of the skull, strongly resemble those of OH 13. The 1813 cranium also shows affinities with the crushed cranium of OH 24 (see page 184): both are of similar overall size and have a short face, small eye sockets, nasal bones and cheekbones placed low on the face, and a projection of the face beneath the nasal opening. The dimensions of the parietal and occipital bones are similar in all three specimens.

Because of these similarities, we have opted to include KNM-ER 1813 in *Homo habilis*, following the classification of paleoanthropologist Bernard Wood. Richard Leakey initially concluded that 1813 was closest to the South African species *Australopithecus africanus*. He has more recently avoided putting a taxonomic label on 1813 other than to say that it and OH 13 should not be called *Homo habilis*, a name he applies to the 1470 cranium. But anatomy supports the idea that 1813 belongs in *habilis* and that this species— on the basis of facial proportions and the shape of the cheekbones and browridge, among other features—strengthens the view that *H. rudolfensis* and *H. habilis* are two distinct species of *Homo*.

***Homo habilis*, KNM-ER 1813** (see also page 6). Similar in size and appearance to OH 24, this cranium from Koobi Fora, Kenya has been placed in the same species. The small face, teeth, and braincase of this specimen distinguish it from other Koobi Fora crania, which belong to two different species of early *Homo*. Actual size. *Photograph by David L. Brill; courtesy of National Museums of Kenya.*

SPECIMEN	LOCALITY	AGE	DISCOVERERS	DATE	PUBLICATION
Partial adult skeleton	Olduvai Gorge, Tanzania	1.8 million years	Tim White	July 21, 1986	Johanson, D.C., F.T. Masao, G.G. Eck, T.D. White, R.C. Walter, W.H. Kimbel, B. Asfaw, P. Manega, P. Ndessokia, and G. Suwa. 1987. New partial skeleton of *Homo habilis* from Olduvai Gorge, Tanzania. *Nature* 327: 205-209

Homo habilis
OH 62

After decades of painstaking research at Olduvai Gorge, Mary Leakey retired from fieldwork to complete scientific publication of her excavations. The untiring efforts of Louis and Mary Leakey assured that no stone would be left unturned in the search for fossil and archaeological clues in this canyon, cut into the flat and generally featureless Serengeti Plain of Tanzania. After each rainy season, the slopes and gullies of the gorge were systematically searched for any new evidence that might have come to the surface as a result of erosion. Owing to the compacted and lightly cemented Olduvai deposits, erosion was slow, and after so many years of surface survey the chances of making an important discovery like *Zinjanthropus* were vanishingly small.

This was the setting when the Institute of Human Origins was invited to assist the Tanzanian Department of Antiquities in reopening work at Olduvai in 1985. The team was well aware of the odds against making a significant find. However, late in the afternoon of July 21, 1986, only three days into the expedition, a hominid fossil was found in an area of the gorge immediately adjacent to a dirt road that had led literally tens of thousands of tourists and scientists into the gorge.

Northern Tanzania, especially Ngorongoro Crater and the Serengeti, draws an endless stream of tourists who come on safari to observe and photograph African animals. On this tourist circuit lies the gem of Olduvai Gorge. After a visit to a small museum detailing the work at Olduvai, guides lead tourists into the gorge to visit localities where the Leakeys made important discoveries, such as the "Zinj" site where a stone plinth marks the exact spot of discovery.

How ironic that sitting on the surface of the ground, a mere 25 meters off the road, was a highly fragmented, 1.8 million-year-old partial skeleton. After recognition of the first fragment of the skeleton, a proximal right ulna, a major effort was put into an arduous search for additional skeletal elements. Because it was the sixty-second hominid to be found at Olduvai, the specimen was given the designation Olduvai Hominid 62, or OH 62. Screening of all

the loose sediments in the immediate area resulted in the recovery of more than 18,000 bone and tooth fragments over an area of about 40 square meters. Most of these were nonhominid remains, but 302 were identified as belonging to OH 62.

Compared with skeletons like Lucy or the Black Skull, OH 62 was scrappy. The teeth were hopelessly splintered into such tiny fragments that, with few exceptions, it was not possible to reassemble a complete tooth crown. It was equally impossible to reassemble the cranial vault fragments into anything even remotely resembling a skull. However, portions of a right arm, including most of a humerus and parts of the ulna and radius, were recovered. Part of the left femur was also recovered, including the femoral neck and some of the shaft. Most important, 32 fragments were successfully glued together to form most of the maxilla, or upper jaw.

It was the maxilla that permitted identification of OH 62. This portion of the facial skeleton most closely resembled specimens from Olduvai and elsewhere that had been assigned to *Homo habilis*. Because of this, as well as the complete lack of robust australopithecine specializations, such as megadont molars, OH 62 was attributed to *H. habilis*.

OH 62 represents the first time that upper and lower limb elements of *H. habilis* were securely associated. The third molar had erupted, and from the heavily worn occlusal surfaces of the teeth it was clear that OH 62 was a relatively old adult. From the diminutive size of the limb bones, especially the femur, which is even smaller than Lucy's, it was postulated that OH 62 was a female. In fact, OH 62 may be the smallest adult fossil hominid ever found, with an estimated stature of about one meter.

The most startling aspect of OH 62 became evident when body proportions were calculated for the upper and lower limbs. Fairly accurate estimates of total limb length for the humerus and the femur, both incomplete, could be calculated. The humerofemoral index of 95 percent indicated that the humerus was 95 percent the length of the femur: a very long arm.

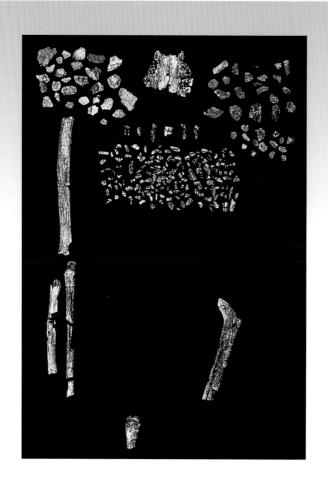

In modern humans this index is roughly 70 percent, while in a quadruped like a chimpanzee it is 100 percent (that is, the humerus and femur are of equal length). Such apelike proportions for *Homo habilis* were unanticipated.

It was previously thought that in *H. habilis*, body proportions would be more modern. The implication was clear: this specimen, OH 62, had limb proportions reminiscent of the primitive condition seen in *A. afarensis* (Lucy). If *H. habilis* was to be considered an ancestor to *H. ergaster/erectus* at 1.6 million years old, then not only would body size have to increase rather considerably, but the relationship between upper and lower limbs would also have to change dramatically. All this would have had to occur over a mere 200,000 years of time. Not an impossibility, but evolution would have had to have been fairly rapid.

OH 62 raises a number of additional important issues. If this is a female *Homo habilis*, as postulated, which other hominids in East Africa represent the males? If the larger femora (KNM-ER 1472 and 1481) from Lake Turkana are from male individuals, then sexual dimorphism is on

***Homo habilis*, OH 62.** Partially pieced together from more than 300 fragments, this tiny skeleton permitted an estimate of body proportions that revealed that this hominid had long arms compared to its legs, like Lucy and African apes. *Courtesy of Donald Johanson, Institute of Human Origins.*

par with that seen in *A. afarensis*—something that had not been anticipated. The discovery of OH 62 has added to the confusion surrounding the diversity in early *Homo* (see page 188). According to Bernard Wood, an anatomist at George Washington University, *H. habilis*, as classically represented by finds at Olduvai Gorge, appears to represent a species distinct from *H. ergaster* and *H. rudolfensis*. Although it appears that early *Homo* may be more speciose than formerly appreciated, the sorting out of the different species will have to await recovery of more complete *Homo* specimens, ideally with cranial and postcranial elements preserved in the same individuals.

SPECIMEN	LOCALITY	AGE	DISCOVERERS	DATE	PUBLICATION
Adult cranium	Koobi Fora, Kenya	1.8 to 1.9 million years	Bernard Ngeneo	August 1972	Leakey, R.E.F. 1973. Evidence for an advanced Plio-Pleistocene hominid from East Rudolf, Kenya. *Nature* 242: 447-450

Homo rudolfensis
KNM-ER 1470   TYPE SPECIMEN

The famous cranium of KNM-ER 1470, which graced the cover of Richard Leakey and Roger Lewin's best-selling book *Origins*, brought Richard Leakey instant fame and vindicated Louis Leakey's belief that a large-brained species of *Homo* inhabited East Africa millions of years ago. Louis, who saw the specimen only four days before he died, carried to his grave the deep-seated conviction of his teacher, Sir Arthur Keith, that our genus had a great antiquity dating back millions of years, and he refused to accept an ancestral role in our lineage for any australopithecine. Although often seen as the epitome of *Homo habilis*, the species Louis helped name and that was dearest to his heart, 1470 may deserve a new identity and reveal a more complicated pattern of evolution in the early days of *Homo*.

In his initial description of the fossil, Richard Leakey chose to label 1470 an indeterminate species of *Homo* rather than include it within *Homo habilis*, primarily because it was thought to be 2.9 million years old, or a million years older than the *habilis* fossils from Olduvai Gorge. This age was based on an estimate of 2.6 million years for the KBS Tuff, a volcanic ash layer at Koobi Fora that lay above the specimen. But it turned out that the age initially calculated for the tuff was made inaccurate by samples contaminated by older volcanic rocks; the real age was only 1.8 million years, and suddenly KNM-ER 1470 became the same age as *Homo habilis*.

The cranium was reconstructed from more than 150 fragments by paleontologist Meave Leakey into two major pieces that constitute the specimen's most striking features: a large braincase and a broad, flat face. Some have pointed out minor features of the face and cranial shape that could link this specimen to *Australopithecus*, but the bulk of its traits and a brain volume of 775 cc support its placement within *Homo*. Cranial capacity estimates on three specimens attributed to *H. rudolfensis* yield a mean of 788 cc, a 70 percent increase over the *A. africanus* mean. The true significance of this increase must be related to body size but none of the crania are associated with postcranial bones.

Compared with the smaller Koobi Fora cranium KNM-ER 1813 (see page 186), for instance, there is only a slight supraorbital torus across the forehead with no sulcus or depression behind it; 1470's face is much longer, with the upper part narrower than the middle, and the upper jaw is squared off rather than rounded, with a very short, shallow palate. The cranium lacks the crests and heavy muscle markings that characterize australopithecine skulls. Although its tooth crowns were not preserved, the remaining roots and sockets reveal that this individual had very large incisors and canines and moderate premolars and molars.

Other features of 1470's anatomy are a marked constriction of the braincase behind the eye sockets (but less than occurs in robust australopithecines), a bulging frontal bone that rises steeply to meet the square parietal bones forming the thin-walled sides of the braincase, and an occipital bone that is smoothly rounded rather than flexed as in *Homo erectus*.

Whether 1470 really belongs within *habilis* has become a matter of heated debate. What features 1470 shares with *habilis* occur in other species of *Homo*, and we have already noted several important differences between 1470 and habilis that apparently are not due to sex, time, or geography. If the smaller specimens from Olduvai and Koobi Fora constitute *Homo habilis*, then perhaps 1470 belongs to a separate species.

In 1986, the scientific name *Homo rudolfensis* was proposed by Valerii Alexeev for KNM-ER 1470. This new species has been identified with a few fossils recovered from Koobi Fora in sediments from the slim slice of time between 1.9 and 1.8 million years ago. Other specimens attributed to *Homo rudolfensis* include the cranial remains KNM-ER 1590 and KNM-ER 3732, and the lower jaw KNM-ER 1802. By acknowledging 1470 as a separate species, we concur with the view that at least two species of early *Homo* lived contemporaneously in East Africa. The evolution of *Homo* was not a simple linear story from *habilis* to *erectus* to *sapiens*, and we must now decide whether *habilis* or *rudolfensis* makes a more likely ancestor for the rest of our lineage.

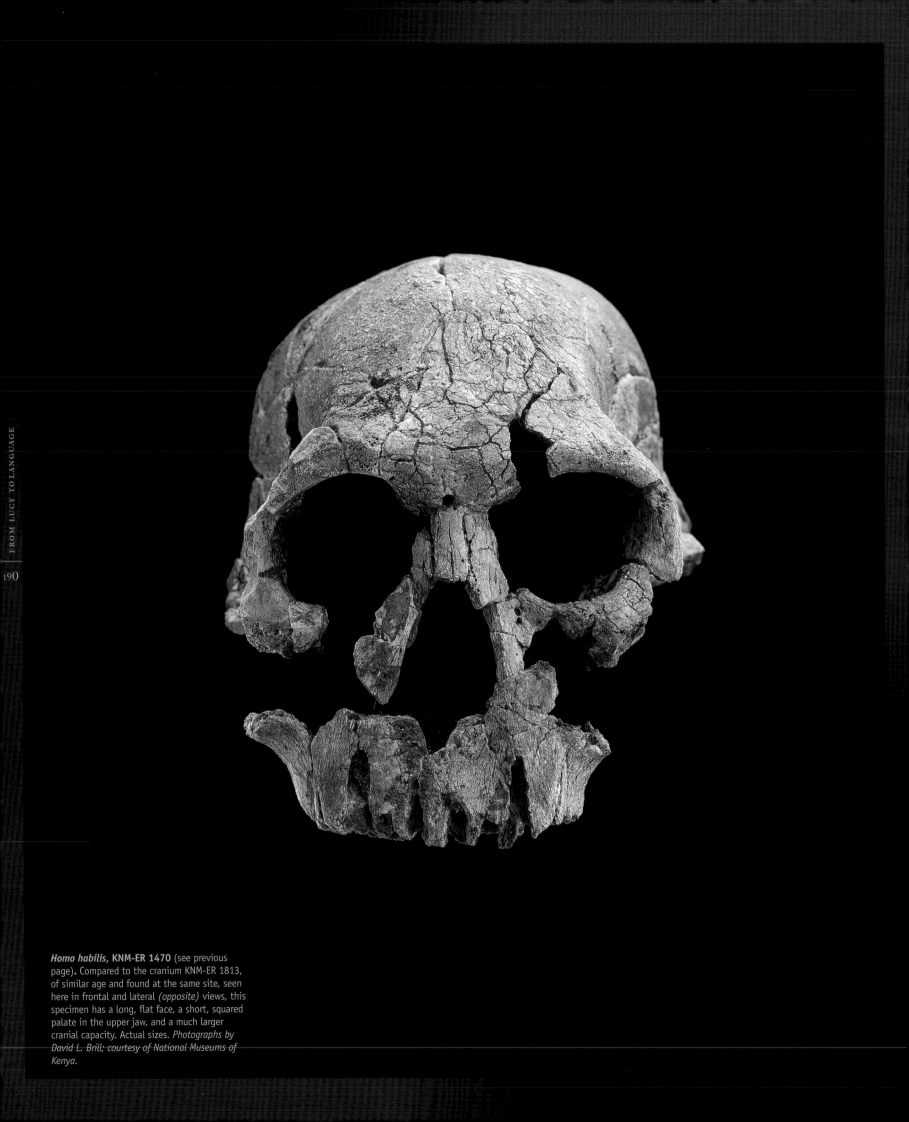

Homo habilis, **KNM-ER 1470** (see previous page). Compared to the cranium KNM-ER 1813, of similar age and found at the same site, seen here in frontal and lateral *(opposite)* views, this specimen has a long, flat face, a short, squared palate in the upper jaw, and a much larger cranial capacity. Actual sizes. *Photographs by David L. Brill; courtesy of National Museums of Kenya.*

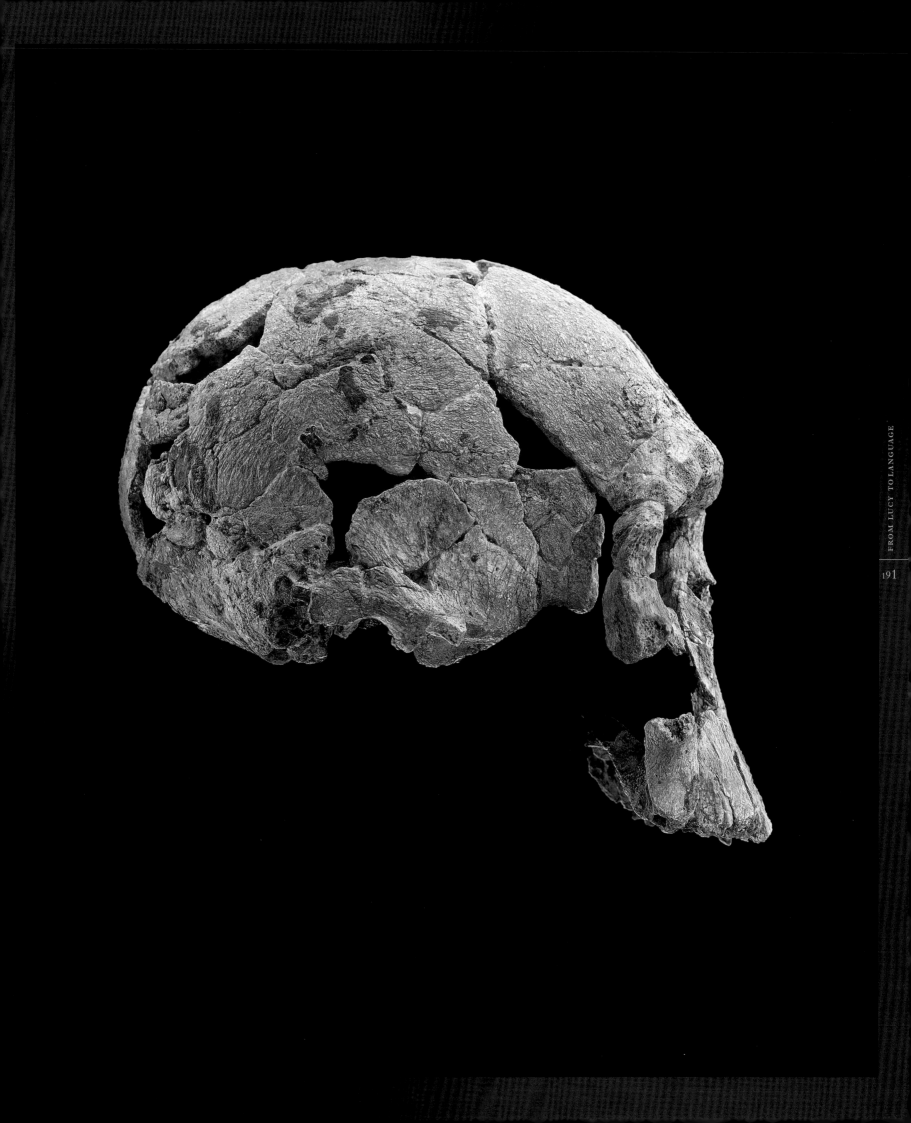

SPECIMEN	LOCALITY	AGE	DISCOVERERS	DATE	PUBLICATION
Adult mandible	Dmanisi, Republic of Georgia	1.8 million years	Gotcha Kiladzé	September 26, 2000	Gabounia, L., M-A de Lumley, V. Abesalom, D. Lordkipanidze and H. de Lumley. 2002. Découverte d'un nouvel hominidé à Dmanissi (Transcaucasie, Géorgie). *C.R. Palevol* v. 1: 243-253

Homo georgicus
D 2600

The two fossil hominid crania recovered from the site of Dmanisi in the Republic of Georgia, located midway between the Caspian and Black Seas, would not look out of place had they been excavated in eastern Africa. Dated to 1.8 million years ago these finds constitute the best incontrovertible evidence that the first "Out of Africa" event significantly predates previous estimates and seriously questions earlier explanations for what triggered exodus from the African motherland.

Initial excavations in the town of Dmanisi, situated on the ancient Silk Road, focused on a medieval citadel, but in 1983 fossilized remains of an extinct rhinoceros were recovered from a grain storage pit in the citadel's basement. It was not until 1991, however, that a hominid mandible was excavated immediately under a skeleton of a saber-toothed cat. The specimen, D 211, was ascertained to have affinities with *Homo erectus*.

The overall anatomy of two crania recovered in 1999—one missing most of the face (D 2280) and the other lacking the midface (D 2282)—supported the earlier contention that the Dmanisi hominids showed resemblances with *H. erectus*. Detailed morphological study, however, emphasized closer affinities to the African form of *H. erectus*, *H. ergaster*, than to the classic material known from China and Southeast Asia. Some of the features supporting a closer affinity with *H. ergaster* are a moderate supraorbital torus, a tall, narrow thin-walled cranial vault, smaller cranial capacities (D 2280, 780 cc; D 2282, 650 cc), a narrow dental arch, and facial proportions.

The Dmanisi crania do show some similarities with Asian *H. erectus* in characters such as a sagittal keel and an angulated occipital (rear) region that in occipital view has a typical *H. erectus* pentagonal outline. This amalgam of anatomical features favors stronger affinity with the African forms, which encouraged the finders to assign the crania to *H. ex gr. ergaster* (meaning from the group of *ergaster*).

A cranium (D 2700) recovered from Dmanisi in 2001, with a small cranial capacity of 600 cc, presents an anatomy more reminiscent of specimens from Africa usually assigned to *H. habilis*. This is seen in the nature of the way the face projects, the lightly built brow ridges, and the rounded curvature of the occipital bone. One of the discoverers, David Lordkipanidze of the Georgian Academy of Sciences, has remarked that the Dmanisi hominids constitute an interesting intermediate between *Homo habilis* and *Homo erectus* (*ergaster*).

The taxonomic identity of the Dmanisi hominids is currently under review because they show affinities with *H. habilis*, *H. erectus*, and even *H. ergaster*. A decision was made, however, to elevate the mandible, D2600, to a new species, *Homo georgicus*. At the moment it is unlear if this taxonomic designation applies to all the Dmanisi hominids or just the D2600 mandible.

Determination of the antiquity of the Dmanisi specimens was achieved using both single-crystal laser-fusion Argon dating and paleomagnetism. Stratigraphically, underlying the horizon from which the hominid specimens derive is a layer of volcanic basalt that has yielded argon dates of approximately 1.8 million years ago. It appears that very little time passed between the deposition of the basalt and the bone bearing horizons. The vertebrate fauna, which contains an abundance of carnivores such as saber-toothed cats (cranium D 2280 may have been a meal for one of these with puncture holes matching inter-canine distance of a saber-toothed cat), hyenas, bears, leopards, and wolves is typical of late Pliocene or early Pleistocene Eurasian faunal assemblages, dating to roughly 1.8 million years ago corroborating the argon and paleomagnetic age estimates.

The paleoecology of the world in which the Dmanisi hominids lived was a mixed woodland with a somewhat warmer and drier climate than today. Associated animals include horses, deer, short-necked giraffe, giant ostriches, gazelles, elephants, and a variety of rodents, all making for a faunally rich world in which to live.

Archaeological remains associated with the Dmanisi hominids are typical of the Oldowan industry of eastern Africa. Made from local raw materials such as quartzite and basalt, the flakes and cores are no more sophisticated than those found in African Oldowan assemblages of same age. Notable is the clear absence of hand axes typical of the Acheulean tradition that apparently did not appear in Africa until roughly 400,000 after hominids populated the Dmanisi area.

Traditional paleoanthropological thought postulated that hominids did not venture out of Africa until about one million years ago when they were large bodied, big brained bipeds capable of manufacturing tools certainly more sophisticated than the Oldowan, and who perhaps even controlled fire to cope with the more seasonal climate. The Dmanisi hominids seriously challenge these assumptions since they were small brained, possessed rudimentary flake and core tools, and, based on fragmentary postcranial remains, were apparently of small stature. So, the question now becomes just what was it that enabled these relatively "primitive" hominids to begin to penetrate into Eurasia?

Part of the explanation may lie in the fact that we, and no doubt our ancestors also, are exploratory creatures, fascinated by what might be beyond the next group of hills, or open patch of grassland. In addition, it appears that some of the animals were familiar African forms and the warmer climate was perhaps not that different from that in the homeland. *H. georgicus* may have simply been part of a larger faunal community that was slowly extending its home range into new geographical areas climatically similar to that of eastern Africa, although the majority of nonhominid species are Eurasian.

Details of the survival strategy of these early hominids are still under investigation. The Oldowan toolkit and associated foraging strategy was apparently sufficiently sophisticated to permit our ancestors to become early pioneers some 1.8 million years ago and significantly expand their home ranges. A few animal bones exhibit cutmarks suggesting some level of meat eating, but how frequent this behavior was and whether it reflected hunting or scavenging is unknown.

The announcement of a skull, which lacked all 32 teeth, suggested to Lordkipanidze that this older male from Dmanisi may have been cared for by other hominids, an act reflecting some degree of compassion. Perhaps the toothless individual subsisted on soft bone marrow, brain, and pounded vegetables.

If the Dmanisi discoveries indicate the first wave of migration out of Africa for our ancestors, they may have been the ancestors for *H. erectus* in Asia. Additionally, they must have undergone a rapid migration to Java where controversial dates indicate that *H. erectus* had made it there 1.8 million years ago. What is clear, however, is that long before early hominids had large brains, controlled fire, or developed sophisticated tools, they migrated out of Africa, something that has come as quite a surprise to anthropologists.

Homo georgicus, **D2700.** This specimen found in the Republic of Georgia is thought to be intermediate between *Homo habilis* and *Homo erectus* (*ergaster*). Actual size. *Photograph courtesy of David Lordkipanidze.*

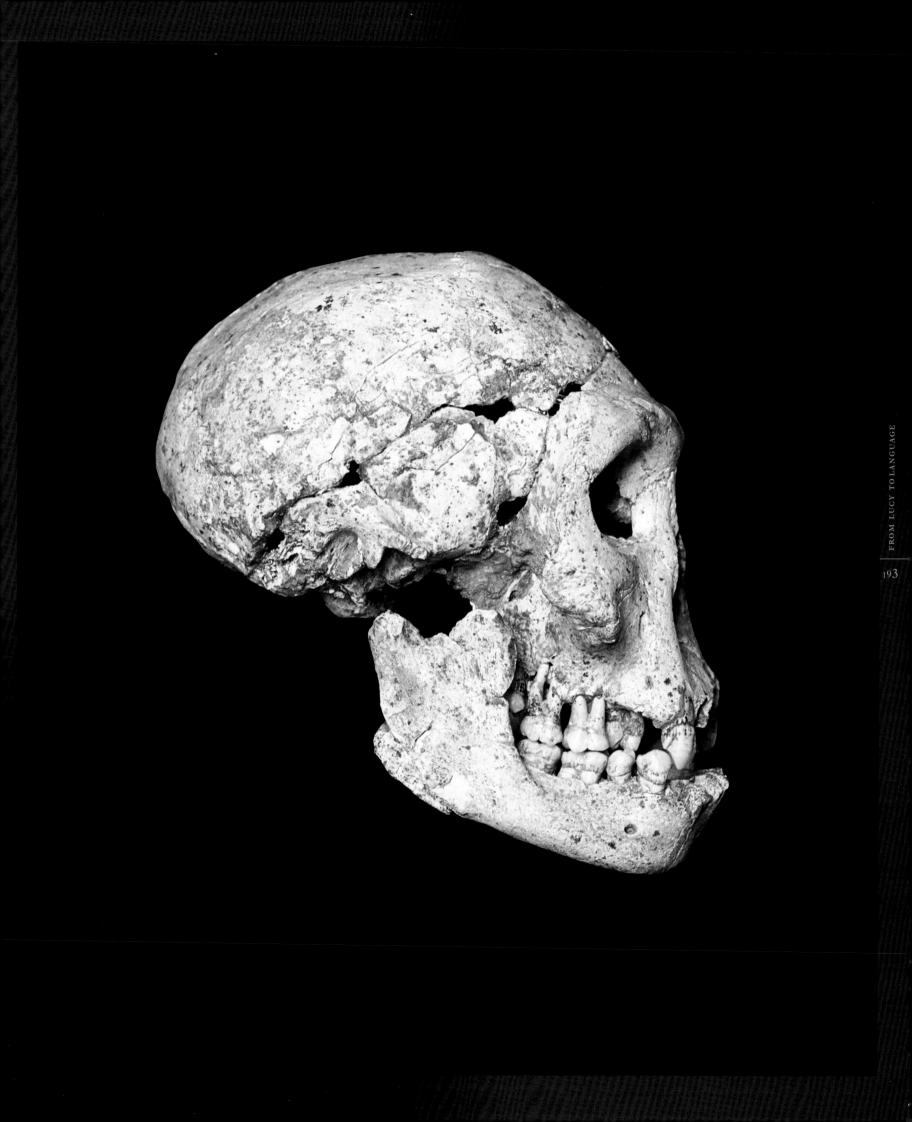

Homo ergaster
KNM-ER 3733

The exquisitely well-preserved cranium of *Homo ergaster*, KNM-ER 3733, confirmed the co-existence of *Homo* with robust australopithecines in East Africa and thus mooted the single species hypothesis for human evolution (see page 53). Until the 1984 excavation of the KNM-WT 15000 skeleton (see page 196), KNM-ER 3733 was the best example of this species. The double-arched browridge atop the eye sockets barely protruded from the ground, but that was enough for Bernard Ngeneo, a member of the "Hominid Gang" of Kenyan fossil finders. The fragmented remains of the face and upper jaw were unearthed nearby, but the lower jaw was never found.

Because the anatomical features of the face are markedly less robust than those of the juvenile male KNM-WT 15000 found on the opposite side of Lake Turkana, the 3733 cranium can be identified as a female with confidence. It had also reached maturity, a conclusion based on the closed sutures between the cranial bones, the degree of dental wear, and the eruption of the third molars.

From the side, KNM-ER 3733 displays the low cranial vault profile characteristic of Asian *Homo erectus*, as well as a swelling or keeling along the midline atop the cranium, a round torus across the occipital bone at the rear, and a cranial base wider than the top of the cranium. In overall size, 3733, with a brain capacity of 850 cc, compares with the Zhoukoudian specimens (see page 202), and the sides of the braincase are flattened instead of arching as in modern humans. 3733 does, however, lack some features that characterize *Homo erectus* in Java and China, including thick skull bones, an angular torus, and an obvious sulcus, or depression, behind the browridge. Such differences may warrant placing this and other African fossils in a separate species, *Homo ergaster*, which was originally named for another fossil from Koobi Fora, KNM-ER 992.

Homo ergaster, **KNM-ER 3733.** Because the face of this cranium appears less robust than that of KNM-WT 15000, the 3733 specimen, seen here in frontal and lateral *(opposite)* views, is considered to be a female. Actual size. *Photographs by David L. Brill; courtesy of National Museums of Kenya.*

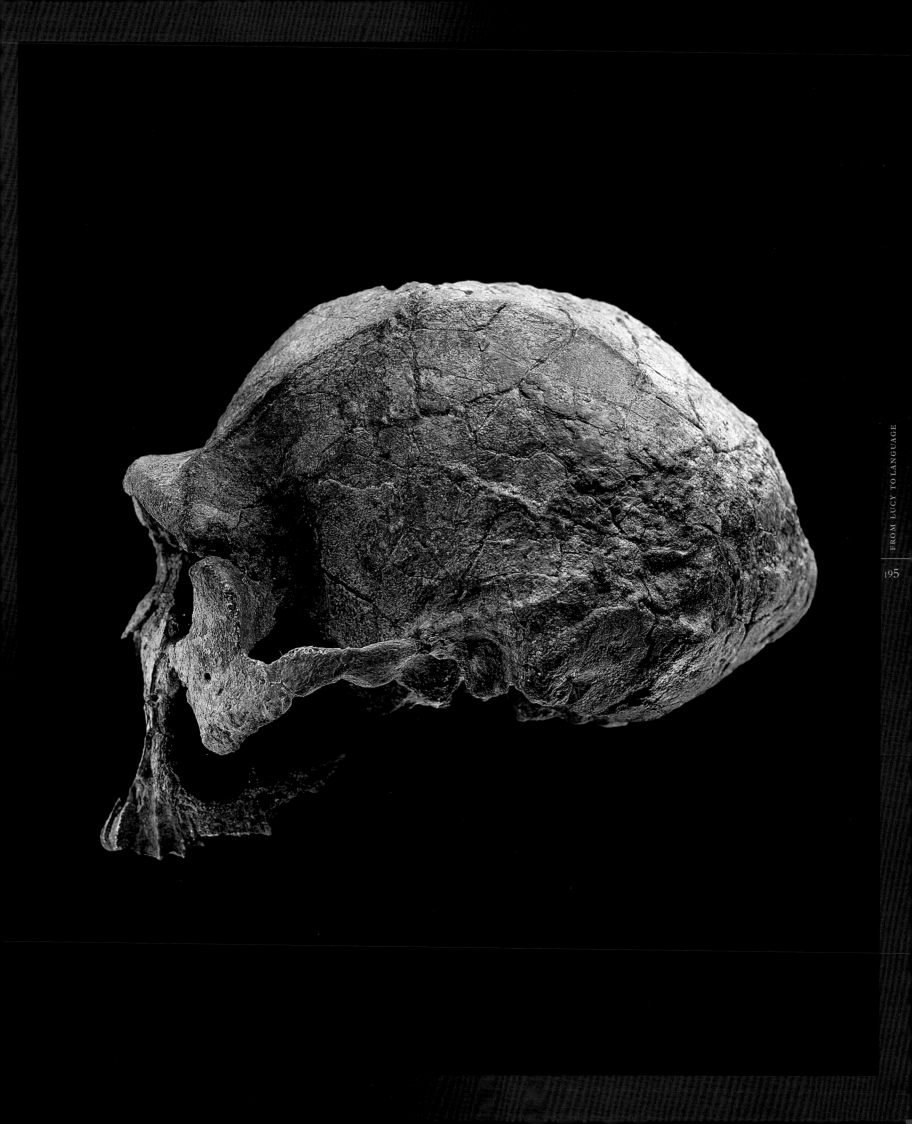

Homo ergaster
KNM-WT 15000

The initial discovery of a tiny cranial fragment on a slope covered with lava pebbles in Nariokotome, Kenya, was a modest start to the eventual recovery of this unparalleled specimen, the most complete skeleton of any early hominid discovered to date and a rich source for research. This is the first early human fossil from before 100,000 years ago with a skull complete enough to precisely measure brain size and with enough of the postcranial skeleton preserved to enable an accurate estimate of body size, and thus the testing of ideas about relative brain size and growth and development. Moreover, the 80 pieces of postcranial bones could help answer questions about limb proportions, locomotion, maturation, gestation, and perhaps whether this hominid had language. Many of the postcranial bones for this species, such as the ribs, were found for the first time with this skeleton from Nariokotome. Experts on all parts of the skeleton have analyzed this specimen in the decades since its unearthing by paleoanthropologists Richard Leakey, Alan Walker, and their team.

The individual was a boy who apparently died in early adolescence (see page 78 for details on estimating his age at death) on the edge of a shallow delta, where his body was covered rapidly by sediments and preserved. Although some large animal, perhaps a hippopotamus, had trampled and scattered some of the bones, the skeleton was remarkably well preserved. Had he lived to adulthood, the Nariokotome boy would have stood over 1.8 meters tall and weighed nearly 68 kilograms. He had already reached a height of more than 1.5 meters and a weight of around 47 kilograms. He was tall and thin, in body shape and limb proportions resembling present-day equatorial Africans. Despite his youth, the boy's limbs nearly matched the mean measurements for white North American adult males. This species was apparently much larger than any earlier hominid (see page 72 for a photograph of the complete skeleton).

The skull was reconstructed from about 70 pieces, not including the teeth, nearly all of which were recovered. The tooth morphology mirrors that of Chinese *Homo erectus* specimens (see page 202).

Features of the skull (a maximum cranial breadth far back, a moderate supratoral sulcus behind the supraorbital torus) recall other *erectus* specimens. Compared to the African specimen OH 9, KNM-WT 15000 has a much smaller supraorbital torus, possibly due to the latter's immaturity. The endocranial volume is 880 cc, slightly larger than that of KNM-ER 3733 (see page 194). Premolar and molar teeth are significantly reduced in size as are the jaws and the areas for insertion for the muscles of mastication, indicating a greatly reduced need for heavy chewing. As a mature adult, the boy's brain size would probably have reached about 909 cc, which would place him between the smaller Kenyan specimens and the larger Chinese *Homo erectus* specimens from Zhoukoudian.

KNM-WT 15000 displays some obvious differences from a modern human skeleton. These differences include longer spines on the vertebrae and a more constricted canal for the spinal cord that limited the number of nerves to the thoracic cavity and may have reduced the ability to regulate air passing from the lungs to the mouth (which in turn could have rendered the boy incapable of speech); a narrow pelvis that may have been more efficiently designed for walking and running than our own (see page 76 for the implications of KNM-WT 15000's narrow pelvis for birth and infant development in the species); and an elongated neck on the femur, a primitive feature also found in australopithecines. Overall, though, KNM-WT 15000 demonstrates that hominids by this time had departed from the apelike australopithecine body plan of our early evolution and were rapidly approaching the body and brain size that characterize modern humans.

Homo ergaster, **KNM-WT 15000** (see also page 72). The boy from Nariokotome, WT 15000, has a skull complete enough to accurately measure brain size, the 80 additional bones recovered from his skeleton allow an estimate of body size and convey much information about this species' biology. *Photograph by David L. Brill; courtesy of National Museums of Kenya.*

SPECIMEN	LOCALITY	AGE	DISCOVERERS	DATE	PUBLICATION
Adult partial cranium	Swartkrans, South Africa	ca. 1.5 million years	Ronald Clarke	July 23, 1969	Clarke, R.J., F.C. Howell, and C.K. Brain. 1970. More evidence of an advanced hominid at Swartkrans. *Nature* 225: 1219-1222

Homo ergaster
SK 847

Not all human fossil discoveries are made in the field. Some are made while reexamining museum specimens in light of discoveries made subsequent to the specimens' collection. The cranium identified as Swartkrans (SK) 847, the best example of early *Homo* from South Africa and perhaps one of the first associated with the use of fire, is a case in point. While inspecting the collections of the Transvaal Museum in Pretoria in 1969, paleoanthropologist Ron Clarke noticed that among the hominid fossil remains from Swartkrans were two pieces of a cranium, a facial fragment and the temporal bone, that looked very different from the robust australopithecine fossils that surrounded them. Clarke found that the two pieces fit together and thus came from the same individual.

Next, Clarke realized that the facial fragment also fit perfectly with part of a maxilla, SK 80, which had been described by Robert Broom and John Robinson in the early 1950s as representing a separate hominid species at Swartkrans from the robust australopithecines, one that they named *Telanthropus capensis* (which Robinson later included within *Homo erectus*). Now Clarke had most of the left side of a face, the cheekbone, and part of the side and base of a cranium from a single individual, and he knew that Broom and Robinson had been right. The cave at Swartkrans has yielded the bulk of the known robust australopithecine remains—bones from more than 85 individuals of *Australopithecus robustus*—but Clarke's serendipity confirmed that a second type of early human had lived at the site. Clarke wrote his doctoral dissertation on this cranium.

SK 847's salient facial features include a relatively short and narrow face, a pronounced browridge or supraorbital torus that is thick both in the middle and to the sides, moderate constriction of the cranium behind the eye socket, an obvious supratoral sulcus, and a sharply sloping frontal bone. In addition, the specimen has delicate, curved cheekbones and rounded, forward-projecting nasal bones. In all of these features, SK 847 contrasts strongly with the anatomy of both *Australopithecus robustus*, such

as SK 48 and SK 79 (see page 160), and *Australopithecus africanus*, such as Sts 5 (see page 146).

Furthermore, SK 847 possesses a short palate and small temporomandibular joint (the area where the lower jaw connects to the cranium) that could only fit a small, short lower jaw and not the typically massive mandible of a robust australopithecine. In fact, although it comes from a different individual, the mandible SK 15 makes a good match for SK 847 and also happens to be the type specimen used to name *Telanthropus*.

Despite all the anatomical evidence suggesting that SK 847 belonged to *Homo*, some paleoanthropologists maintained, largely by invoking the single species hypothesis (see page 170), that all the hominids from Swartkrans were of the same species and that this specimen was simply a small robust australopithecine. But the facial anatomy of SK 847 can be clearly distinguished from that of a presumed female robust australopithecine, KNM-ER 732 (see pages 73). A much more favorable comparison, however, can be made between SK 847 and another Koobi Fora specimen, KNM-ER 3733 (see page 194), which has also been classified as *Homo ergaster* or African *Homo erectus*.

The presence of *Homo* at Swartkrans raises an interesting question. Should we assume that this hominid—rather than the robust ones whose bones constitute at least 95 percent of the hominid fossils from the site—was responsible for the stone tools and the evidence of controlled fire (see page 96) found in the cave? Despite the statistical odds, it seems a reasonable conclusion. Just as there is no clear association at any site of stone tools and robust australopithecine bones, even though the hands of *Australopithecus robustus* had the necessary musculature and precision grip to flake stone, so there is no evidence elsewhere linking this species to the creation or curation of fire.

Homo ergaster, **SK 847.** South African cave sites are known for preserving large numbers of australopithecine fossils, but this is one of the much rarer examples of early *Homo*. Actual size. *Photograph by David L. Brill; courtesy of Transvaal Museum.*

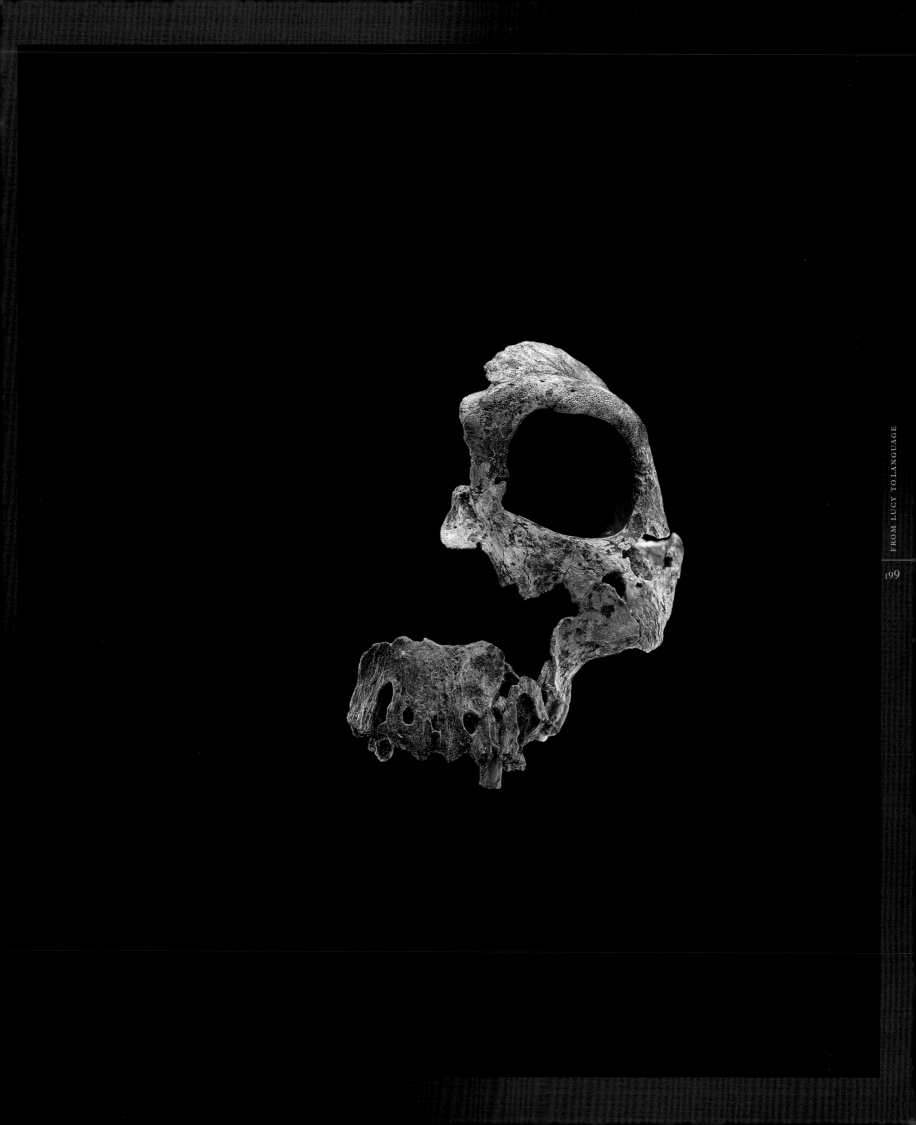

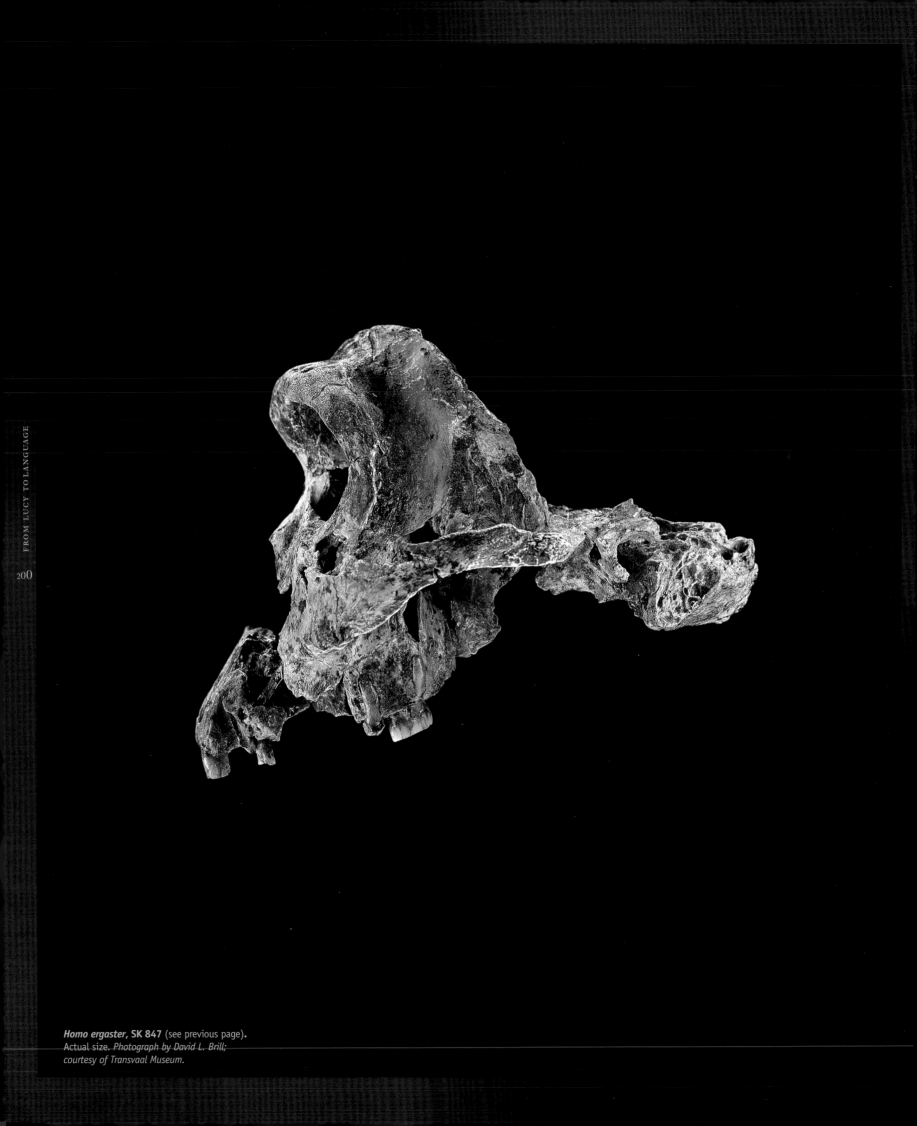

Homo ergaster, SK 847 (see previous page).
Actual size. *Photograph by David L. Brill;*
courtesy of Transvaal Museum.

Homo erectus
TRINIL 2 TYPE SPECIMEN *JAVA MAN*

If the discovery of a skullcap and assorted bones in the Neander Valley initiated paleoanthropology as an endeavor, the specimen known as Trinil 2 was the first fossil to suggest our ancestors' antiquity and geographic dispersion. Its discoverer, Eugène Dubois, set out from Holland to find the missing link between apes and humans and succeeded in making the first hominid discovery outside of Europe. Trinil 2, or Java Man, became the holotype specimen of *Homo erectus*.

Inspired by British naturalist Alfred Russel Wallace's conviction that clues to our origins lay in southeast Asia, Dubois enlisted as an army surgeon in the Royal Dutch East Indies Army and became an avocational fossil hunter in Sumatra. He had little luck there, but 1890 he continued his search on Java, along the bank of the Solo River.

Success came with the unearthing of this thick, mineralized cranium with a cranial capacity of 940 cc, consisting of a flat frontal bone, much of both parietals, and the upper portion of the occipital bone. The pronounced supraorbital torus and the constriction behind it are striking features of this specimen. Much of the surface anatomy has worn smooth, but the extremely flat forehead has an obvious keel along the bone's midline. A pathological femur subsequently found upstream from the cranium may or may not belong to the same individual and has a remarkably modern-looking form.

In 1894, Dubois named his Java Man *Pithecanthropus erectus*, taking the genus name coined by biologist Ernst Haeckel for a hypothetical human ancestor. Since the 1950s, when evolutionary theorist Ernst Mayr proposed that the Javan

hominids and the roughly contemporaneous fossils of Peking Man could be placed in a single species, the fossils have been identified as *Homo erectus*. Further discoveries in Java have turned up remains from a total of about 40 individuals of this species, and an equal number have come from China, mainly from the site of Zhoukoudian (see page 202). Dubois was ridiculed by his contemporaries who thought his find was really that of an ape. Dubois, however, must be recognized as the first scholar to ask if a fossil was more closely related to an ape or to a human rather than relegating such finds to a long-extinct race of modern humans.

No tools have ever been found associated with the fossils of *H. erectus* in Java. This is curious, because elsewhere the species is associated with the conspicu-

ous and ubiquitous heavy-duty butchering artifacts of the Acheulean industry. Some have proposed that this hominid in Indonesia and East Asia relied on cutting tools made from abundant bamboo. Alternatively, in light of the very ancient dates recently announced for two *H. erectus* specimens from Modjokerto and Sangiran (see page 205), it may well be that this species left Africa before the Acheulean had even evolved there. Acheulean tools spread outside of Africa with subsequent emigrants but never penetrated far into Asia. Despite Dubois's early success and that of subsequent fossil hunters there, Asia has not turned out to be the cradle of humankind. Darwin, not Wallace, had it right that we should seek our ultimate origins in Africa.

Homo erectus, Trinil 2.
The first hominid fossil found outside of Europe, Trinil 2 focused the search for our origins in Asia until older fossils began to be discovered in Africa. Actual size.
Photograph by John Reader, Science Source/ Photo Researchers.

Homo erectus
PEKING MAN

The cave of Zhoukoudian, or Dragon Bone Hill, about 40 kilometers south of Beijing, China, has yielded the largest collection of *Homo erectus* fossils from any one place. Accounting for a third of the known fossils from this species, the bones from 40 individuals of the same species as Peking Man provided a rich trove for research: five skullcaps, several cranial and facial fragments, 11 mandibles, and 147 teeth. The fossils met a tragic and mysterious end: all of the original fossils, save for two teeth sent to Sweden in the 1920s, vanished in December 1941 after being packed for shipment to the United States in the wake of the Japanese invasion of China during World War II. Postwar excavations turned up a few more teeth, cranial, jaw, and limb bone fragments, but the rest were gone. Fortunately, in anticipation of the invasion, exacting casts of each specimen had been made.

Paleoanthropologists have just put a new face on this famous old fossil. The skull shown here is a recent reconstruction by Gary Sawyer and Ian Tattersall of the American Museum of Natural History, New York, which departs significantly from the original reconstruction made by German anatomist Franz Weidenreich and his assistant Lucille Swan in the 1930s. The original reconstruction used a cranium and lower jaw from two females and a probably female maxilla. The new reconstruction takes advantage of some previously ignored bones from males, and thus finally affords a look at Peking Man himself.

Bones from four individuals were used in the new reconstruction. The skullcap from Skull XII, which included nasal bones and the rim of the left eye socket, a partial maxilla from Skull XIV, and two facial fragments from Skull X created a nearly complete left side of the face. The mandible was reconstructed using the left and right fragments GI and GII and following the bone's bilateral symmetry to make a mirror image for missing parts from the opposite side. The complete lower jaw could then guide reconstruction of the midface so that teeth and other points of contact were aligned correctly. Isolated teeth were added to complete the dentition in both jaws.

The result is a larger cranium with a more massive and projecting face overall and a broader, taller nasal region. Peking Man now more closely resembles the anatomy of *erectus* specimens from elsewhere.

The Zhoukoudian fossils were excavated beginning in the 1920s and given the name *Sinanthropus pekinensis* by Canadian physician Davidson Black based on a few isolated teeth (all the fossils were later lumped within the species *Homo erectus*). The first skullcap was found in 1929, and as more fossils accumulated in the following decade, Weidenreich made extensive studies of them that contributed to his multiregional view of human evolution by his recognition of 12 anatomical features that he believed Peking Man shared with modern Chinese. His series of monographs on Zhoukoudian, published between 1936 and 1943, remain definitive, and we can thank his foresight in bringing all of the master molds and casts—made before the original fossils became lost—with him to the American Museum of Natural History, where they are still kept today.

The five skullcaps have a mean cranial capacity of 1,043 cc, about the same as that for the Olduvai Gorge cranium OH 9. The supraorbital torus of Peking Man is smaller than that of OH 9 or Sangiran 17 from Java (see page 205). Occipital bones from Zhoukoudian are strongly flexed (note the back of the skull in lateral view) with a broad torus across the bone's width. These specimens are also characterized by flat, thick, rectangular parietal bones, massive facial bones, and bulky mandibles.

Zhoukoudian contains a rich record of fauna and archaeology that attests to the cave's occupation for more than 200,000 years. Hominids seem to have alternated with carnivores at the site, perhaps occasionally competing for food and shelter at the same time. Bones belonging to 97 species of mammals and 62 kinds of birds have been excavated. At least 17,000 stone tools were made by the inhabitants, predominantly from quartz, with smaller percentages of rock crystal, sandstone, and flint artifacts. Zhoukoudian also preserved what until recently was the earliest evidence of the human use of fire (see page 96).

Homo erectus, **original cast of Peking Man** (see also pages 7 and 177). Franz Weidenreich's original cast of cranium XII from Zhoukoudian, China provided the braincase and upper face for a new reconstruction of Peking Man's skull. Actual size. *Photograph by David L. Brill; courtesy of American Museum of Natural History.*

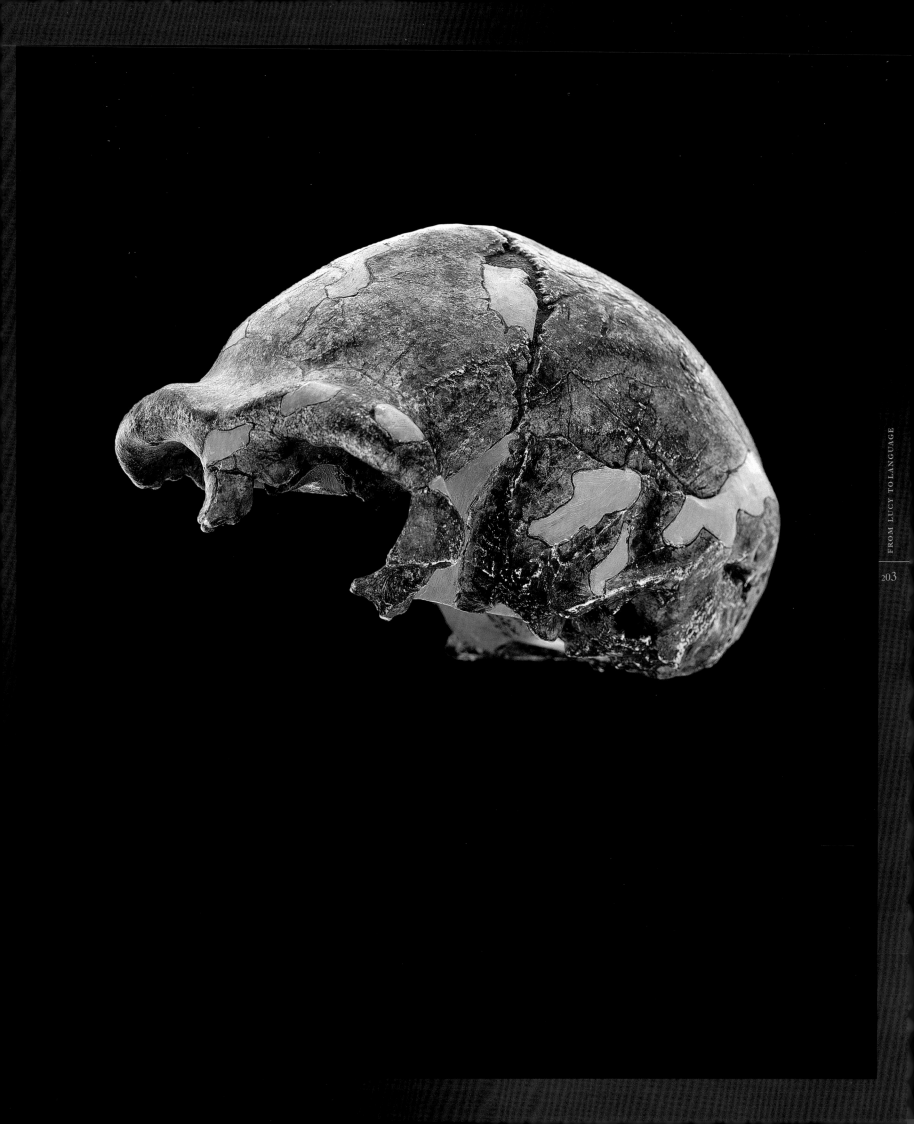

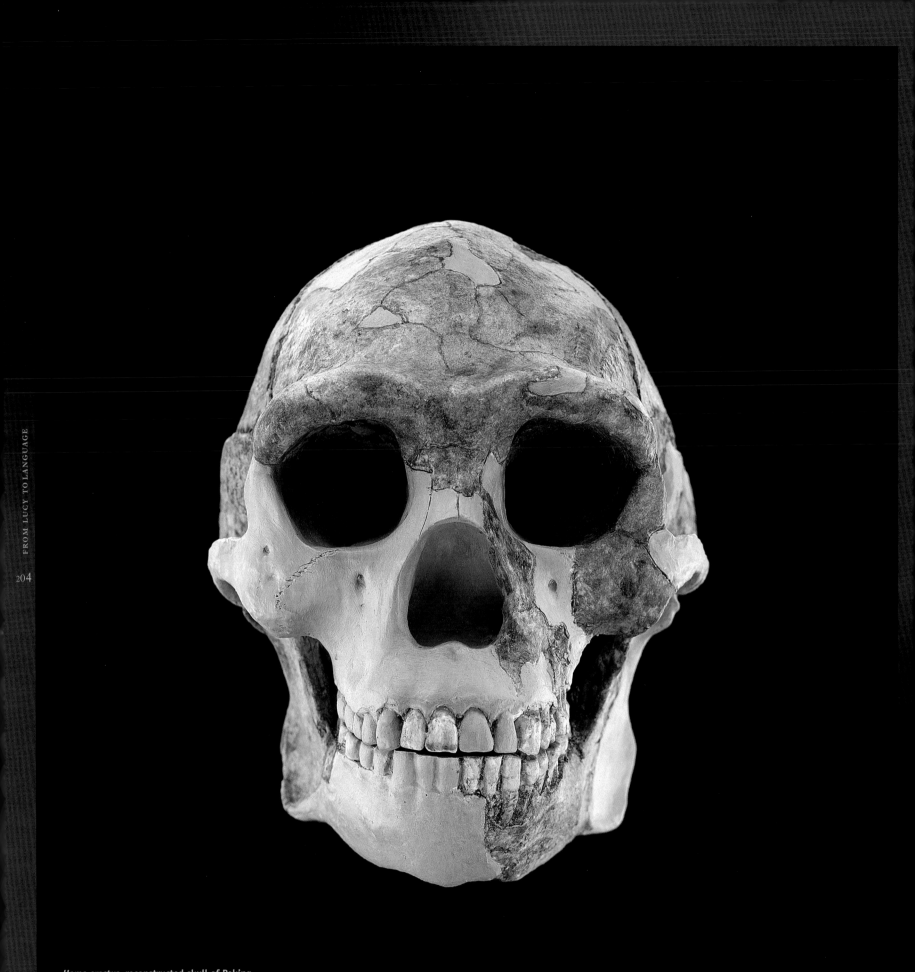

***Homo erectus*, reconstructed skull of Peking Man** (see previous page). This revised reconstruction gives Peking Man a more massive, projecting face and broader nose than earlier versions and appears more similar to *Homo erectus* fossils from elsewhere in Asia. Actual size. *Photograph by David L. Brill; courtesy of American Museum of Natural History.*

SPECIMEN	LOCALITY	AGE	DISCOVERERS	DATE	PUBLICATION
Adult male cranium	Sangiran, Java, Indonesia	ca. 800,000 years	Mr. Towikromo	September 13, 1969	Sartono, S. 1971. Observations on a new skull of *Pithecanthropus erectus* (*Pithecanthropus* VIII) from Sangiran, Central Java. *Proceedings of the Koninklijke Nederlandsche Akademie van Wetenschappen, Amsterdam*, series B, 74: 185-194

Homo erectus
SANGIRAN 17

Eighty years elapsed after Eugène Dubois discovered his famous fossil skullcap at Trinil (see page 201) before paleoanthropologists could finally gaze into the face of Java Man. Although other fossil hunters subsequently found complete crania, skullcaps, and cranial fragments from a dozen individuals, all lacked the facial skeleton. Then, in 1969, a farmer cultivating his land struck the side of a skull embedded in the sandstone soil. His blow opened a large hole in the fossil, but the fragments were collected, along with the more intact bone, which was heavily mineralized and had a rusty brown color from iron oxide. Once it was reconstructed, Sangiran 17 constituted the best-preserved hominid cranium from Java and the only known adult male *Homo erectus* from anywhere. But by providing the first face of Indonesian *Homo erectus*, the specimen took on a central role in the multiregional model for the origin of *Homo sapiens*.

This model hypothesizes that *Homo erectus* evolved in place in Java and accumulated anatomical traits that were then passed on to its descendant species, *Homo sapiens*. These traits can be traced through time in various fossils in Java and even to a population of modern humans that migrated from Java to Australia perhaps 50,000 years ago. A separate regional lineage is said to link erectus and sapiens fossils in China, including specimens from Zhoukoudian (see page 202) and Dali (see page 250), and it is speculated that another migration took members of our species from China into Australia at around the same time.

In Java, this direct line of evolution allegedly begins with the Sangiran specimens and continues with specimens from Sambungmachan and the much younger crania from Ngandong, all the way to present-day Javanese. Some of the cited stable anatomical traits include a long, relatively flat frontal bone; a projecting face with massive, flat cheekbones; a distinct ridge, called a zygomaxillary tuberosity, at the base of the cheekbone; a rounded edge to the bottom of the eye sockets; and the lack of a clear demarcation between the nasal region and the lower face. Similar features supposedly carry over to much younger Australian fossils such as the partial cranium WLH 50 from Willandra Lakes and the skeletons from Kow Swamp (see page 263), the latter being only 10,000 to 14,000 years old. Many of these facial features are adaptations for heavy chewing power.

Other characteristics visible in Sangiran 17 that distinguish Indonesian *Homo erectus* from other populations of the species and that figure in the multiregional model include the very thick bones of the braincase, which is flattened along the sides; a maximum breadth at the base of the cranium; reduced development of the frontal and parietal lobes of the brain (with a cranial capacity of 1,029 cc, Sangiran 17 is toward the high end of the range in brain size); and prominent muscle markings along the sides and back of the cranium.

Originally the multiregional model suggested that *erectus* had been evolving in China and Java for a million years. But new dates from Javan fossils—including some hominid skull fragments found at Sangiran in the 1970s, now estimated by the argon-argon technique to be 1.66 million years old, although Sangiran 17 is a much younger specimen—would push back the beginning of this species' isolation in Java to nearly 2 million years ago. Since the exact geological layer where the Java specimens were found is known, there is serious doubt about their association with 1.66 million year old volcanic rocks. Multiregional evolution proponents, however, are adapting their ideas to the new dates, but 2 million years of local evolution is too long a period to consider this *erectus* population as a plausible modern human ancestor. No other example from hominid evolution exists to support the idea that this population evolved in place, with little or no gene flow from elsewhere, for nearly 2 million years.

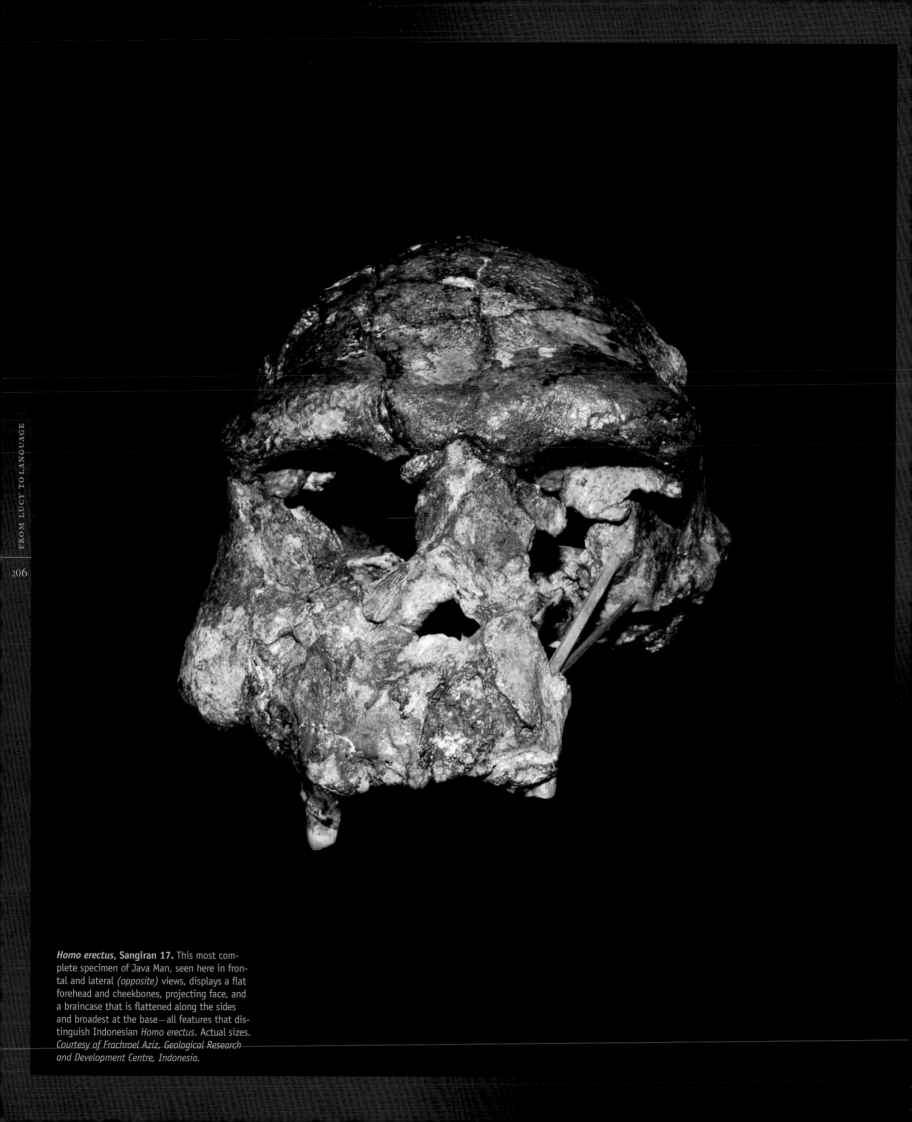

Homo erectus, **Sangiran 17.** This most complete specimen of Java Man, seen here in frontal and lateral *(opposite)* views, displays a flat forehead and cheekbones, projecting face, and a braincase that is flattened along the sides and broadest at the base—all features that distinguish Indonesian *Homo erectus*. Actual sizes. *Courtesy of Frachroel Aziz, Geological Research and Development Centre, Indonesia.*

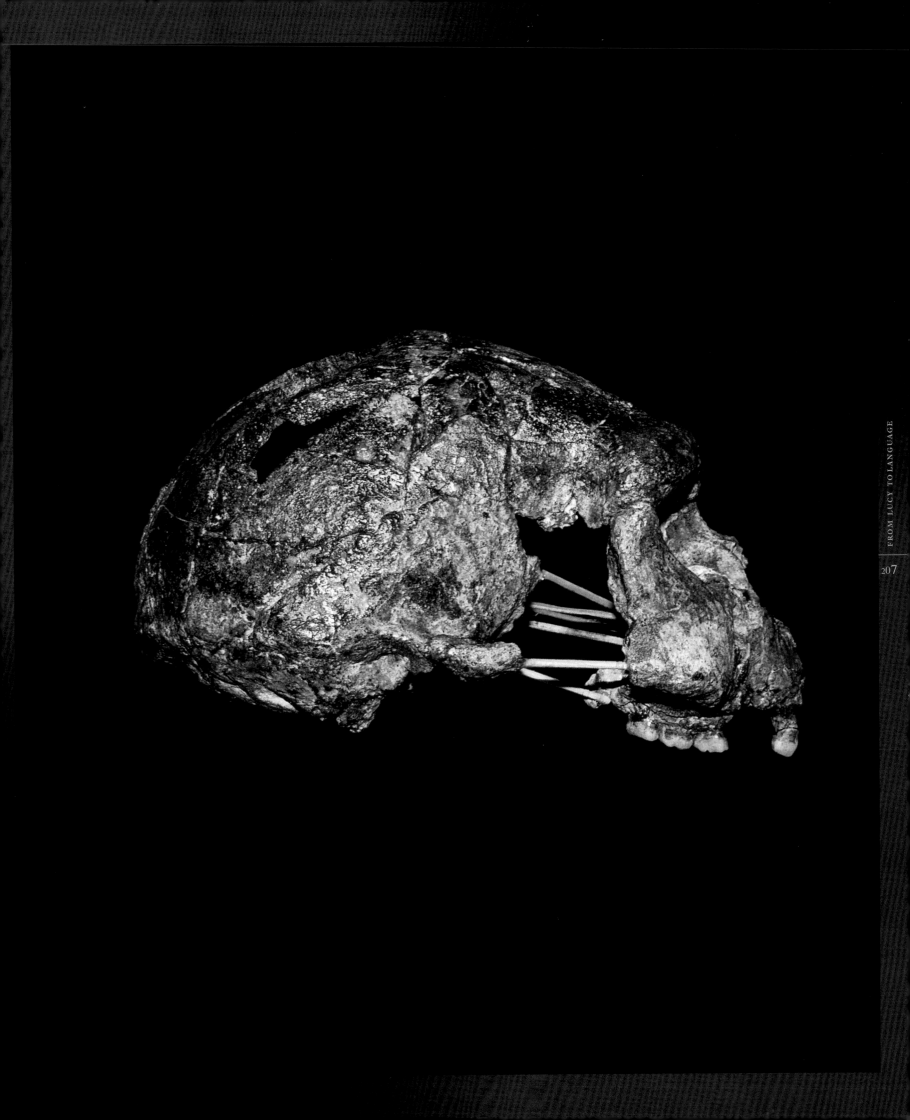

Homo heidelbergensis
BODO CRANIUM

Creationists love to argue that paleoanthropologists lack transitional fossils to show that one hominid species evolved into another. The specimen known as Bodo, after its place of discovery in Bodo d'Ar, Ethiopia, refutes that argument. It possesses a mosaic of anatomical features typical of different species of *Homo* and as such does not fit easily into any one species. Even as this specimen captures a biological transition in progress, the stone tools collected at Bodo reflect an unfolding cultural transition.

Bodo sports the biggest face known in the human fossil record. In 1981, five years after this massive cranium was found, the left lower rear corner of a parietal bone from the skull of a second individual was recovered from Bodo d'Ar, and even this small fragment reveals the robust skull structure of early humans at this site. Part of an upper arm, or humerus, found at Bodo d'Ar in 1990, however, is smaller than that of many modern humans, perhaps reflecting size differences between ancient males and females in this population.

Despite its broader, more massive face, Bodo bears a striking resemblance to the younger specimen from Kabwe, or Broken Hill, commonly called Rhodesian Man (see page 222). Both have a broad face and a thick, prominent ridge, or supraorbital torus, over the eye sockets. Like many *Homo erectus* specimens, Bodo has a bony midline keel atop the cranium. Its *sapiens*-like features include the nasal bones, the mandibular fossa (a shallow depression in front of the ear hole where the lower jaw attaches), and the forehead or frontal bone. More important for proponents of the view that modern humans evolved first and only in Africa before spreading around the world, Bodo has anatomical affinities with the Petralona specimen from Greece (see page 214).

When the specimen was discovered, its age was thought to be only 350,000 years, contemporaneous with other European specimens of *Homo heidelbergensis*. New dates reported in 1994 for the Bodo d'Ar site have increased the estimated age of this skull to around 600,000 years, within the time frame of *H. erectus*. But

Bodo's anatomy has advanced beyond that of its *erectus* contemporaries in Asia, including those at the Peking Man site of Zhoukoudian.

The site at Bodo d'Ar provides an extremely rare snapshot of simultaneous biological and cultural evolution. Near where the cranium was found, stone tools—Acheulean hand axes and cleavers—litter the ground, along with the bones of hippos, baboons, and antelopes. The numerous lava tools recovered from the same river sands as the human fossils document a technological shift from the more simple single-faced cores and flakes of the Developed Oldowan industry to the bifacial, heavy-duty artifacts of the succeeding Acheulean industry. Curiously, this transition at Bodo d'Ar occurred very late, since elsewhere in Ethiopia's Middle Awash region, at Konso-Gardula, Africa's earliest Acheulean tools have been found, dating to about 1.4 million years ago.

Humans may have been butchering hippos at some of the archaeological sites near Bodo, but some stone tools were apparently used on Bodo Man himself. Distinctive cut marks discovered on several parts of the cranium after the bone was cleaned indicate that Bodo Man was the victim of defleshing and butchery, either before his death or shortly thereafter (see page 93).

Homo heidelbergensis, **Bodo** (see also page 94). This immense specimen, with the largest face in the human fossil record, possesses a mix of primitive anatomical features, such as the thick browridge and cranial wall, and more modern ones, including the shape of the forehead and nasal bones. Actual size. *Courtesy of Donald Johanson, Institute of Human Origins.*

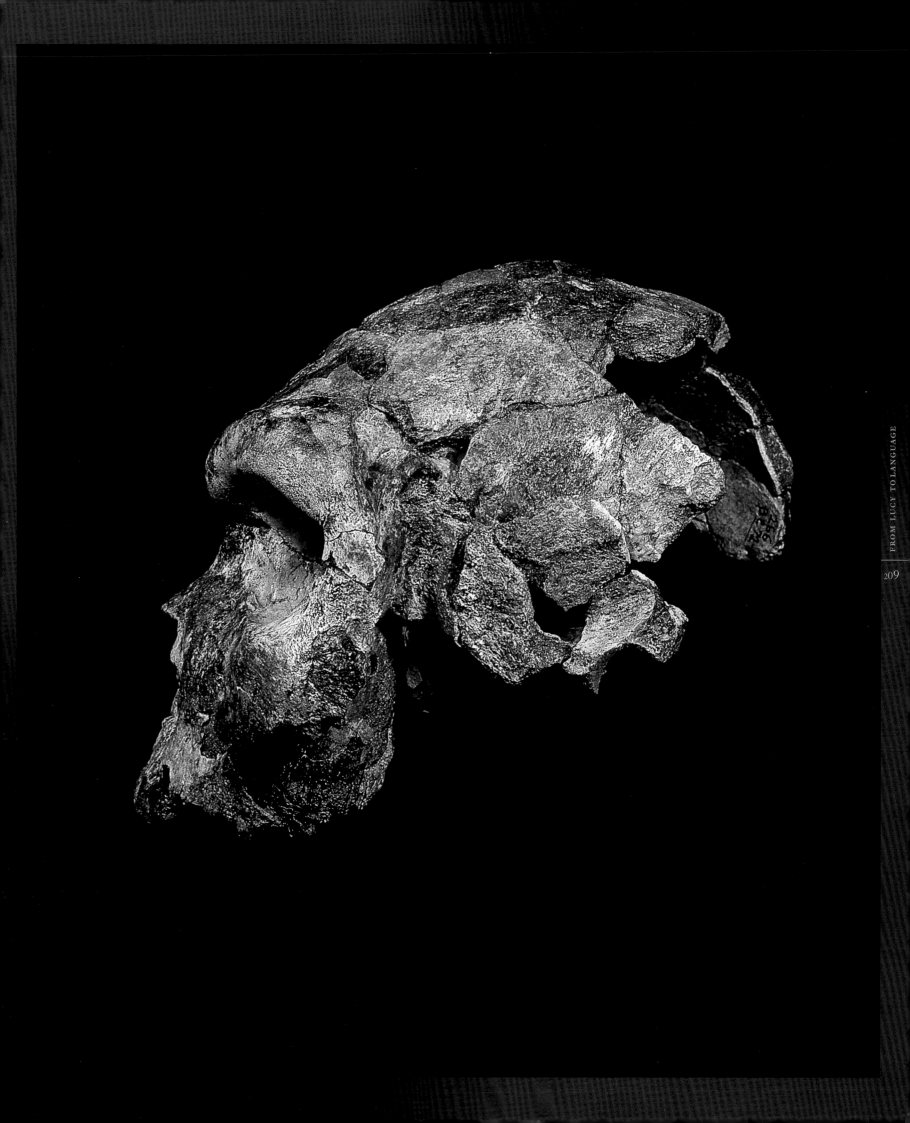

SPECIMEN	LOCALITY	AGE	DISCOVERERS	DATE	PUBLICATION
Adult mandible	Mauer sand pits, Germany	ca. 400,000 to 500,000 years	Workman for Joseph Rosch	October 21, 1907	Schoetensack, O. 1908. *Der Unterkeifer des* Homo heidelbergensis *aus den Sanden von Mauer bei Heidelberg,* 1-67 (Wilhelm Engleman, Leipzig)

Homo heidelbergensis
MAUER 1 TYPE SPECIMEN

Until the 1990s, this robust jawbone had been the oldest known hominid fossil in Europe. Although the recently discovered mandible from Dmanisi, Georgia, and fossils from Gran Dolina, Spain (see pages 192 and 218), eclipse the Mauer specimen in age, this fossil remains one of the earliest Europeans. Also known as Heidelberg Man, the mandible is the type specimen of *Homo heidelbergensis*, a scientific name that has recently gained currency and for most anthropologists refers to the species at the base of the lineages that led to *H. sapiens* and the Neandertals.

When this fossil was found near the village of Mauer, a few kilometers southeast of Heidelberg, it validated the long-held conviction of paleontologist Otto Schoetensack. The river sands deposited at the Mauer quarry were known to contain the remains of extinct mammals—rhino, elephant, bear, bison, deer, and horse among them—and were thought to date deep into the Pleistocene. Schoetensack believed that the gravel pits at Mauer might yield the remains of some Pleistocene human, so he made frequent train trips to the site from Heidelberg. He had to wait 20 years, but finally a worker discovered this mandible in a layer that had been buried by nearly 24 meters of deposits. Schoetensack completed and published his extensive monograph on the mandible just a year later. Five years after the discovery he died.

Although Schoetensack created the name without justifying it by describing the unique anatomical features of this species, there are features that seem to set *H. heidelbergensis* apart from *Homo erectus*, Neandertals, and modern humans. In the case of Mauer 1 the anatomy is clearly more primitive than that of either Neandertals or modern humans. There is no projecting chin, and the symphysis at the front of the massive mandible slopes down and back from the teeth. The specimen possesses a surprisingly broad ramus, the vertical portion that connects the jaw to the cranium, which anchored strong chewing muscles. Other than taurodontism, a combination of enlarged pulp cavities and fused roots, the completely preserved teeth show no typical Neandertal traits, and the large molars are small for *H. erectus* teeth but overlap those of some early modern humans.

Although no unambiguous tools were found with Heidelberg Man (some bone fragments and pebbles have been argued to bear signs of hominid use), he was thought to be a contemporary of the people who made the Acheulean tools that have been found elsewhere in western Europe, and to have lived during the warm period between the first two glaciations of Pleistocene Europe. The exact age of the specimen, however, remains somewhat uncertain.

In a bit of hyperbole, Schoetensack wrote that apart from the teeth of this fossil, "Even an expert could not be blamed if he hesitated to accept it as human." Echoing Schoetensack's sentiments, English anatomist Sir Arthur Keith later wrote, "From the very first, anatomists have been struck by the apparent discrepancy between the 'humanity' of the teeth and the massive power—almost bestiality—of the jaw itself." These two authorities considered Mauer to be an early and primitive type of Neandertal, a view that has been strengthened by the recovery of subsequent fossils in Europe.

Homo heidelbergensis, **Mauer 1.** The type specimen for this species, the mandible of Heidelberg Man dates to nearly half a million years ago and remains one of the oldest human fossils from Europe. Note the absence of a chin in front and the very broad ascending ramus at the rear of the bone. Actual size. *Photograph by John Reader, Science Source/Photo Researchers.*

SPECIMEN	LOCALITY	AGE	DISCOVERERS	DATE	PUBLICATION
Adult cranium	Caune de l'Arago, Tautavel, France	ca. 400,000 years	Henry de Lumley	1971	de Lumley, H., and M.-A. de Lumley. 1971. Découverte de restes humains anténéandertaliens datés du début de Riss á la Caune de l'Arago (Tautavel, Pyrénées-Orientales). *C.R. Acad. Sci. Paris* 272: 1729-1742

Homo heidelbergensis
ARAGO XXI

An impressive limestone cave in the eastern Pyrenees, perched above the Verdouble River, Arago has yielded about 60 human fossils since excavation began in 1964. These have included 14 teeth and some finger bones, but the most spectacular fossils are a large hip bone and the cranium known as Arago XXI. Often referred to as Tautavel Man, the cranium has traditionally been considered that of a young male due because of its size and robust facial features.

Despite its obvious distortion, Arago XXI is important for representing the robust end of anatomical variation in the early humans of Europe during the Middle Pleistocene (from about 700,000 to 200,000 years ago). Arago XXI possesses a fairly complete face, with five molar teeth, and part of the braincase. A prominent browridge frames the slightly projecting face. Behind the browridge, a deep sulcus, or depression, separates the face from the long, flattened forehead, and the cranium narrows behind the eye sockets (a feature called postorbital constriction). The sides of the cranium, the parietal bones, display an angular torus, a trait usually associated with *Homo erectus*.

This specimen has in fact been classified by its discoverers to that species. The hip bone from Arago also shows similarities to bones belonging to *H. erectus* times from Zhoukoudian, in China, Koobi Fora, in Kenya, and the Olduvai Gorge, in Tanzania. Two chinless lower jaws from Arago, with distinct sizes that could reflect differences between males and females, have also been compared with *H. erectus* specimens.

But there is some doubt whether *Homo erectus* ever entered Europe (see page 46), and other aspects of Arago XXI's anatomy argue against its placement in this species. For instance, the frontal, or forehead, bone of Arago is broader than that of a typical *erectus*, has less postorbital constriction, and shows no sagittal keel along the midline of the bone. Arago has a more straight-sided braincase, and its estimated brain size of 1,166 cc exceeds the known range for *erectus*.

Like Petralona 1 (see page 214) or the Steinheim cranium (see page 216), the Arago specimen can instead be considered part of the "archaic" European population preceding the Neandertals, a population that would be expected to retain some primitive traits from its *erectus* ancestors. In many respects, the Steinheim cranium resembles a smaller version of the Arago cranium, with a similar frontal bone and facial anatomy, although the Steinheim cranium has a more projecting upper jaw and hollowed cheekbones. We therefore include Arago XXI in *Homo heidelbergensis*, a species showing early signs of adapting to the Pleistocene glacial environments of western Europe by evolving the incipient signs of a Neandertal physique, especially in the face.

Homo heidelbergensis, **Arago XXI.** Arago XXI represents the high end of variation in size and robusticity evident in European crania after 400,000 years ago and before the appearance of Neandertals. Note the degree of cranial constriction behind the eye sockets. Actual size. *Photograph by David L. Brill; courtesy Musée de l'Homme.*

SPECIMEN	LOCALITY	AGE	DISCOVERERS	DATE	PUBLICATION
Adult cranium	Katsika Hill, Petralona, Greece	300,000-400,000 years	J. Malkotsis, J. Stathis, B. Avaramis, C. Sarijanides, and C. St. Hantzarides	September 16, 1960	Kokkoros, P., and A. Kanellis. 1960. Découverte d'un crane d'homme paléolithique dan peninsule Chalcidique. *Anthropologie* 64: 132-47

Homo heidelbergensis
PETRALONA 1

In a cave of vaulted galleries full of hanging stalactites and climbing stalagmites, the cranium of Petralona 1 was found hanging suspended in a stalagmite 23 centimeters above the ground. Early reports that the rest of the skeleton lay nearby are apocryphal, for no skeleton was ever recovered. This is a shame, for the remarkably well-preserved cranium has proved to be one of Europe's most enigmatic human fossils, its identity and geological age as uncertain as the stories surrounding its discovery. This specimen underscores the difficulty of trying to pin a species designation—an inherently dynamic, biological concept—to a static and singular fossil.

Petralona 1 is very transitional in form, with some Neandertal-like features and others that are much more primitive. The cranium looks as though a Neandertal face had been grafted to the braincase and rear of a cranium from some other species. In fact, it was originally classified as a Neandertal but later was claimed to be *Homo erectus*. The best match for Petralona's odd anatomy, though, comes from other European specimens of *Homo heidelbergensis* such as Arago XXI (see page 212) and the recently discovered Atapuerca crania (see page 218).

The Neandertal-like features of Petralona's face include the double-arched browridge rimming the upper eye sockets like a pair of spectacles and the broad nasal opening, which became exaggerated in later "classic" Neandertals. The cheekbones are inflated like those of Neandertals, whereas *Homo sapiens* has distinctive, hollowed-out cheekbones (see Cro-Magnon 1, page 260). An important difference from Neandertals is that the middle of Petralona's face does not project forward. Overall the face, especially the upper half, is massive and broad compared to any Neandertal, and the upper jaw has a wide palate that would be too broad to fit even the lower jaw from Mauer (see page 210). As for the rest of the cranium, the thick occipital bone at the rear, with its prominent transverse torus, or bar, for anchoring muscles resembles what is seen in a typical *Homo erectus*. But the expanded braincase is quite unlike *erectus*, and the estimated cranial capacity of 1,220 cc compares with the largest *erectus* brains and smallest Neandertal brains.

Petralona 1 was originally thought to be only 70,000 years old, as young as many Neandertal remains. Later estimates increased its antiquity tenfold, but the actual age most likely falls in the middle of these extremes. A pair of analytical techniques, electron spin resonance and uranium series dating (see page 26), have been tried to determine the specimen's age by dating the calcite that covered the cranium and other stratigraphic layers in the cave. The results give a minimum age for Petralona 1 of 200,000 years, but the specimen's primitive morphology indicates that its true age is closer to twice that figure.

Homo heidelbergensis, **Petralona 1** (see also page 8)**.** Another transitional form like Bodo, Petralona 1 possesses a Neandertal-like face, including a double-arched browridge and full cheekbones, but lacks Neandertal traits on the braincase or rear of the cranium. Actual size. *Photograph by David L. Brill; courtesy Paleontological Museum, University of Thessaloniki.*

Homo heidelbergensis
STEINHEIM

A curious mosaic of primitive and advanced traits, the Steinheim cranium is the odd man (or, more likely from its size and anatomy, woman) out among Middle Pleistocene crania. It lacks the robust cranial characters of other *H. heidelbergensis* specimens like Arago XXI (see page 212) and Petralona 1 (see page 214), yet it is not Neandertal either. Unlike the Petralona 1 cranium, with its Neandertal-like face and primitive rear, Steinheim shows just the opposite pattern by possessing Neandertal traits only at the back of the cranium, such as the depression in the occipital bone called a suprainiac fossa.

Steinheim figured in the pre-*sapiens* hypothesis, first proposed by anatomists Arthur Keith and Marcellin Boule and defended by their students, according to which modern humans must have had a very ancient ancestry, perhaps dating back millions of years, and the Neandertals and other archaic fossils had nothing to do with it. The remains of Piltdown Man fitted nicely into this scheme until they were exposed as a hoax. The Steinheim individual in turn was seen as the base of a relatively late-branching Neandertal stem that became an evolutionary dead end. The pre-sapiens view has since been discredited: there are no ancient anatomically modern humans in Europe.

Although Steinheim can be considered a plausible Neandertal ancestor—and we have opted to include it in *Homo heidelbergensis*—in some features it rather anticipates our own species. It retains some primitive features, including the marked supraorbital torus beneath the forehead and a cranial capacity of around 1,100 cc. (Steinheim was smaller brained than its approximate contemporary, the partial cranium from Swanscombe, England, and a cranial capacity of 1,100 cc is quite small for a Neandertal.) In certain traits of the face and the shape of flexion of the cranial base, the Steinheim cranium more closely resembles *Homo sapiens* than *Homo neanderthalensis*. Although warped and distorted from postmortem damage, the face clearly appears flatter than a Neandertal's and has a canine fossa, the slight depression beside the nasal opening, a trait not found in Neandertals. Other modern traits include the large

frontal sinus, the straight, expanded parietal bones, and small third molars. The mastoid process, the round protuberance at the base of the skull behind the ear canal, lacks the pronounced crests typically found on the mastoids of Neandertals as a result of the heavy forces that Neandertals placed on their front teeth. No other parts of the skeleton or any stone tools that could provide further clues to its identity accompanied the cranium.

***Homo heidelbergensis*, Steinheim.** In contrast to the Petralona specimen, seen in frontal *(opposite)* and lateral *(above)* views, Steinheim has some Neandertal traits on the rear of the cranium combined with the flatter face, hollowed cheekbone, and small third molars of a modern human. *Courtesy of Staatliches Museum für Naturkunde.*

Homo heidelbergensis
ATAPUERCA 5

Atapuerca 5 is the most complete premodern skull in the human fossil record. It is even more complete than the Cro-Magnon remains, which are nearly ten times younger. More important, this skull and more than 4,000 other human fossils from this site give the first comprehensive glimpse of Neandertals in the making.

Located at the bottom of a 15-meter shaft, the Sima de los Huesos, or "pit of the bones," is a remarkable site. This tiny, low-ceilinged chamber has yielded the remains of at least 32 individuals, including children, adolescents, and adults. Interestingly 15 of the individuals were younger than 18 years of age, and only three were older than 30 years when they died. Within the sample of 32 persons it is possible to identify nine males and nine females with the remainder being of indeterminate sex. Every bone in the skeleton is represented, even the fragile hyoid, previously found only with the Kebara Neandertal skeleton (see page 232). More than 70 percent of all human postcranial bones from the entire Middle Pleistocene (the period spanning 700,000 to 200,000 years ago) come from here, so Atapuerca offers an unprecedented opportunity to study variation due to age and sex in a single early human population from this time.

Speleologists stumbled upon the chamber in 1976 while mapping the cave system. They were followed by a Spanish excavation team, which spent six years removing tons of sediment from the chamber and searching for human bone fragments among thousands of pieces of cave bear bones before they hit paleontological paydirt. Such small finds as bones from the tips of fingers convinced the crew that the pit contained complete human skeletons.

Working under wet, cold, cramped conditions in the chamber for only a few hours at a time, five people suspended above the ground on wooden planks carefully carved away wet clay to expose the delicate, easily shattered bones. The fossils were put in boxes, which were then placed inside padded bags and packed in backpacks to be carried out of the cave, where they were allowed to dry and harden.

In 1992 the complete calvaria of Cranium 4 was found, cause for a champagne celebration in the cave. On the last day of fieldwork that year, the team uncovered the calvaria of Cranium 5 nearby. Returning to the site one last time to close up for the season, the team decided to dig a little more and soon found the face that fit Cranium 5. In 1993 they found a lower jaw for Cranium 5 and completed the skull.

Atapuerca 5, a presumed male, has a broad, bulky face with the widest nasal opening of the human fossil record. The cranium is relatively small and round. Its brain capacity, 1,125 cc, is smaller than that of any adult from Atapuerca (and smaller than almost every hominid brain from the Middle Pleistocene of Africa or Eurasia), but the upper canine from this skull is the biggest in the collection, indicating a wide range of size variation among individuals. Another presumed male, Cranium 4, in contrast, may have the largest known brain capacity, 1,390 cc, of any Middle Pleistocene human fossil. These brains far surpassed those of *Homo erectus* in size and were close to the size of Neandertal brains. The degree of size difference between Cranium 4 and Cranium 5 is comparable to that between the crania from Steinheim (see page 216) and Petralona (see page 214).

Certain traits in the Atapuerca crania foreshadow the features of "classic" European Neandertals, including the prominent double-arched browridge above the eyes, the projecting middle face (particularly noticeable on Cranium 5), a well-developed nuchal torus, or bar, on the rear of the braincase, worn front teeth, and, in Cranium 5, a retromolar space, the gap behind the third molar that comes from the middle of the face being pulled forward. Cranium 5 lacks a canine fossa (the hollowing of the cheekbone above the canine that occurs in modern humans but not in Neandertals, giving the latter an inflated look to the cheekbones). Another specimen in the collection does possess a canine fossa.

In other traits, however, the Atapuerca population shows little tendency in the Neandertal direction. The crania recovered from the site are high and round in side view, rather than long and low like a typical Neandertal cranium. In occipital view (back view) Neandertal skulls narrow inferiorly while the Atapuerca skulls are broadest at the base. The mastoid process, the bump behind the ear opening, is prominent and projecting in adult Atapuerca specimens, unlike Neandertals where the mastoid is a minor feature. Typical features of the occipital bones seen in Neandertals such as a protruding occipital bun and an elliptical depression, called the suprainiac fossa, are absent in the Atapuerca fossils.

Judging from the postcranial bones, the Atapuerca people were very strongly built and no doubt led physically demanding lives. The long bones exhibit extremely thick cross-sections characterized by massively thick cortical bone occupying nearly the entire cross-section, leaving little room for the spongy medullary canal. Employing the long bones for stature estimates suggests that males averaged 1.75 meters while females only slightly shorter at 1.7 meters, indicating essentially a lack of sexual dimorphism in stature. Body weight in larger individuals is estimated to have been 90 kilograms.

Apparently the anterior dentition was used extensively and is heavily worn, and in older individuals (30 years old) entire tooth crowns are missing. Perhaps the Atapuerca people used these anterior teeth for holding objects, while working animal hides. The intense use of the anterior dentition subjected the hinge joint (temporomandibular join) of the mandible where it articulates with the skull base to considerable stress, producing arthritis.Cranium 5 (Miguelón) in addition to exhibiting a series of 13 lesions due to impact or infection, evidenced signs of pathology in the facial skeleton consistent with septicemia. This generalized infection that appears to have begun in the upper jaw and progressed upwards into the eye socket may have been the cause of death of this large male. Cranium 4 (Agamemnon) must have been deaf during life since his ear canals are almost entirely blocked by a bony growth typical of some of ear infection.

The Atapuerca fossils capture an early stage of isolated evolution on the European continent that would culminate in the emergence and extinction of Neandertals. Although the bones were not found in association with tools or other evidence of habitation and were not deliberately buried, they may have entered the pit by human hands, perhaps foreshadowing another Neandertal behavioral trait, the burying of their dead.

Homo heidelbergensis, **Atapuerca 4.** Unlike the relatively small cranial capacity of Cranium 5 (1125 cc), Atapuerca 4 has the largest brain capacity, 1,390 cc, of any Middle Pleistocene human fossil. When it was found, the excavators celebrated with champagne in the cramped cave. Actual size. *Photograph by Javier Trueba, Madrid Scientific Films; courtesy of Juan-Luis Arsuaga.*

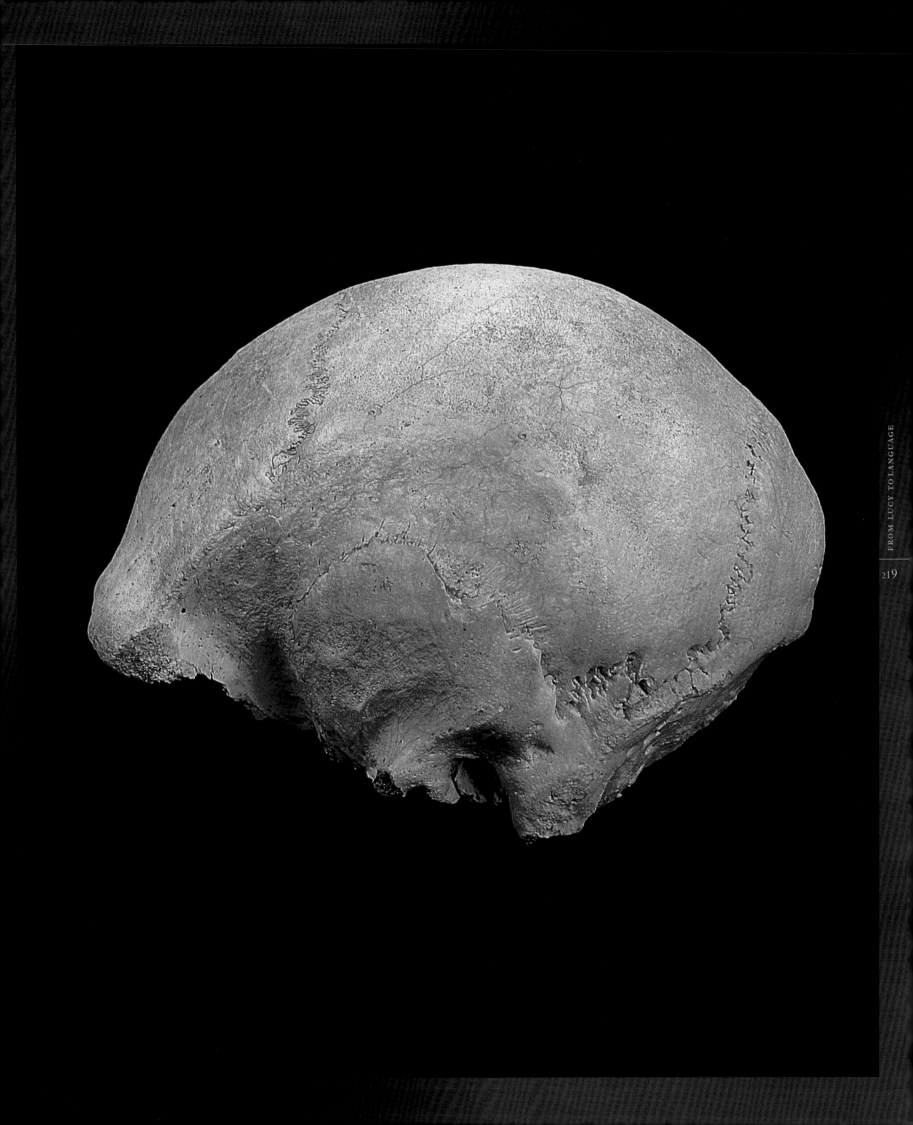

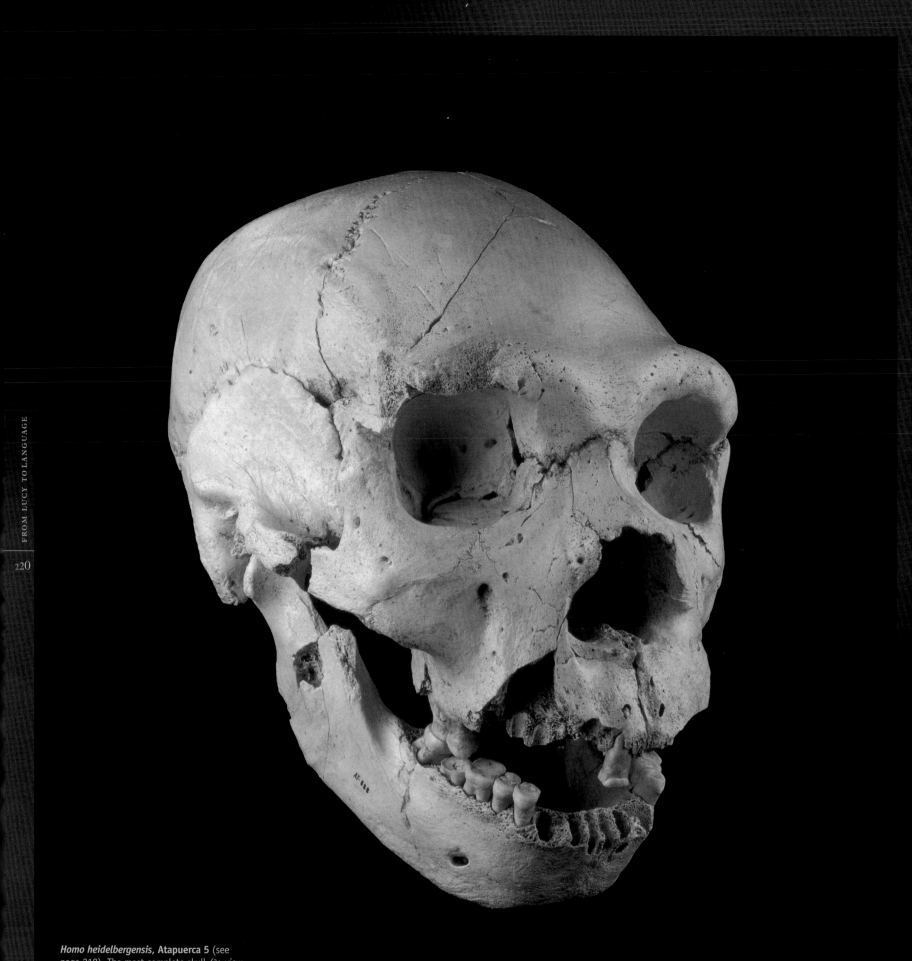

Homo heidelbergensis, **Atapuerca 5** (see page 218)**.** The most complete skull *(3/4 view, left, and lateral, right)* from a rich Spanish cave site, Atapuerca 5 displays such incipient Neandertal traits as a projecting mid-face, double-arched browridge, inflated cheekbones, and a gap behind the third molar. Actual sizes. *Photographs by Javier Trueba, Madrid Scientific Films; courtesy of Juan-Luis Arsuaga.*

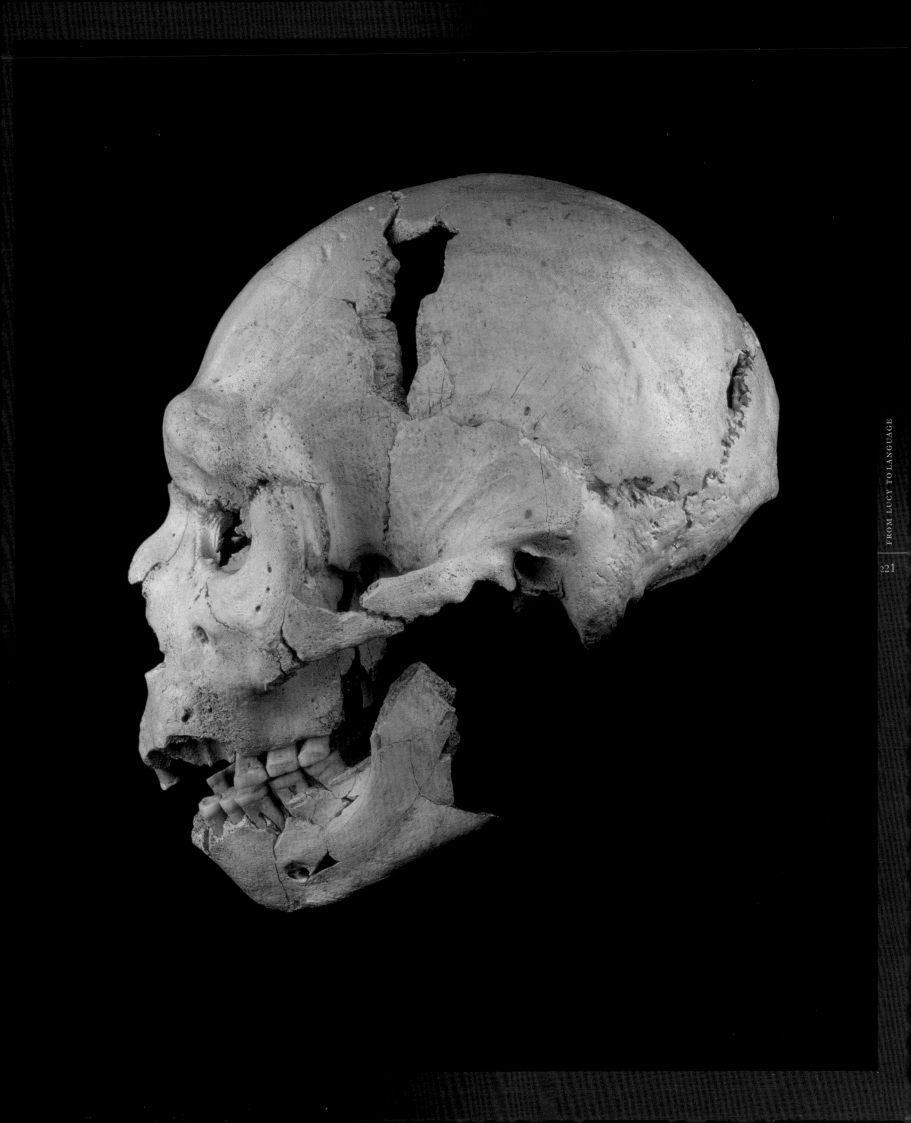

Homo heidelbergensis
BROKEN HILL 1

The quest for lead and zinc in a vast cave beneath a limestone kopje (small hill) led to the discovery of this famous specimen, the first fossil of any human ancestor to be found in Africa. As such, it marked the start of paleoanthropologists' continuing quest for human origins on that continent. Tucked away in the wall at the base of the cave, 27 meters beneath the surface, the specimen known as Broken Hill 1 was brought to light in 1921 by miner Tom Zwigelaar, who reportedly hoisted the hefty cranium onto a pole as a way of prodding his men to work harder. A later search for the lower jaw turned up some modern-looking limb bones that came from different, possibly less ancient individuals. Soon the well-preserved fossil was delivered to anatomist Arthur Smith Woodward, in London, who described it in the journal *Nature* as a new species, *Homo rhodesiensis*, or Rhodesian Man.

Ever since the discovery of the Broken Hill cranium, several "archaic" human specimens have turned up in Africa, and we choose to include these with several European specimens in the species *Homo heidelbergensis*. Like Bodo (see page 208) and other fossils, Broken Hill 1 may be an individual from the ancient African population from which our species first evolved.

The Broken Hill, or Kabwe, specimen was once thought to be only 40,000 years old, based on an assessment of the associated stone tools (that assessment was later revised). Thus, American anthropologist Carlton Coon, who constructed an ambitious but flawed categorization of the world's races (see page 56), used the cranium in his 1962 book *The Origin of Races* as evidence that Africans remained at a *Homo erectus* level of evolution at a time when *Homo sapiens* had already appeared in Europe. But evidence from other vertebrate fauna at the site suggests that Broken Hill 1 is at least 125,000 years old, and probably much older, so its primitive features make sense, in light of its obvious antiquity. Rather than being the contemporary of the very modern-looking Cro-Magnon specimens (see page 260), Broken Hill 1 is closer in age to the Petralona 1 cranium (see page 214).

Woodward noted some similarities between the large, heavy face of Broken Hill 1 and that of the Neandertal found at La Chapelle-aux-Saints, but he considered the shape of the Broken Hill specimen's braincase to be "much more ordinarily human." It has a low, sloping forehead and a long cranium that held a brain approaching 1,300 cc in size (significantly larger than the average for *Homo erectus*). In 1925, Woodward's colleague, Sir Arthur Keith, was impressed enough by this specimen to devote to it two chapters in the second volume of his revised *The Antiquity of Man*. He wrote that Broken Hill 1 revealed that "the wild dreams of the Darwinists have a solid basis in fact" and provided "for the first time a glimpse of our ancestral state."

Keith thought that the strikingly massive browridge over the eyes could not be explained as merely an adaptation to heavy chewing and speculated that, like the large canines of male gorillas, this was a secondary sex characteristic for males to attract mates. But there is no evidence that this was the case.

The specimen displays some disease and wounding. Ten of the 16 upper teeth contain cavities, and abscesses formed in the jaw. A partially healed, quarter-inch-diameter wound above and in front of the left ear opening may have been inflicted by a sharp instrument or by carnivore teeth.

Homo heidelbergensis, **Broken Hill 1.**
The first hominid fossil to be found in Africa, Broken Hill 1 resembles the Bodo cranium with its robust, primitive facial features. Ten of its teeth contain cavities, and a partially healed wound penetrated the bone by the left ear. *Photograph by John Reader, Science Source/ Photo Researchers.*

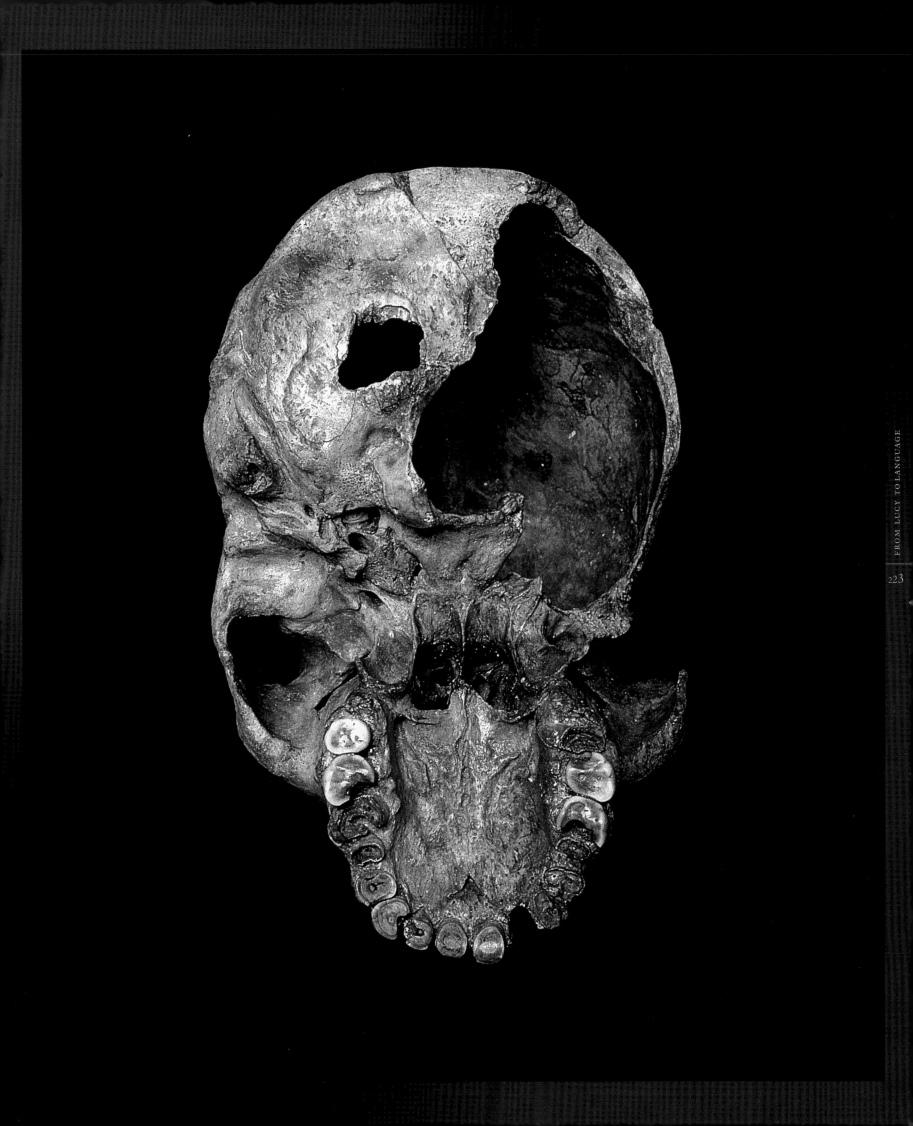

224

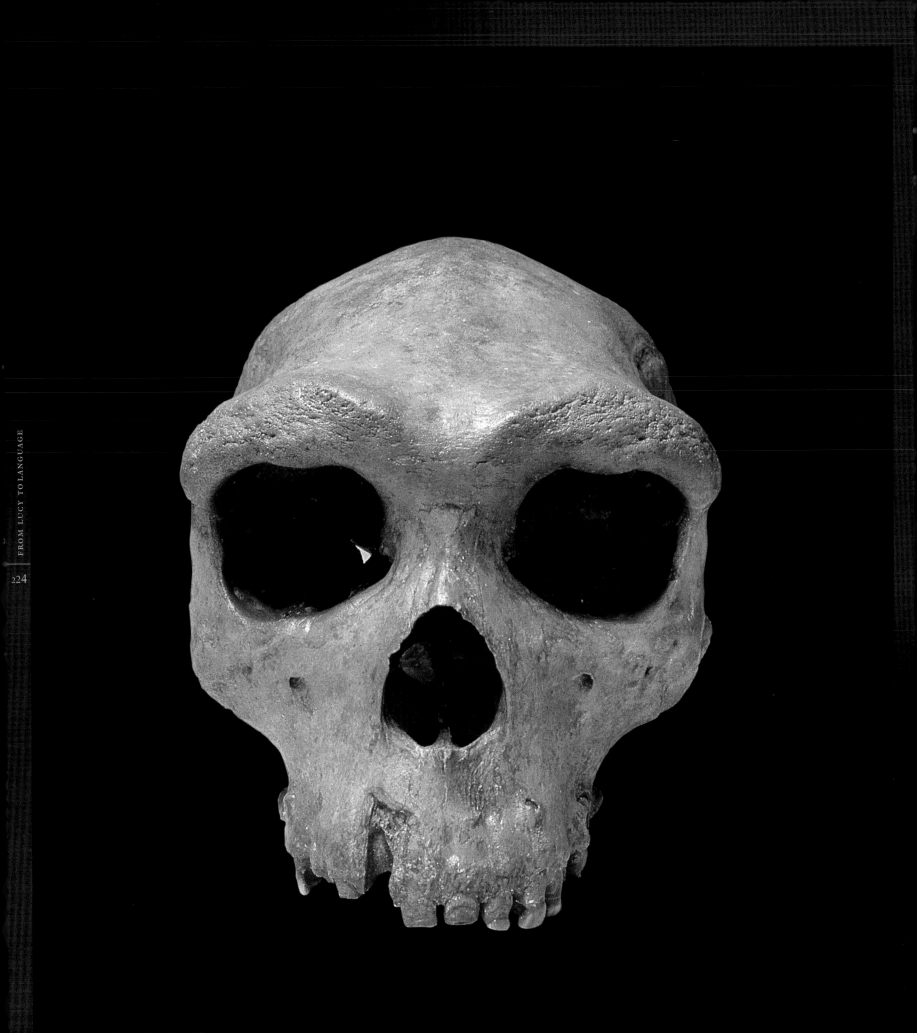

Homo heidelbergensis, **Broken Hill 1**
(see previous page). Actual size. *Photograph by
John Reader, Science Source/Photo Researchers.*

SPECIMEN	LOCALITY	AGE	DISCOVERERS	DATE	PUBLICATION
Adult female partial cranium	Krapina cave, Croatia	130,000 years	Karl Gorjanovic-Kramberger	September 1899-July 1905	Gorjanovic-Kramberger, K. 1906. *Der diluviale Mensch von Krapina in Kroatien* (C.W. Kreidels Verlag, Wiesbaden)

Homo neanderthalensis
KRAPINA C

Paleoanthropologists typically work with scattered pieces of skulls and skeletons. Sometimes a nearly or fully complete skeleton is found. On very rare occasions, one site yields an entire ancient population for study. That is the case at Krapina, a limestone cave in the mountains of Hrvatsko Zagorje, Croatia, where the largest known sample of Neandertal remains has been found. Males and females, children to adults, and almost every part of the skeleton are represented among the more than 850 human fossils from up to 80 individuals, most of whom died between the ages of 16 and 24 years. Three thousand animal bones and 1,000 stone tools were also recovered. Because many of the hominid bones are broken into small pieces, determining exactly which bones go with which individuals has been a mostly insurmountable challenge since the turn-of-the-century excavations, but five crania have been partially reconstructed—Krapina C being the most complete.

As should be expected from such a large sample, the Krapina fossils exhibit considerable variation in size and anatomy, particularly in the teeth, lower jaws, and certain cranial features. The sample of 279 teeth includes front teeth with large crowns and molars that show the Neandertal trait of taurodontism, or an enlarged pulp chamber and fused roots. Such cranial traits as the projecting midface, a retromolar space in the lower jaw, and an occipital bun align the Krapina sample with other, more recent European Neandertals, as do postcranial bones revealing the Krapina people to have been barrel-chested with muscular forearms and powerful hands.

In general, however, the Krapina fossils lack the classic anatomy of later Neandertals from further west, such as La Ferrassie (see page 240). Neither are they quite like the Near East Neandertals. Compared to "classic" Neandertals, the Krapina individuals tended to be more lightly built but to have large faces, widely separated eye sockets, and broad crania. Some had higher foreheads and limb bones that closely resemble those of modern humans. The intermediate nature of the anatomy has been cited as evidence that the Krapina Neandertals evolved directly into the modern human populations of Central

Europe, but evidence is lacking from elsewhere in Europe for such a transitional scenario.

The fragmentary state of human fossils from Krapina and another Croatian cave site, Vindija, as well as the presence of stone tool cut marks on certain bones strongly suggests that some of these Neandertals were butchered and possibly cannibalized (see page 93). Debate on this issue has continued ever since the initial excavator, Karl Gorjanovic-Kramberger, concluded that the pattern of bone breakage and the burning of some bones provided evidence of cannibalism. He wrote, "These men ate their fellow tribesmen, and what's more, they cracked open the hollow bones and sucked out the marrow. . . ."

Some of the breakage may have come from falling rock debris, but falling rocks cannot explain why only the elbow ends of the humeri are present and not the shoulder ends of this marrow-rich upper arm bone. Meaty and marrow-filled femora are missing from the sample, except for one shaft fragment from a child. And there is no sign of carnivore chewing on the bones that remain. In addition to obvious cut marks, some bones show concoidal scars, the rippling fracture pattern that is a distinctive signature of blows from a human stone artifact, further evidence that human bones were actively processed for the nutrients they contained.

Homo neanderthalensis, **Krapina C** (see also page 95). Krapina C is the most complete of five crania from this Croatian cave, which preserved highly fragmentary remains from a population of about 80 individuals, from children to adults. This specimen shows the typically broad, lightly built face of these particular Neandertals. Actual size. *Photograph by David L. Brill; courtesy of Geolosko-Paleontologi Musej.*

SPECIMEN	LOCALITY	AGE	DISCOVERERS	DATE	PUBLICATION
Adult female cranium	Saccopastore quarry, Rome, Italy	ca. 120,000 years	Mario Grazioli	May 13, 1929	Sergi, S. 1929. La scoperta di un cranio del tipo di Neanderthal presso Roma. *Rivista di Antropologia* 28: 457-462

Homo neanderthalensis
SACCOPASTORE 1

With its lightly built cranium, the specimen known as Saccopastore 1, found in the Saccopastore quarry in Rome, resembles early modern humans almost as much as it does the "classic" Neandertals that appeared in Europe 60,000 years later. That fact gave Saccopastore 1 a pivotal place in the pre-Neandertal hypothesis developed by Italian anthropologist Sergio Sergi, who first studied and described this specimen. Sergi proposed that the evolutionary line leading from an anatomically generalized "pre-Neandertal" type of early human split into two branches, one leading to later Neandertals and the other to modern *Homo sapiens*. He based this idea on the resemblance of Saccopastore to specimens from Steinheim (see page 216), Swanscombe, Fontéchevade, and Ehringsdorf. The Swanscombe and Steinheim individuals formed the pre-Neandertal base, with the specimens from Saccopastore, Krapina, and Ehringsdorf occupying the Neandertal branch and the cranial fragments from Fontéchevade on the branch to modern humans. This idea, which came to dominate the previously popular pre-*sapiens* hypothesis (see page 43), made it acceptable for archaic-looking early humans to be viewed as modern human ancestors, although the western European Neandertals were still seen as an evolutionary dead end. This scenario is partially right: Neandertals were the end product of isolated evolution in Europe from *Homo heidelbergensis*, but it is not clear that modern *Homo sapiens* descended from any pre-Neandertal stock.

Missing only its browridges, Saccopastore 1 preserves the most complete evidence for an early Neandertal from the Riss-Würm interglacial period (between 127,000 to 115,000 years ago), making it comparable in geological age to the large but highly fragmented Neandertal population from Krapina (see page 225). Six years after this female cranium was found with Mousterian tools at a gravel quarry along a tributary of the Tiber River, a companion specimen was found by anthropologists Alberto Blanc and Abbé Henri Breuil during a visit to the site. Saccopastore 2 is a more robust male represented by the cranial base and pieces of a face that more closely resembles the Petralona 1 cranium (see page 214) than

a typical Neandertal. Both specimens reveal that early Neandertal crania contained many generalized features.

Yet they are unmistakably Neandertals. Saccopastore 1 has a low cranial vault and a projecting midface with a broad nasal opening. It exhibits all the typical Neandertal features from the rear and base of the skull—including a suprainiac fossa and occipitomastoid crest—except the occipital bun. Unlike most Neandertals, however, Saccopastore 1's occipital bone appears rounded, and the cranial base appears more flexed than in such "classic" Neandertal specimens as those from La Chapelle-aux-Saints and La Ferrassie. This latter feature may have implications (see page 106) for Neandertals' capacity to use spoken language.

Homo neanderthalensis, **Saccopastore 1.**
A female member of an early Neandertal population, this specimen, seen here in frontal view, and lateral, next page, has a typically broad nasal opening and low forehead, as seen in frontal view. From the side, the low cranial vault and projecting mid-face are clearly visible, along with the flexed cranial base. Actual size. *Courtesy of Giorgio Manzi, Museo di Antropologia of the University of Rome "La Sapienza."*

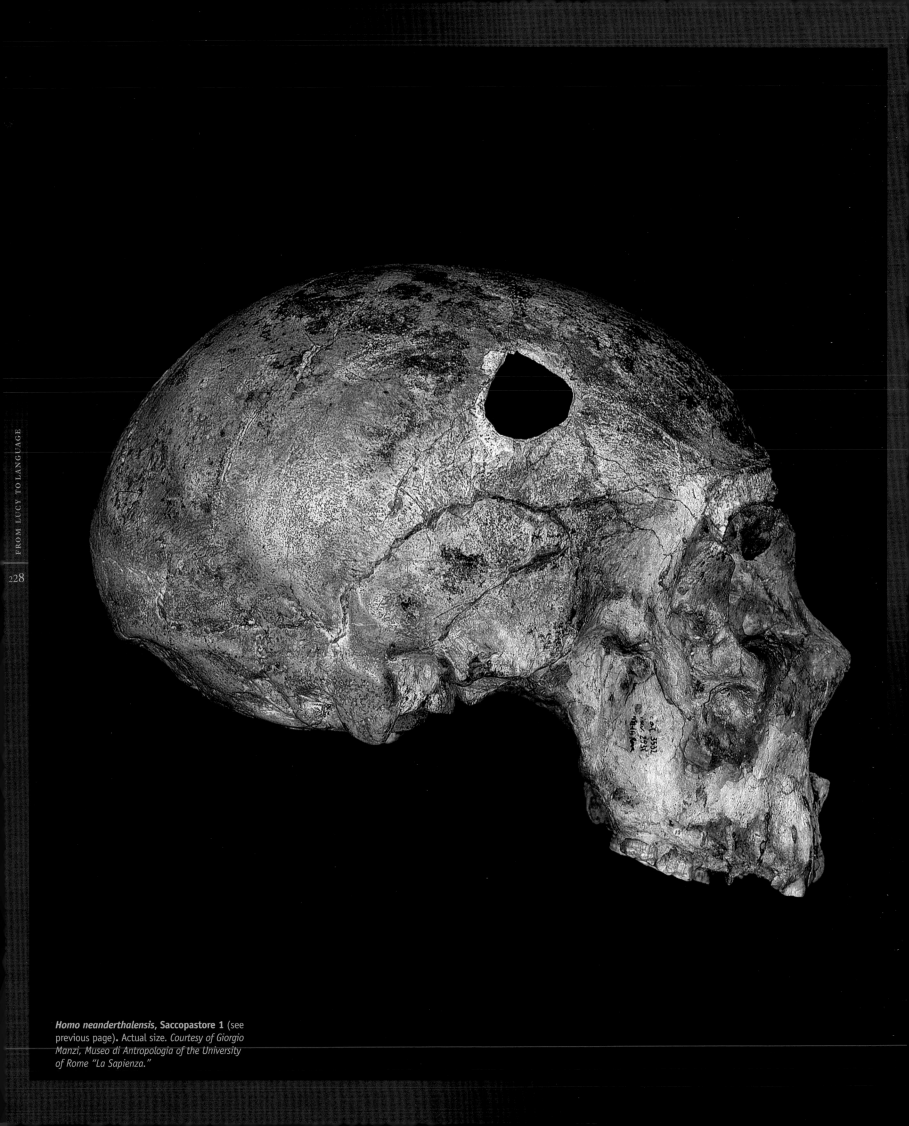

228

***Homo neanderthalensis*, Saccopastore 1** (see previous page)**.** Actual size. *Courtesy of Giorgio Manzi, Museo di Antropologia of the University of Rome "La Sapienza."*

Homo neanderthalensis
TESHIK-TASH

The skeleton known as Teshik-Tash, after its place of discovery in Uzbekistan, defines the easternmost extent of the Neandertals' known geographic range. Some 3,200 kilometers from the nearest Neandertal in Europe and 1,600 kilometers from his conspecifics in the Shanidar cave of Iraq, this young boy lay in a cave due south of Samarkand at an altitude of nearly 1.6 kilometers in the rugged and remote Gissar mountain range. The Teshik-Tash specimen demonstrates that Neandertals inhabited isolated and extreme environments.

Teshik-Tash was the first Paleolithic site in Central Asia to be excavated, but other nearby cave sites indicate a long human history in the region. Discovery of the skeleton, its bones yellowed by the surrounding sediment, came at the beginning of a two-year excavation. The cave contained five separate occupation layers and a dozen hearths, some of which had clusters of broken bones and wild goat horns and stone flakes and core tools nearby.

The skeleton had apparently been buried at the base of the thickest and archaeologically richest occupation layer, along the cave's western wall, with the feet pointed toward the entrance. The shallow pit in which it lay protected the skeleton from falling rocks that shattered bones in the layer immediately above the corpse, but the pressure of overlying sediments flattened the skull. Fortunately, the cranium was well preserved and could be reconstructed from about 150 pieces.

The Teshik-Tash boy died at around age nine, but he had reached sufficient maturity to have developed some of the distinctive features of Neandertal anatomy. These included a large face and nasal region, a developing browridge, a receding forehead, a long cranium, and a lower jaw lacking a bony chin. His cranial capacity was around 1,500 cc, high for the boy's age by modern human standards.

Various parts of the postcranial skeleton, including one neck vertebra, several ribs, a humerus, the clavicles or collarbones, a femur, a tibia, and the fibulae (the smaller of the lower leg bones), lay scattered near the cranium. The ends of both the humerus and femur had been gnawed off, and a coprolite (fossilized scat) lay near the skeleton, suggesting that a hyena or other carnivore had dug up the burial and removed some of the bones. No artifacts had been included with the body.

The most striking feature of the grave was the presence of six pairs of large bony horn cores from Siberian ibex, or mountain goat, which had been placed point down in a circle surrounding the skull, and a few other bones. A small fire had been briefly lit beside the body. These clues have made Teshik-Tash a frequently cited example of Neandertal ritual burial (see page 100).

It is difficult to know whether these horns were intentionally placed around the corpse in some symbolic gesture. Perhaps the horns were tools that were used to bury the boy and were then discarded, serving no ritualistic purpose. Or the proximity of the horn cores to the skeleton may be fortuitous. Ibex horns and bones are very common throughout the site; of 769 bones found in the cave from mammals other than rodents, 761 of these belonged to ibex. Judging from such prevalence, ibex was the animal of choice hunted by the Neandertal inhabitants, and villagers in this remote region still relied on goat hunting for subsistence as recently as the late 1930s, when this skeleton was found. Whether or not the Neandertals of Teshik-Tash practiced a cult of the dead and a cult of wild goats, they depended on these agile, majestic mammals to survive here.

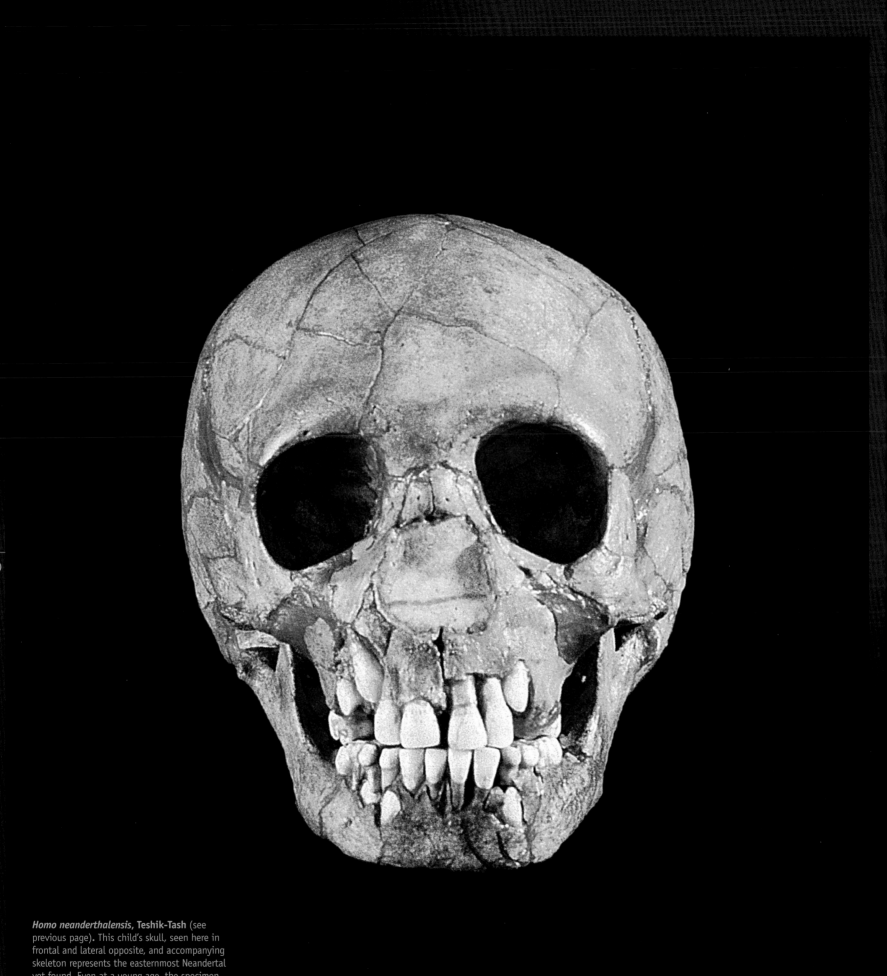

***Homo neanderthalensis*, Teshik-Tash** (see previous page). This child's skull, seen here in frontal and lateral opposite, and accompanying skeleton represents the easternmost Neandertal yet found. Even at a young age, the specimen exhibits such distinctive Neandertal features as the long, low cranium, large nasal opening, a developing browridge, and a lower jaw lacking a chin. Actual sizes. *Photographs by Andrei Mauer; courtesy of Institute and Museum of Anthropology, Moscow State University.*

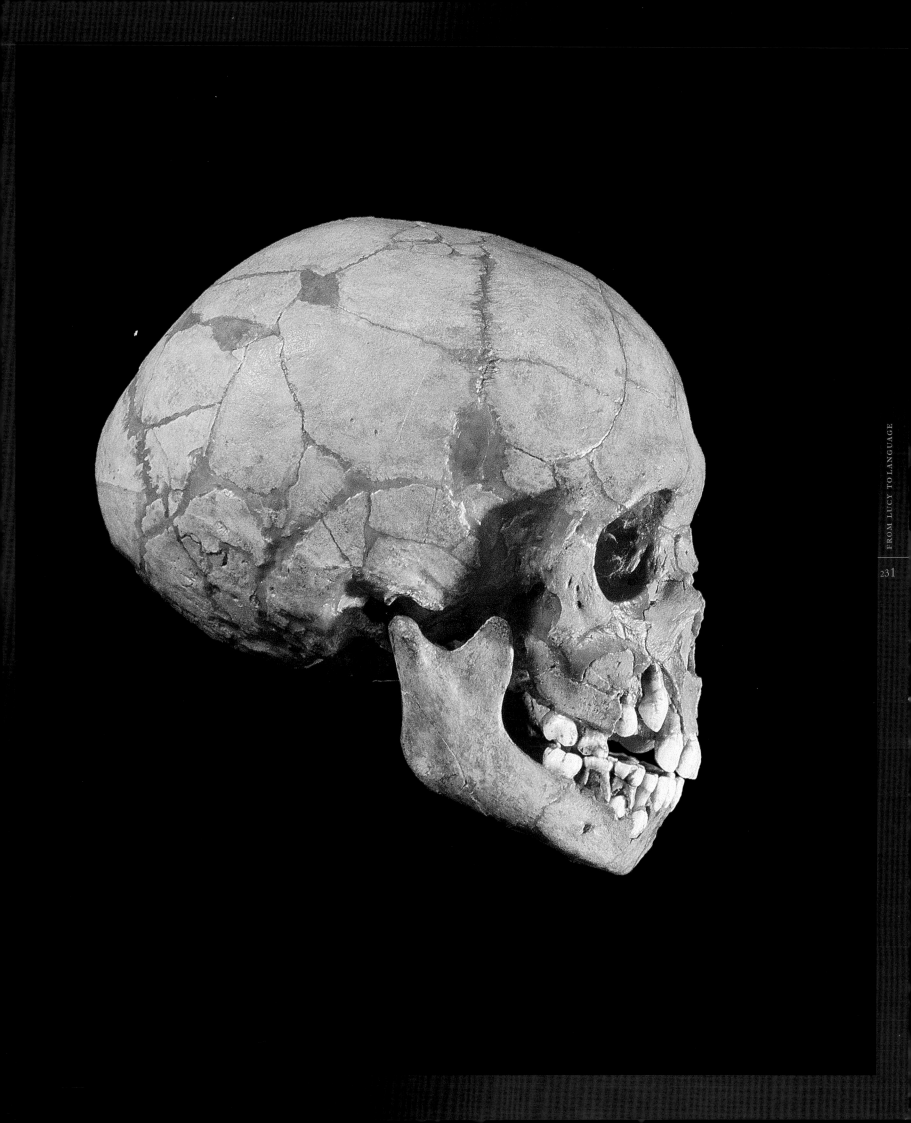

SPECIMEN	LOCALITY	AGE	DISCOVERERS	DATE	PUBLICATION
Adult male skeleton	Kebara Cave, Israel	ca. 60,000 years	Lynne Schepartz	October 1983	Arensburg, B., O. Bar Yosef, M. Chech, P. Goldberg, H. Laville, L. Meignen, Y. Rak, E. Tchernov, A.-M. Tillier, and B. Vandermeersch. 1985. Une sépulture néanderthalien dans la grotte de Kebara (Israel). *Comptes Rendus de l'Académie des Sciences* (Paris), Série II, 300: 227-230

Homo neanderthalensis
KEBARA 2

Nicknamed "Moshe" in honor of Moshe Stekelis, a previous excavator of this cave site, the Kebara 2 specimen with its intact mandible is the most complete trunk skeleton of a Neandertal discovered. It includes the first complete set of ribs, vertebrae, and pelvis ever found.

Another preservation first for Kebara 2 is the hyoid bone, the only bone in the human body that connects to no other bones. Rooted in the cartilage surrounding the larynx, the hyoid anchors throat muscles necessary for speech. The discovery in this specimen of a hyoid bone identical to that of modern humans has important implications for the Neandertals' capacity for language (see page 106).

Moshe's skeleton was discovered on the edge of a test pit dug during initial excavation during the 1960s (which revealed the fragmentary infant skeleton labeled Kebara 1). In an apparent burial, he had been placed on his back inside a shallow pit, with the right arm laid across his chest and the left arm resting on his abdomen. He was probably between the ages of 25 and 35 when he died, and there are no clues in the bones to death by violence or disease. Mysteriously, Moshe's cranium has vanished, possibly carried off by a carnivore or buried elsewhere in the cave. Only one upper third molar was found. The right leg and the lower half of the left leg are also missing.

Although we cannot regard his face, Kebara 2 has clear similarities to skeletons from Amud, Tabun, and Shanidar, but it is more robust than all of them. At 1.7 meters, the individual was taller than a typical European Neandertal. The massive mandible, with a complete set of teeth, possesses the retromolar space common to Neandertals and lacks a chin. Other parts of the skeleton, such as the hyoid and vertebrae, are indistinguishable in size and shape from those of modern humans. The fifth lumbar vertebra, at the base of the spinal column, for instance, has the same modern modifications for bipedalism found in our own spines.

The complete pelvis also speaks to this hominid's posture and locomotion. With a longer distance between the hip joints on the sides and the joint of the pubic bones in front of the pelvis, Neandertals had their center of gravity shifted forward and may have lacked the degree of cushioning and shock absorption found in a modern human pelvis.

Besides this spectacular fossil specimen, the Kebara Cave preserved a rich archaeological record that offers a detailed look at the lives of Near East Neandertals, who occupied the site from before 60,000 years ago until at least 48,000 years ago. (Less frequent human occupation occurred during the Upper Paleolithic, and consequently carnivore remains are more common from this time period.) More than 25,000 stone tools at least 2.5 centimeters in size were collected from the site. These represent the Levallois industry typical of this time period and generally associated with Neandertal remains. The Neandertals preferred fine-grained flints, mostly collected within a few miles of the cave, for their tools, and they used efficient flaking techniques to make the most of their raw material. Half of the triangular points and flakes that have been analyzed for signs of wear bear impact fractures, suggesting that these tools had been hafted to wooden shafts and then heaved as projectiles. Other tools have wear traces associated with woodworking and cutting. Curiously, the Kebara artifacts closely resemble stone tools excavated from the Israeli cave of Qafzeh, but the more ancient skeletons buried at Qafzeh (see page 255) are clearly modern humans rather than Neandertals. Why two distinct populations, different species, would share the same culture remains a mystery.

Numerous round and oval hearths were excavated in the cave, along with layers where grassy and woody vegetation was burned. Pieces of burned flint retrieved from the hearths proved crucial in dating the site and the skeleton by the thermoluminescence technique. Locally obtained oak wood fueled the hearth fires, which were apparently used to parch wild peas and perhaps to roast gazelle and deer meat. Garbage dumps full of broken animal bones and waste from stone tool manufacture accumulated along the cave's north wall, away from the central living area. The picture that emerges from this broad range of archaeological evidence suggests that the dwellers of Kebara were far more technologically capable and sophisticated than Neandertals have often been portrayed.

Homo neanderthalensis, **Kebara 2.** No cranium was found with this skeleton buried in an Israeli cave, but Kebara 2 is nonetheless the most complete Neandertal specimen known and includes the first set of ribs, vertebrae, and pelvis ever found. *Photograph by David L. Brill; courtesy of Sackler School of Medicine, Tel Aviv University.*

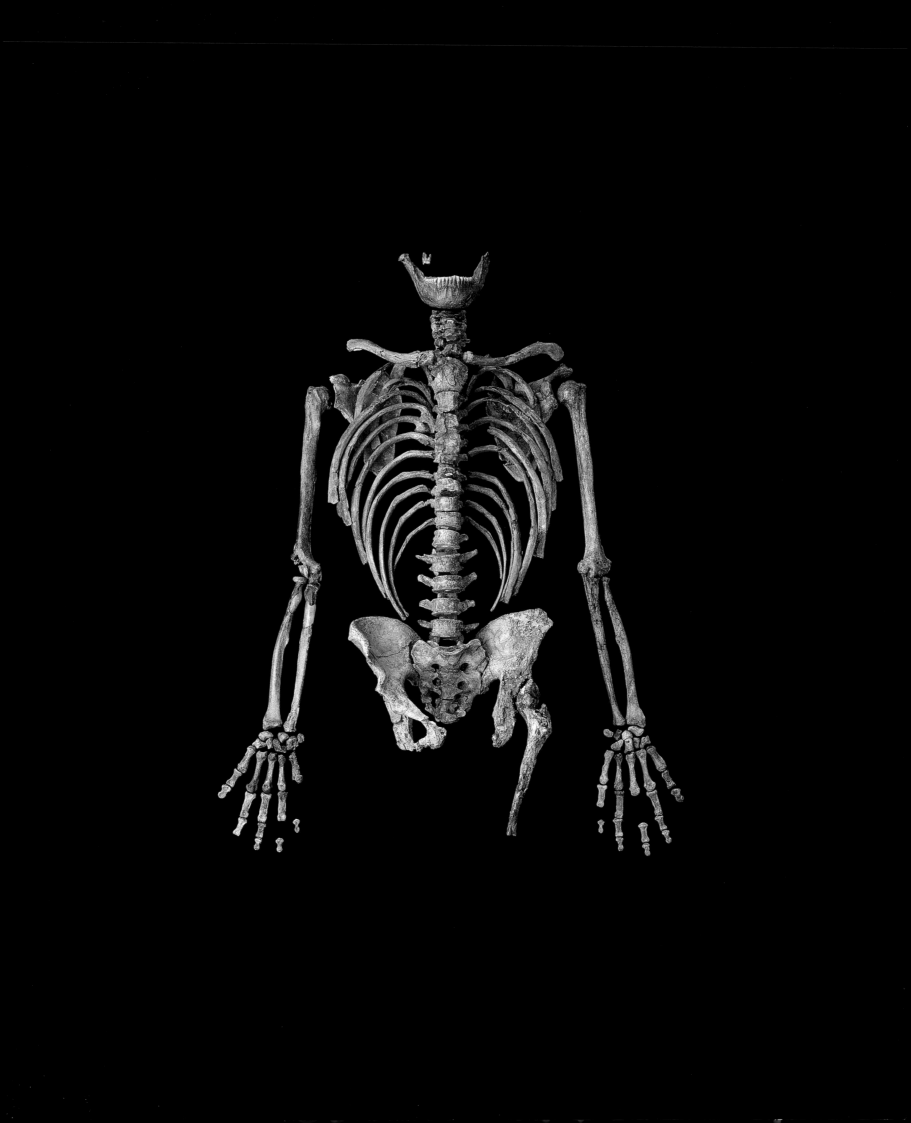

SPECIMEN	LOCALITY	AGE	DISCOVERERS	DATE	PUBLICATION
Adult male skull	Amud Cave, Israel	40,000–50,000 years	Hisashi Suzuki	June 28, 1961	Suzuki, H., and F. Takai, editors. 1970. *The Amud Man and His Cave Site*. Tokyo: University of Tokyo Press

Homo neanderthalensis
AMUD 1

The adult male hominid skeleton found in Amud Cave, Israel, can be described by two superlatives. At more than 1.8 meters, the individual is the tallest Neandertal known, and the skull enclosed the largest brain—a whopping 1,740 cc—of any known fossil hominid. Judging from the degree of closure in the sutures between the cranial bones, he probably died at around age 25.

The shattered skeleton lay on its left side with its limbs flexed, at the top of the Middle Paleolithic level, surrounded by younger Upper Paleolithic tools and even younger pottery. The pale gray bones had not completely fossilized and proved fragile in the cave's soft soil. The skull, which had been crushed from the side, was found first and was so encrusted with limestone rubble that it was nearly mistaken for a chunk of fallen cave ceiling. Unfortunately, the palate and much of the diagnostic facial region is missing, but the enormous lower jaw is intact. The jaws contain a complete set of 32 teeth, which are small for such a large individual. Many parts are less complete: the vertebrae, pelvis, and right lower limbs, for instance, were badly damaged or missing.

Amud 1's anatomy presents an eclectic and peculiar moasic of features. Although clearly a Neandertal, it is not a "classic" Neandertal such as the most famous European specimens. Amud 1 shows closest affinity to the Neandertals of Shanidar Cave in Iraq (the skull was reconstructed using Shanidar I as a guide) and Tabun, Israel, but it also has some similarities to the Skhul and Qafzeh populations of early modern *Homo sapiens*.

Although the large, long, narrow face falls within the range of European Neandertal dimensions, there are differences in the details. The browridge above the eyes is more slender and angles backward on each side of the face. The eye sockets have sharply defined margins, as in modern humans, rather than the typically rounded edges of Neandertals. The maxilla appears less inflated in the cheeks than can be seen in the Neandertal from La Chapelle-aux-Saints or even in specimens from Shanidar, and there is a more marked chin at the front of the mandible than is typical of Neandertals.

The mix of Neandertal and modern features continues on the cranium. The cranium is wide and long; it is longer than in any modern human and longer than in the biggest European Neandertal, from La Ferrassie. The mastoid process of the temporal bone, which protrudes like a knob beneath the ear opening, approaches modern human size and is larger than the mastoid process in the Neandertals at La Chapelle-aux-Saints and Gibraltar. In rear view, the cranium is perfectly round, like a bowling ball—a typical Neandertal trait—but the occipital bone combines a Neandertal-like torus, or bar, of bone across its width with a sharply curved overall shape that resembles a modern human cranium.

Amud is Hebrew for "pillar" and refers to a stone pinnacle near the cave entrance. The cave was perched about 30 meters above the streambed of the Wadi Amud, through which water flows to the Sea of Galilee. A nearby perennial spring may have also attracted Neandertals to the site. Initial attempts to date the Neandertal occupation of the cave by radiocarbon isotopic decay methods fell far short of the real age; more recent estimates using the electron spin resonance technique on a mammal tooth suggest an age of between 40,000 and 50,000 years, relatively late in the Neandertals' time on Earth. Perhaps the combination of a relatively young age and peculiar anatomy indicate that Amud 1 is a more evolved form of Neandertal than its immediate European predecessors; or it may simply be a variant due to geographic separation.

Homo neanderthalensis, **Amud 1** (see also page 178). Though it lacks the important midface region, this specimen can be identified as a Neandertal, resembling those from other Near East sites. Note the retromolar gap behind the tooth row in the lower jaw. Actual size. *Photograph by David L. Brill; courtesy Israel Antiquities Authority, Rockefeller Museum.*

SPECIMEN	LOCALITY	AGE	DISCOVERERS	DATE	PUBLICATION
Parial infant skeleton	Amud Cave, Israel	50,000 to 60,000 years	Tina Hietala and Yoel Rak	June 9, 1992	Rak, Y., W.H. Kimbel, and E. Hovers. 1994. A Neandertal infant from Amud Cave, Israel. *Journal of Human Evolution* 26: 313-324

Homo neanderthalensis
AMUD 7

Renewed excavation at Amud Cave in the early 1990s unearthed the remains of four humans. The bones indicated that these individuals had all died in the first year of life except for a child of about eight years, represented by a single tooth. The most complete specimen, a tiny newborn's skeleton catalogued as Amud 6, had been buried just on the edge of a Middle Paleolithic layer, but radiocarbon dating of Amud 6 determined that this infant was a modern human who had died between 680 and 880 A.D. and whose grave had been dug into older layers of the site.

Amud 7, found the following year in 1992, had been buried even lower in the stratigraphic sequence of the cave than the male Neandertal skeleton of Amud 1 (see page 234) that had been discovered in 1961. Amud 7, therefore, dates without question to Neandertal times, specifically between 50,000 and 60,000 years ago. Based on the degree of tooth eruption, the infant died when it was just ten months old.

Although the collapsed cranium of Amud 7 includes only the occipital, parietal, and temporal bones, the lower jaw, or mandible, is relatively complete, and the ribs and vertebrae are well preserved. Found lying on its right side in a niche on the cave's north wall, the skeleton's articulation, especially that of the intact hand and foot bones, suggests that this was also an intentional burial. Intriguingly, a red deer's upper jaw leaned against what remained of the infant's pelvis. Two early modern human burials at the nearby cave sites of Qafzeh and Skhul also had associated animal bones (see Skhul V, page 258), but if the deer jaw had been placed beside the Amud infant, it would mark the first known occurrence in the Levant of grave goods associated with a Neandertal.

Analysis of the Amud 7 skeleton has revealed some previously unrecognized features that appear to be unique to Neandertals and should prove useful in identifying other fossil remains to this species. For instance, the foramen magnum, the hole in the base of the cranium through which the spinal cord passes, is extremely elongated and oval shaped in Amud 7, whereas modern humans retain the more primitive round shape (shared with most primates and many other mammals). A similarly elongated and oval foramen magnum can be seen in juvenile Neandertal crania from Engis, La Quina, and Teshik-Tash (see page 229).

Three more distinctive Neandertal traits occur in the mandible of Amud 7. First, there is no bony chin as in modern humans, so, when viewed from below, the front of the Amud baby's jaw appears squared off rather than pointed. Second, the U-shaped notch on top of the mandible's vertical ramus ends in the middle of the condyle, the knob that articulates the jaw with the cranium, and not on the side of the condyle as in all other hominids. Third, a prominent lipped tubercle or bump on the inner surface of the ramus marks the attachment of the medial pterygoid muscle. This muscle aids in chewing, and the uppermost muscle fibers appear to have been particularly well developed in Neandertals, as indicated by the rugose muscle marks on this side of the ramus that end at the medial pterygoid tubercle, which is absent in other early humans.

That such features appear even on the skull of a young infant indicates that they are genetically determined traits and reaffirms the many anatomical differences between Neandertals and modern humans. If babies can be so easily distinguished by the traits, it is additional evidence that Neandertals indeed constitute a separate species from modern humans.

Homo neanderthalensis, **Amud 7.** Remains of a ten-month-old infant excavated from Amud cave contain some features unique to Neandertals, including an oval foramen magnum at the base of the cranium. *Photograph by David L. Brill.*

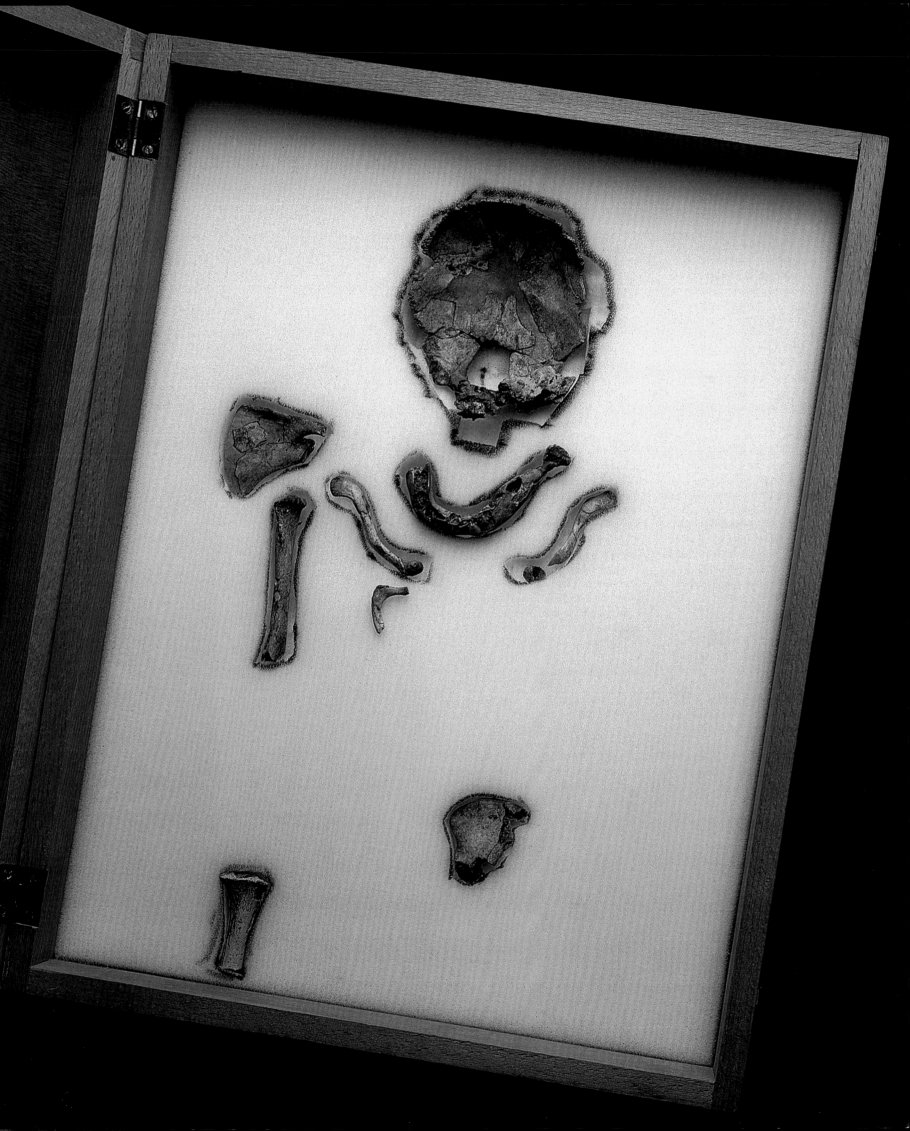

Homo neanderthalensis
LA CHAPELLE-AUX-SAINTS

The most complete Neandertal specimen known when it was found, the skeleton of a nearly toothless old man of La Chapelle-aux-Saints came to symbolize Neandertalness, but in the errant view of Neandertals as shuffling, brutish cavemen—a complete misrepresentation of their anatomy, gait, and intelligence. Marcellin Boule's in-depth study of this skeleton, published between 1909 and 1912, singlehandedly created this lingering and damaging image. Boule correctly noted that a large suite of skeletal features distinguished the Neandertals, but his unfortunate description resulted in Neandertals being considered more closely related to apes than to modern humans. By effectively making the La Chapelle-aux-Saints individual a reference specimen for its species, he left no place for Neandertals in our evolutionary lineage. This influential and persuasive study maligned Neandertals for decades and helped put the pre-*sapiens* hypothesis (see page 43) at center stage in the emerging debate over intellect and brain size in the hominid family tree.

Boule misinterpreted several pathological aspects of the skeleton and other features that reflect adaptation to a cold environment as being primitive and ape-like. Among the pathological parts are a deformed left hip, a crushed toe, severe arthritis in the neck vertebrae, a broken rib, and a damaged kneecap. So unwavering was Boule in his bias that he noted the immense brain capacity of the La Chapelle individual—1,625 cc, well in excess of the modern human mean—but nonetheless discounted Neandertals' mental aptitude in the face of so many other "primitive" traits.

Reevaluation of this skeleton by anatomists William Straus and A.J.E. Cave in the 1950s showed how the deforming arthritis had skewed Boule's view of Neandertal gait. Neandertals were suddenly embraced and became all too readily included as a subspecies of human. But this Neandertal renaissance has since given way to a growing consensus that this was a separate species, a successful and highly derived—not primitive—one that ultimately went extinct. There was not enough time for Neandertals to have evolved into modern humans in Europe; they could not have been our immediate ancestors.

More recently, the La Chapelle-aux-Saints specimen has figured prominently in a debate over whether Neandertals possessed the equivalent of modern spoken language. It was argued that because the cranium had an unusually flat base, the larynx would have sat too high in the throat of Neandertals for them to pronounce the vowels a, i, and u. The base of a modern human cranium, once past infancy, becomes flexed, which lowers the larynx in the throat and permits a broader range of sounds to pass from the pharynx. A more recent reconstruction of the cranium of the La Chapelle-aux-Saints individual resulted in greater base flexion than Boule allowed in his reconstruction, but it still appears to have a flatter cranial base than in modern humans. Exactly what conclusion can be drawn from this feature remains unclear, but perhaps it is time to refrain from reading further Neandertal generalities into the diseased and damaged specimen from La Chapelle-aux-Saints.

Whenever a new anatomical region of a Neandertal skeleton is targeted for study, it turns out to have distinctive anatomy. Computerized-tomography scans (CT scans) were used to examine the bony labyrinth located inside the petrous bone in the inner ear in modern humans and fossil hominids. The three semicircular canals are associated with balance and body movement, and therefore presumably with locomotion. Fred Spoor and colleagues were able to distinguish a human (bipedal) and ape (quadrupedal) configuration. In early hominids *Australopithecus* the configuration was apelike and in early *Homo*, modernlike.

In mammals and birds the size of the canals relates to the degree of agility. Those creatures more agile have larger canals. When the inner ear was scanned in some 15 Neandertals, including La Chapelle-aux-Saints, the canals were much smaller than those of modern humans, but also of more ancient ancestors. The conclusion is that Neandertals were presumably less agile than other hominids, perhaps a reflection of the fact that Neandertal postcranial bones suggest a short, stocky body-type that was built more for power rather than agility.

Homo neanderthalensis, **La Chapelle-aux-Saints.** The aged, pathological skeleton from La Chapelle formed the basis of a pervasive, but errant, view of Neandertals as shuffling, stupid brutes. The resorption of tooth sockets in both jaws, arthritis, and broken bones are just a few of the ailments that afflicted this individual. Actual size. *Photograph by John Reader, Science Source/Photo Researchers.*

Homo neanderthalensis
LA FERRASSIE 1

The adult male skeleton catalogued as La Ferrassie 1 best exhibits the "classic" Neandertal anatomy that evolved in glacier-covered western Europe midway between this unusual species's appearance and extinction. All the defining Neandertal features appear in the cranium of La Ferrassie 1: the receding forehead, the long, low vault of the braincase, the globular shape when viewed from behind, the prominent, double-arched browridge, the projecting midface and backward-swept cheekbones, the weakly developed chin, the heavily worn front teeth, the retromolar gap behind the tooth row (resulting from the teeth having been pulled forward with the face), and the huge brain capacity (in this case more than 1,600 cc), among other traits. The limbs show the Neandertal tendency for stout, thick bones with large joints.

Classic Neandertals constitute the clearest example in human evolution of a distinctive and unique set of anatomical features characterizing a population. Although seemingly trivial when considered individually, these shared features compose a consistent pattern that helps to define the Neandertals as an evolutionary unit—a distinct species from modern humans and other hominids. Some of the features visible in La Ferrassie 1 include, in lateral view, the mastoid tuberosity behind the ear opening and above the knoblike mastoid process at the base of the skull, and the occipitomastoid crest immediately behind the mastoid process; and, in rear view, an occipital torus, a horizontal bar of bone, on the cranium, and just above this torus an elliptical depression called the suprainiac fossa.

Of all the Neandertal traits present in this specimen, the teeth tell a particularly interesting story. In addition to abscesses and bone recession in the jaws, the teeth of La Ferrassie 1 are dramatically worn down, to the point that many of the enamel crowns have disappeared and dentine or pulp cavities are exposed, especially on the lower left teeth. In the upper jaw, the teeth on the right side are most severely worn. Grooves along the back of the teeth

indicate that the lower jaw moved in a predominantly horizontal chewing motion rather than a vertical one, and the teeth were required to grind coarse, gritty foods.

Much of the damage to the front teeth, however, may stem from their use as extra hands. A comparison of the incisors of La Ferrassie 1 with teeth from gorillas (strict herbivores with a tough, fibrous diet) and Eskimos (who eat the least vegetation of any human population or nonhuman primate) for microscopic surface damage showed that tooth wear in the Neandertal closely resembled tooth wear in Eskimos. La Ferrassie's incisors have rounded edges with superficial flakes and gouges and deeper pits down to the dentine, as well as several fine, linear scratches. Eskimos use their teeth as tools, specifically to hold everything from harness lines to sealskin boots in a tight, viselike grip. Apparently Neandertals also used their teeth to clamp down on bulky objects, and to an even greater extent than modern Eskimos.

In addition to the nearly complete skeleton of a middle-aged male, the La Ferrassie rockshelter has yielded a largely intact adult female skeleton and the remains of five children ranging in age from prenatal to ten years: in composite, the largest set of juvenile human fossils from any site in France. The skeletons occupied six graves. The adult male and female were found lying head to head, and the skull from one child's skeleton had been removed and buried separately beneath a stone slab marked with curious depressions. Doubts had persisted about the reported Neandertal burial from La Chapelle-aux-Saints (see page 238), found a few years earlier, but the collective evidence from La Ferrassie confirmed that Neandertals buried their dead.

Soon after their discovery, Marcellin Boule used the La Ferrassie 1 and 2 skeletons to study missing bones for his infamous analysis of the La Chapelle-aux-Saints specimen. No one thoroughly studied the collection of La Ferrassie fossils until Jean-Louis Heim completed the first detailed descriptions, which were published as a monograph in 1976. Thanks to Heim's

work and subsequent studies on this group of specimens, we can assemble a few facts about Neandertal life.

A half-size model of the La Ferrassie Neandertal suggested that this 2.71 meter tall male, who weighed some 85 kilograms, possessed some 2.1 square meters of skin. Employing equations developed for modern humans to estimate the basal metabolic rate of an individual, calculations for the La Ferrassie Neandertal estimated a daily caloric requirement of nearly 5,000 kilocalories. This stands in contrast to Inuit Eskimos who requires some 3,000 to 4,000 kilocalories per day.

Analyses of the isotopes found in Neandertals points conclusively to a diet that consisted almost entirely of meat. Such a dietary specialization, requiring a high level of kilocalories suggests that Neandertals would have consumed some two kilos of meat per day. Such a diet it is assumed would have demanded a high level of oxygen uptake, perhaps explaining the enlarged chests and presumably large lungs of Neandertals.

Such a specialized diet requiring significant daily caloric intake may have made Neandertals prone to periods of starvation as is indicated in the preponderance of growth defects evident on Neandertal teeth that are called hypoplasia. Furthermore, when behaviorally modern humans entered Europe around 40,000 years ago with a very sophisticated hunting strategy, Neandertals may have not been able to sustain such high daily caloric intake, making them even more vulnerable to starvation and perhaps ultimately to extinction.

Homo neanderthalensis, **La Ferrassie 1** (see also page 9)**.** Extreme, sloping wear on this male specimen's front teeth, some with dentine exposed, reflect a horizontal chewing motion and the use of teeth as an extra pair of hands to grip animal skins or other objects like a vise. Actual size. *Photograph by David L. Brill; courtesy Musée de l'Homme.*

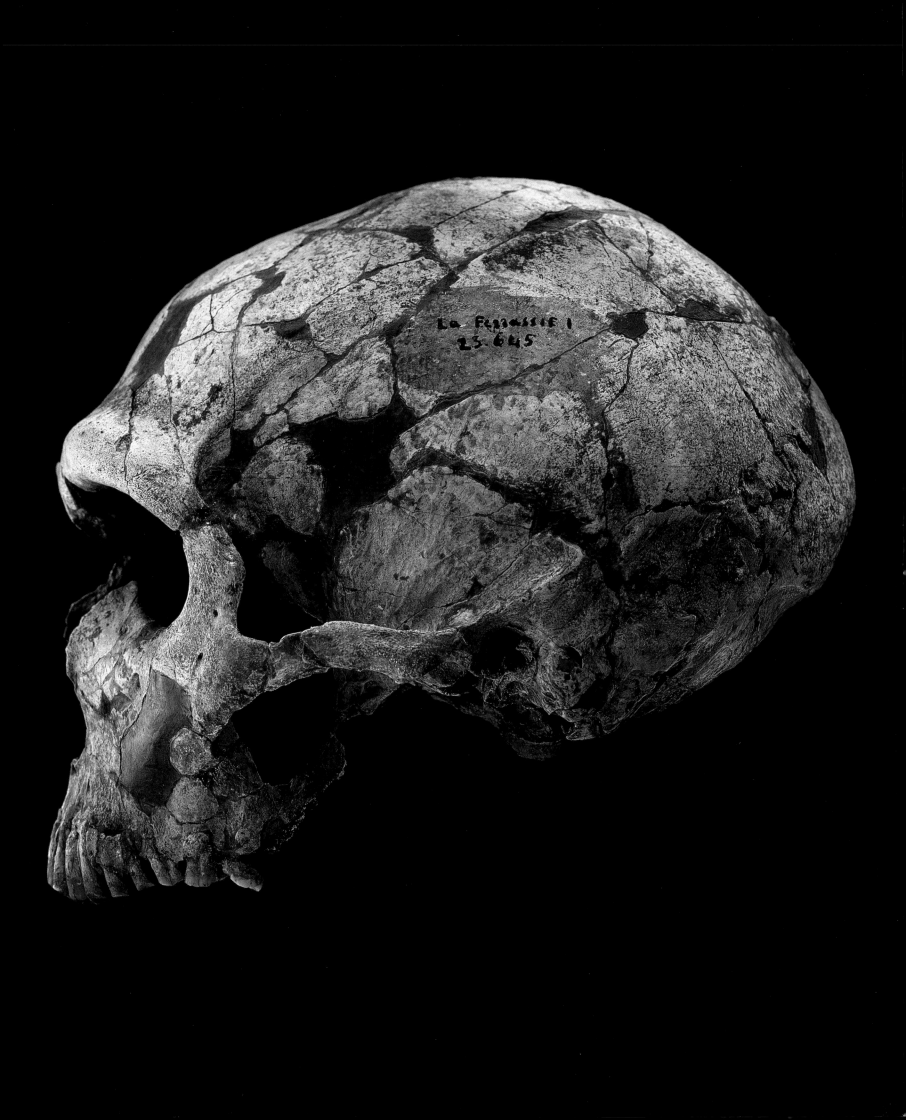

Homo neanderthalensis

NEANDERTAL 1 TYPE SPECIMEN

The discovery and study of a partial adult hominid skeleton from the Neander Valley in Germany marked the start of paleoanthropology as a science and the beginning of efforts to understand our evolutionary past. Although a few other Neandertal specimens, such as those from Engis and Gibraltar (see page 244), had been found before limestone quarry workers uncovered this skeleton in 1856, the Neandertal 1 individual was the first to be recognized as something truly different. It was so different, in fact, that it became the first fossil hominid species to be named, when, in 1863, geologist William King proposed *Homo neanderthalensis* at a meeting of the British Association. King even stated in print that Neandertals might deserve their own genus. No one thinks so today, but the appellation has stood the tests of time and subsequent evidence. Today paleoanthropologists differ over whether *neanderthalensis* should be treated as a separate species or merely as a subspecies of *Homo sapiens*.

The name Neandertal comes from the valley near Düsseldorf where these bones were found. The valley had been named in memory of a 17th-century composer and vicar, Joachim Neumann, who took the adopted name Neander, meaning "new man." Thal (with a silent h) means "valley" in German, but the spelling changed early in the twentieth century to tal. Neandertal is still spelled both with and without the h. Since the International Code of Zoological Nomenclature governs the naming of a species, the "h" must remain in the formal species designation *Homo neanderthalensis*.

When the workmen's shovels struck bone at Feldhofer, they uncovered first the top of a cranium. Then they found the two femora, the three right arm bones, two of the left arm bones, part of the left ilium, and fragments of a shoulder blade and ribs. Perhaps more of the skeleton lay in the cave mud, but only the largest bones were collected and saved for a local teacher and amateur naturalist, Johann Carl Fuhlrott. It is sobering to think that the opportunity to recover a complete Neandertal skeleton may have been missed, but the workers thought the bones came from a cave bear. Fuhlrott, however, suspected

that the Feldhofer fossils represented unique pieces of the human past. He left their description to anatomist Hermann Schaaffhausen, and their joint announcement appeared in 1857, two years before Charles Darwin published his revolutionary work, *On the Origin of Species*.

Schaaffhausen noted the thick, well-developed muscle markings and ridges on the Neandertal bones. Even more striking was the oval shape of the skullcap, with its low, receding forehead and prominent, curved browridge. Because this appearance was unlike that of any human skull, the Neandertal cranium was labeled "apelike."

In 1997 scientists, using archival records, located the pile of clay deposits that had been removed from the Feldhofer Grotte where the Neandertal remains were originally interred. Careful excavation of these deposits yielded Pleistocene faunal remains, Middle and Upper Paleolithic artifacts and 62 specimens of human bones. Several of the newly found fragments were fitted onto the original Neandertal 1 partial skeleton. For example a fragment of temporal bone and a cheekbone fit onto the skullcap and a chunk of leg bone fit perfectly onto the distal end of the femur. The additional human bones suggest the presence of another adult Neandertal and a subadult individual that may or may not be Neandertal.

Renewed interest in the Feldhofer Neandertal led to the application of radiocarbon dating for the specimen. Three samples were taken from two of the newly discovered bones and one from the original 1856 partial skeleton. The three dates clustered very tightly around 40,000 years, providing a more precise age for the Feldhofer Neandertals.

A small sample of bone was taken from the right humerus of the second adult Neandertal from Feldhofer with the goal of sequencing the ancient DNA. Successful sequencing of the mtDNA of Neandertal 2 shows some minor nucleotide differences with the mtDNA sequence initially derived from Neandertal 1. Since the mtDNA is inherited only along the maternal line, this suggests that the two Feldhofer Neandertals did not share the same "mother".

Overall the mtDNA sequence derived from Neandertal 2 clusters with those already known from Neandertal 1, as well as specimens from Vindija in Croatia and Mezmaiskaya in Russia, all of which are distinct from modern human mtDNA sequences.

If Neandertal 1 signaled the birth of paleoanthropology, it also sparked the sort of intense controversy that has come to characterize this science. Prominent German scientists attributed the Neandertal's bowed thigh bones to habitual horse-riding and a case of rickets. The browridges, it was thought, resulted from prolonged frowning in pain from a poorly healed fracture of the left ulna. Such spurious arguments framed Neandertal 1 as a diseased modern human, perhaps a Cossack cavalryman, rather than some hitherto unknown primitive species. But reason prevailed, and although the validity of Neandertal Man was challenged in the country of his discovery, acceptance was more forthcoming in England. Thomas Henry Huxley commented on the Neandertal bones in his 1863 essays on man's place in nature, but it was King in the same year that dared go out on a limb and give these fossils the identity that we still recognize today.

The discovery of Neandertals sparked the longest lasting controversy in paleoanthropology: Where did they fit on the human family tree and were they ancestral to modern humans? Recognized as "peculiar" looking by scientists in the 19th Century many, including T.H. Huxley, saw them as being evolutionarily close to modern humans. Other thought the distinctiveness of Neandertal anatomy justified their being placed in a separate species from us. Those who wanted a compromise placed them in *Homo sapiens neanderthalensis* and modern humans in *Homo sapiens sapiens*, recognizing their closeness but also providing them with a subspecies distinction.

In the 1960s two distinctly unique and quite different schools of thought emerged. One school stressed continuity and concluded that there was a Neandertal phase in our evolutionary past and modern humans evolved from these ancestors. A contrary school emphasized the idea of replacement. This school contended that Neandertals and we were different species,

and when *H. sapiens* moved into Europe some 40,000 years ago and encountered the Neandertals, they eventually triumphed and Neandertals went extinct and did not contribute genes to modern humans.

Neandertals appear to be distinct from us anatomically, behaviorally, and genetically. They may represent the descendents of more ancient, later Middle Pleistocene European hominid populations (*H. heidelbergensis* or *H. antecessor*) that became isolated in glacial and peri-glacial environs. According to the replacement scenario, Neandertals evolved separately, in a classic Ernst Mayr speciation event, and developed their characteristic anatomy in response to living in a cold climate. This would include large brains which reflected increased blood supply to keep the brain warm, enlarged nasal openings and maxillary sinuses to moisten and warm incoming air, and short, squat bodies that performed as heat conservers with a small surface areas to body mass ratio.

With the appearance of much more sophisticated Upper Paleolithic hunters, *H. sapiens*, moving into Europe, Neandertals were out competed for food and/or territory. It may also be that the end of the ice age also had a negative impact on the Neandertals. Specific details of the disappearance of Neandertals may never be known, but the increasingly widespread view among anthropologists is that they went extinct and have left no genes in modern humans. A number of eminent scholars, however, still champion the view that Neandertals were indeed in our direct ancestry.

Homo neanderthalensis, **Neandertal 1.** Although it was not the first Neandertal fossil found, discovery of this skullcap in 1856 eventually led to the recognition of Neandertals as an ancient, extinct relative and began the ongoing debate about their place in our ancestry. Actual size. *Courtesy of Rheinisches Landesmuseum Bonn.*

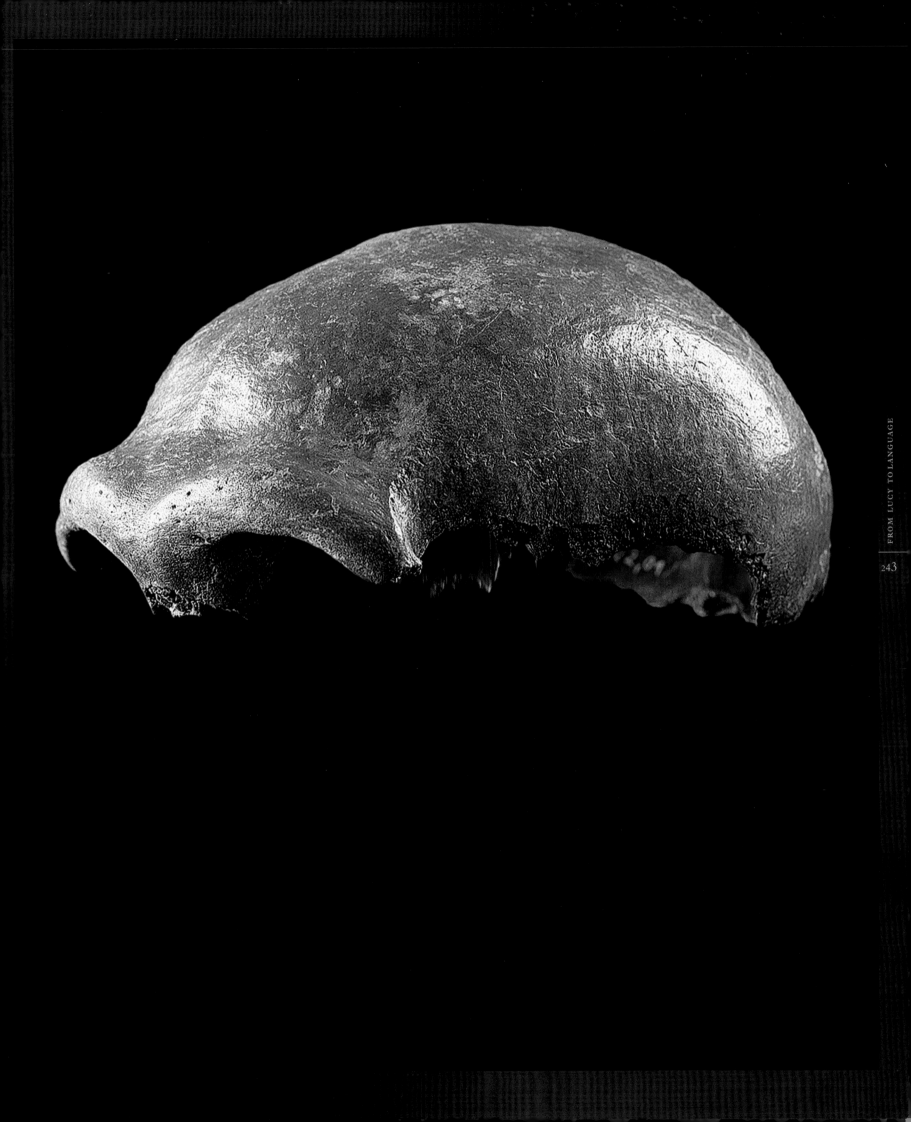

Homo neanderthalensis
GIBRALTAR 1

The individual represented by a female cranium found on Gibraltar in 1848 could be called the forgotten Neandertal. The impressive and remarkably complete cranium was found eight years before the discovery of the skullcap and limb bones from Feldhofer, Germany (see page 242), which gave us the name Neandertal. Because no one knew what to make of it at the time, the Gibraltar specimen languished for 16 years before its significance was recognized. Even then, it received little attention until 1907, when it was finally described in detail, but by a geologist rather than an anatomist.

The exact location and circumstances of its discovery are unclear. A Captain Brome, who ran the military prison on Gibraltar, was an amateur fossil collector and employed prisoners to explore the local caves (a transgression that cost Brome his position). Apparently this cranium turned up in the course of construction on the Mediterranean island sometime before March 3, 1848, when brief mention of the specimen appeared in the minutes of the Gibraltar Scientific Society. There were no accompanying stratigraphic, archaeological, or faunal data that might help determine its geological age.

Fortunately, in 1863, the cranium was sent to George Busk, the London zoologist who in 1861 had translated Hermann Schaaffhausen and Johann Fuhlrott's initial description of the Feldhofer Neandertal finds into English. Despite the stony matrix that covered the cranium, Busk recognized the Gibraltar specimen as a Neandertal. Most notably Busk was the first to draw attention to two of the most distinctive features of the Neandertal face: midface prognathism and expanded maxillary sinuses. He collaborated with paleontologist Hugh Falconer in presenting the find to the British Association in 1869. That same year, French anatomist Paul Broca examined the Gibraltar specimen and pronounced it a Neandertal. Falconer described Gibraltar as "a very low type of humanity—very low and savage, and of extreme antiquity—but still man" and enthusiastically suggested several names for the specimen, including *Homo calpicus*, from Calfe, an ancient name for Gibraltar.

Gibraltar 1 was not the first Neandertal specimen ever found. That distinction goes to the child's skull from Engis Cave, Belgium, found in 1829 or 1830, the significance of which also went unrecognized, in part because influential French paleontologist Georges Cuvier dismissed the antiquity of Engis. Gibraltar 1 was, however, the first complete adult cranium. At the rear and base of the cranium, Gibraltar 1 displays such Neandertal traits as a projecting occipital bone with a suprainiac fossa and a distinctive crest behind the round mastoid process of the temporal bone. This specimen also had the face that the Feldhofer specimen lacked, and many distinctive Neandertal features occur in the face. It is remarkable that, without a face to study, William King recognized the Feldhofer fossils as different enough to warrant the species name neanderthalensis. Had King waited to publish his description, he might have been made aware of Busk's possession and thus saved this female Neandertal from enduring obscurity.

Homo neanderthalensis, **Gibraltar 1.** The first complete adult Neandertal cranium found, this female specimen has the large projecting face and receding cheekbones that distinguishes this species. Actual size. *Photograph by John Reader, Science Source/Photo Researchers.*

SPECIMEN	LOCALITY	AGE	DISCOVERERS	DATE	PUBLICATION
Partial adult skull	Pierrot's Rock, Charente-Maritime, France	ca. 36,000 years	François Lévêque	July 27, 1979	Lévêque, F., and B. Vandermeersch. 1980. Les déscouvertes de restes humains dans un horizon castelperronien de Saint-Césaire (Charente-Maritime). *Bull. Soc. prehist. francaise* 77: 35

Homo neanderthalensis
SAINT-CÉSAIRE

Once the sole contender for "last Neandertal," this specimen now must relinquish that distinction to the Neandertal lower jaw recovered from Zafarraya, Spain, in the early 1980s but recently dated to 30,000 years ago and to a 34,000 year-old Neandertal temporal bone from Arcy-sur-Cure, France. The Saint-Césaire skeleton nonetheless holds special significance in the debate surrounding the ultimate fate of Neandertals.

Excavation of the rockshelter began in 1976 after stone tools turned up in the course of widening the road for a mushroom-growing operation in the limestone caves of Pierrot's Rock. Many mammal bones and some stone tools from both Middle and Upper Paleolithic technologies were uncovered, along with the Neandertal.

The skeleton was found flexed into a small oval burial posture. The remains consist of the right half of a skull, some ribs, a shoulder blade, two robust arm bones, and fragments of the kneecap and shin bones. The skull reveals typical Neandertal traits, including the absence of a fossa, or shallow hollowing, above the upper canine, a gap in the tooth row behind the third molar (called a retromolar space), and the lack of a chin on the lower jaw.

Although the age of the site and the specimen had initially been estimated by comparisons of the fauna and archaeology with other sites of known age, Saint-Césaire generated excitement in 1991 after burnt flints associated with the skeleton had been dated by the thermoluminescence technique to 36,300 years, plus or minus 2,700 years. The implication of this date and the even younger ones from Zafarraya and Arcy-sur-Cure is that Neandertals and *Homo sapiens*, the Cro-Magnons, overlapped in western Europe for up to 10,000 years. The nature of their co-existence—whether it was peaceful or violent, intimate or distant—has been the subject of much dispute and speculation. Perhaps any interactions played out differently in different regions.

Although there may have been ample time for cultural convergence, this overlap appears too brief for Neandertals to have evolved the modern morphology of Cro-Magnons. It has been suggested that the reduced amount of midface projection in the Saint-Césaire individual as compared to earlier western European Neandertals could indicate hybridization with *Homo sapiens*. Despite the difficulty of recognizing species—not to mention hybrids—from fossils alone, judging from their clearly disparate anatomies these groups were too biologically distinct to have shared anything more than culture. There are differing views on the question of whether Neandertals and modern humans would have exchanged genes. If we accept them as separate species, *H. neanderthalensis* and *H. sapiens*, then following Mayr's biological definition of a species—"a group of interbreeding natural populations that are reproductively isolated from other such group"—we must conclude there was no gene exchange. There are some anthropologists, however, who accept a species distinction but contend that there was some, but very limited gene exchange.

The 1998 uncovering of a 24,500 year-old burial of a four year old child at Lagar Velho, Portugal was announced as a hybrid between *H. sapiens* and *H. neanderthalensis*. The specimen, unfortunately damaged by road-working equipment, possessed what the discoverers described as an intriguing amalgam of morphological features of both modern humans and Neandertals. The modern features include a chin and some details of the teeth. Neandertal affinities are seen in the reduced length of the tibia, relative to the femur. Serious reservations about the possibility of this the Lagar Velho child being a modern/Neandertal hybrid have surfaced. Foremost among these is the observation that the fossil evidence in Europe suggests that by around 30,000 years ago Neandertals had become extinct and it is extremely unlikely that Neandertal features could have been maintained in modern humans for more than 5,000 years during which time they would have been swamped by modern human genes. Perhaps an even more potent objection is that when two species are artificially bred, such as a lion and tiger, the hybrid is a mixture of the two cats, not an offspring with the head of a lion and body of a tiger. It seems more reasonable that the Lagar Velho child is a *H. sapiens*, evincing some cold adapted anatomy as seen in the relatively short tibia.

The Saint-Césaire cave provides intriguing evidence about the late culture of Neandertals. This is the only site where identifiable human remains have been found in association with distinctive Châtelperronian tools, including points and backed blades. The Châtelperronian industry found in France and Spain displays features of both the preceding Mousterian technology of the Middle Paleolithic, generally associated with Neandertals, and the later Aurignacian industry, usually viewed as the first stage of Upper Paleolithic technology and associated with modern *Homo sapiens*. Before the discovery of the Saint-Césaire individual, Châtelperronian tools had often been attributed to the hands and minds of modern *Homo sapiens*, but apparently such tools were the product of Neandertal industriousness.

One interpretation of this distinctive industry sees it as the Neandertals' attempt to mimic and master the newfangled technology that arrived in Europe with modern human immigrants. It remains unclear if the development of the Châtelperronian industry occurred in an atmosphere of cooperation or competition, but Saint-Césaire captures part of the dynamic physical and cultural transition among human populations in Europe between 30,000 and 40,000 years ago.

Homo neanderthalensis, **Saint-Césaire.**
Note the absence of a depression in the cheekbone above the canine, indicating that this skull had the Neandertal trait of inflated cheekbones. Other Neandertal traits are clearly visible in a lateral view (see page 109) of the mandible, including a retromolar gap by the ascending ramus and the absence of a chin. Actual size. *Photograph by David L. Brill; courtesy of Université de Bordeaux I.*

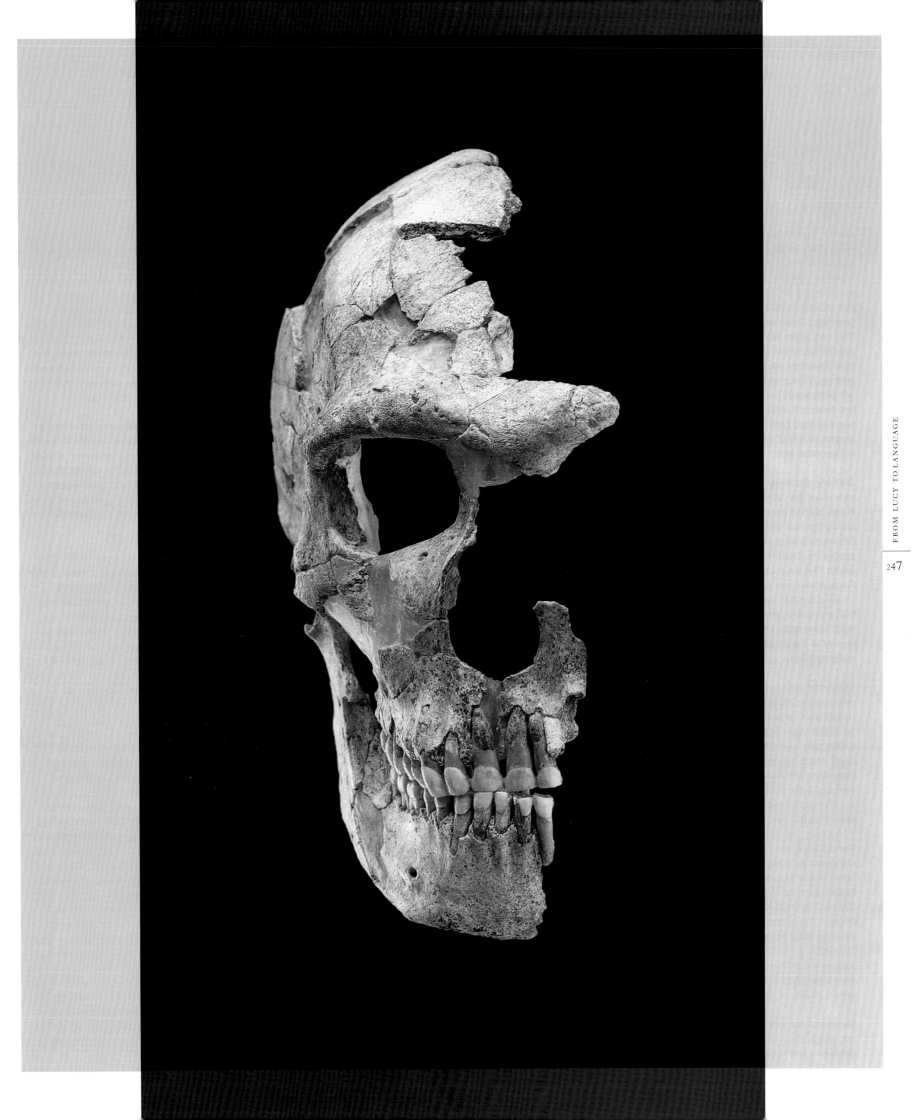

Homo floresiensis

LB 1 TYPE SPECIMEN

When is fact stranger than fiction? Answer: When an excavation turns up a three and half foot tall fossil hominid with a brain the size of a softball, who seemingly made fairly sophisticated stone tools, hunted and cooked, but lived only 18,000 years ago. These are the facts strange as they may seem, but when this completely unanticipated and unpredicted discovery from Indonesia was announced it stunned anthropologists and stimulated whole new scenarios about our ancestry. After all, the last time such little hominids with small brains walked Earth they were australopithecines that lived more than three million years ago.

Until the discovery of these little people, nicknamed the "Hobbits," on the island of Flores, we had been pretty secure in our knowledge that since the demise of the Neandertals some 30,000 years ago our species has been the lone human on the planet. Not only was the anatomy of the Flores hominid perplexing, but now the observation that here was another human that co-existed with us after we had already penetrated into nearly all habitable regions of the world cried out for deep contemplation.

The partial skeleton of a petite hominid excavated in 2003 by an Australian/Indonesian team came from sediments dated to 18,000 years by carbon-14. Found in a limestone cave called Liang Bua (Cool Cave), the specimen, LB1, consists of a fairly complete skull, major portions of the legs, parts of hands and feet, as well as other postcranial bone fragments, and most importantly the left half of a pelvis.

The stature of the LB1 specimen was calculated by using the length of the femur in formulas developed from modern human pygmies. With a femur length of 280 mm, interestingly identical to that of Lucy's femur, the Flores individual apparently stood slightly over a meter tall (106 cm; 3.5 feet). An adult status for the specimen is attested to by the presence of permanent third molars and fusion of the growth plates on the long bones.

Aspects of the pelvis, such as a wide sciatic notch, imply that LB1 was a female. Detailed comparisons of the left pelvic bone (the innominate) with a modern human female pelvis reveal strikingly similar anatomy. Hence, there is no question that the Flores specimen was bipedal, a conclusion well supported by details of femoral anatomy.

The most astonishing observation, however, about the find is that the cranial capacity of LB1 is about 417 cc! This is dramatically below the modern human mean of roughly 1,400 cc. and within the range of variation for chimpanzees and early hominids like *Australopithecus afarensis*. For the moment the cranial capacity for the Flores specimen is among the smallest recorded for a hominid (with the exception of the *S. tchadensis* cranium; see page 116) and well below the brain size of any specimen previously admitted into the genus *Homo*.

A computer-generated model of the skull of LB1 has provided insight into the architecture of the brain. In spite of having such a small size, aspects such as enlarged temporal and frontal lobes as well as the presence of deep, complex convolutions on the surface of the brain (inferred from impressions left on the internal surface of the brain case), all point to a brain that is reorganized in the direction of modern humans. The brain, however, is not identical in its proportions to either a modern human or *Homo erectus*, but seems to have its own unique configuration.

The dentition is small, within the range of *Homo erectus* and modern humans, and much reduced compared to that of earlier hominids, especially *Australopithecus*. The face is fairly lightly built and lacks the heavy buttressing and prognathism of the significantly older *Australopithecus*. The mandible bears no chin and is attached to a long, low cranium that is extremely small compared with other *Homo* species, though it resembles *H. erectus* in various cranial indices. In rear view the contour of the brain case, being widest very low down and with a reduced cranial height, also resembles *H. erectus*. In the place of the characteristic bar-like brow ridges of *H. erectus*, the Flores cranium has much smaller swollen arches over each orbit.

Excavation recovered stone artifacts and a variety of animal remains including Komodo dragons and dwarf elephants (Stegodon), as well as a rat the size of a rabbit. The tools range from simple flakes to points, blades and perforators, and even microblades. It appears that the hominids at Liang Bua were technologically capable in spite of possessing amazingly small brains. While relatively few stone artifacts were found directly associated with the skeleton, thousands were found in the cave at stratigraphic levels apparently equivalent to LB1. Interestingly, 17 individuals of juvenile dwarf elephants were excavated, some in association with tools, insinuating that perhaps the early inhabitants of the cave hunted and consumed these creatures.

The taxonomic conundrum offered by the Liang Bua hominid prompted the discoverers to announce their find as a new species, *Homo floresiensis*. This species with a mix of advanced and seemingly primitive traits is unique in human evolution, totally unexpected by anthropologists, and has major implications for our understanding of relatively recent human prehistory. Furthermore, geological evidence suggests that *H. floresiensis* existed in Indonesia between 74,000 roughly 13,000 years ago. This is remarkable since *H. sapiens* had spread to all continents prior 13,000 years ago, yet here was an isolated enclave of tiny hominids living on Flores.

A minority of anthropologists suggested that the Flores skull was really a microcephalic abnormality, perhaps a pituitary dwarf. On the contrary, the proportions of the skull and brain show that this was no medical oddity, but a member of a very short hominid species. With likenesses to *H. erectus* some have asserted that LB1 is a dwarfed version of that species, while others have drawn attention to resemblances between LB1 and *Homo habilis* (especially KNM-ER 1813; see page 186)) from eastern Africa.

All indications are that the Flores hominids were very small people indeed and that being the case an explanation must be sought for the significantly reduced body size, something we see in modern Pygmies, for example. The reason for dwarfing might be sough in a process termed "insular dwarfing" whereby there is strong natural selection for smaller bodied animals in isolated environments (as on a small island) with few predators and intense competition for limited food resources. A smaller body size would certainly reduce the caloric needs of an organism such as these early hominids, thus giving smaller body size a selective advantage over a larger one.

Island dwarfism has been identified in the paleontological record in places like Sicily where over a brief period of some 5,000 years, four meter tall elephants were transformed into dwarf elephants one-fourth that height. Perhaps this was the process by which the Flores hominids became small.

If the controversial evidence for stone tools on Flores dating to approximately 840,000 years ago is supported, possibly the people who made these tools were larger bodied *H. erectus*. In this case, then natural selection would have had a long stretch of time to craft a small-bodied *H. floresiensis* hominid, leading some authorities to subsume the Flores fossils under the title *Homo erectus*, miniature variety.

While short stature may have found an explanation, the diminutive brain of LB1 still requires elucidation. One of the major trends characterizing human evolution has been that of brain enlargement, encephalization. But in Flores we apparently see an incredibly small-brained creature capable of modern human behavior. Why such a small brain? Pygmies may be short, but they have the same brain size as non-Pygmy humans. The brain is known to be the hungriest organ in the body, and maybe natural selection favored smaller, less needy brains as a way to further economize on energy requirements. If this is the case then the general correlation of brain size and "intelligence" must be revisited, giving more importance to how the brain is organized and wired because getting smaller brained and brighter at the same time is the reverse of what has always been believed.

For those interested in encountering an alien from another world, traveling to Flores just 20,000 years ago would have been an incredible opportunity. On one hand, perhaps there is some truth to local folklore of an elfin-like, bipedal creature with an insatiable appetite and an in-comprehensible language that, according to some, still lives on Flores. Known as ebu gogo, "the grandmother who eats everything," these diminutive people may have actually survived until about 13,000 years ago when cataclysmic volcanic erup-tions, and perhaps the peopling of the island by *Homo sapiens*, brought about their extinction as it did to other animals like the dwarf elephants. But on the other hand, maybe the Flores people were simply diminutive pathological modern humans...more discoveries will resolve this interesting debate.

Homo floresiensis, **LB1.** This presumed female skull is associated with a partial skeleton found on the island of Flores, Indonesia. Although only 18,000 years old, it has a cranial capacity of only 417 cc. Some inventigators believe the specimen is a pathological *H. sapiens*. Actual size. *Photograph courtesy of Peter Brown.*

Homo sapiens [archaic]
DALI

Outside of Africa, the adult male cranium found at Dali, in China, is the best candidate for ancestor to modern humans. This specimen plays a crucial part in the multiregional model for the origin of *Homo sapiens*. According to this model (see page 46), the Dali cranium and an older Middle Pleistocene cranium and skeleton from Jinniushan in China's Liaoning Province constitute a link between the earlier *Homo erectus* fossils of Zhoukoudian (see page 202) and modern Chinese and illustrate longterm regional continuity in evolution.

In most cranial measurements, Dali falls between *H. erectus* and modern *H. sapiens*. Like *H. erectus*, Dali has a low, long, thick-walled cranium with a large and very thick browridge over the eyes. The rugged nuchal region at the rear of the cranium has a prominent torus, and on the top of the cranium is a slight sagittal keel.

Dali differs from *H. erectus* in being broadest at a higher point on the cranium and having less constriction behind the eye sockets. The estimated cranial capacity of the Dali specimen, 1,120 cc, is somewhat greater than that for the Zhoukoudian crania. The face presages the small, flat faces, and prominent cheekbones of modern Chinese. Unfortunately, the absence of front teeth precludes checking for shovel-shaped incisors, a common trait in Asian populations today. Its forward-facing cheekbones gave Dali a flatter face than many archaic humans from Europe (such as *Homo heidelbergensis*) or Africa, and certainly much flatter than a Neandertal face. Although the face is relatively squat, deformation of the lower face during burial makes it difficult to measure the facial height accurately. In life, this face was longer and more projecting.

Despite these progressive features, a straightforward ancestral scenario with the Dali specimen a connecting link between earlier and later Chinese hominids cannot be proved. In some respects, it is just a typical Middle Pleistocene face, with resemblances to the modern humans of Qafzeh (see page 255). Analyses of skull shape reveal distinct differences between Dali and both more recent modern humans in China (including the Upper Cave crania

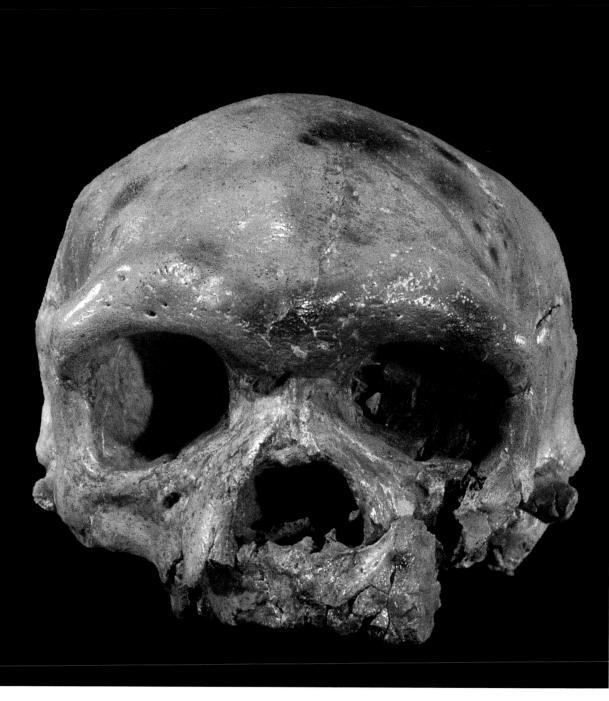

Homo sapiens, **Dali.** A short, flat face surmounted by a massive browridge reveals Dali's blend of traits found in both *Homo erectus* and *Homo sapiens*. In lateral view *(opposite)*, the low cranial vault and torus at the rear recall erectus, but the cranial capacity exceeds that of Peking Man. Actual sizes. *Courtesy of Wu Xinzhi, Chinese Academy of Sciences.*

from Zhoukoudian) and contemporary Chinese populations. In fact, in skull shape the Dali specimen more closely resembles modern Europeans than Asians and shares similarities with late Middle Pleistocene fossils from Africa (including those from Omo, see page 252). Rather than prolonged regional evolution in China, the shape of the Dali skull could reflect the spread of *Homo heidelbergensis* into Asia from Europe or Africa. If its age, estimated by the uranium series technique applied to ox teeth as 209,000 years plus or minus 23,000 years, is accurate, then Dali probably preceded the emergence from Africa of modern *Homo sapiens*.

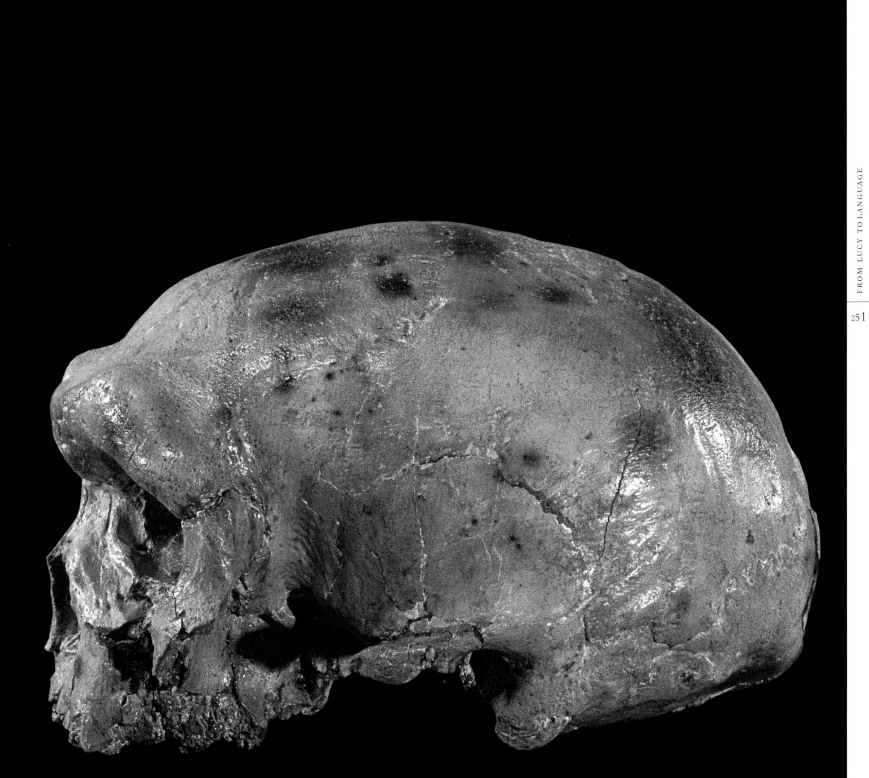

Homo sapiens
OMO I AND OMO II

In 1967 paleoanthropological field research was just beginning in Ethiopia when three research groups, a Kenyan team, under the direction of Richard Leakey, a French team led by Yves Coppens and Camille Arambourg, and an American team directed by F. Clark Howell began explorations in the lower Omo River valley in the southwest corner of Ethiopia, just north of Lake Turkana (Lake Rudolf at that time). A friendly rivalry ensued and every day the teams would call each other by radio to compare progress. One night Leakey's team reported the discovery of two skulls and parts of a skeleton from an area called Kibish.

Catalogued as Omo I, the first specimen was undeniably modern in its anatomy and it presumed great antiquity quickly made it a critical piece of evidence for those who argue that *Homo sapiens* evolved first, and relatively, recently, in Africa before spreading into the rest of the world. Geological dating proved to be quite a challenge because it falls outside the range of time adequately addressed by the radiocarbon method. A less precise method, Uranium-Thorium dating, was applied to associated oyster shells which yielded an equivocal date of approximately 130,000 years.

In addition to the skull, the associated postcranial bones from a shoulder, arm, hand, ribs, spine, legs, and foot display fully modern human anatomy. A small number of stone tools and broken animal bones, as well as a complete buffalo skeleton, were also found nearby.

The obvious modern traits of the Omo I skull include the long and curved parietal bones of the expanded braincase, where the cranium reaches its maximum breadth, coupled with a short, broad face and high forehead. Its prominent browridge tapers at the sides instead of forming a consistently thick bar as in more archaic humans. The facial bones are fragmentary, but when the upper jaw is pieced together it reveals a modern-looking, U-shaped palate. The lower jaw possesses a chin, and the pair of worn teeth that survived appear modern in size and shape. Little of the cranial base was

preserved, but a fortuitous break in the petrous region of the temporal bone—the hardest bone in the human skull, located internally from the ear canal—allowed a cast to form of the cochlea and semicircular canals of the inner ear.

Another specimen of presumably comparable age, Omo II, is a faceless cranium that displays some markedly different, more archaic anatomy. More heavily built than Omo I, this cranium has rugged muscle markings, a recessed forehead, and a conspicuous occipital torus, or bar, across the back. It shares, however, with Omo I the modern traits of long, arched parietal bones along the sides of the braincase and of being broader at the top than the base. The estimated brain size for Omo II is 1,435 cc. Omo I, although more difficult to measure because of its incomplete preservation, had a brain at least as large, and well within the modern human range. But the striking morphological differences between Omo I and Omo II raise the question of whether either specimen was artificially intruded into the layer from which they were unearthed. They may not sample a single, contemporaneous population. The third human specimen found at Omo consists of a few cranial fragments that are closer in form to Omo I.

In the late 1990s and early 2000s a group of researchers, including archaeologist Zelalem Assefa, anthropologist John Fleagle, geologist Frank Brown, and volcanic rock dating specialist Ian McDougall, returned to the important site of Kibish to further explore the possibility of ascertaining a more accurate date for the fossil finds. Brown had been a member of the original American contingent to the area in 1967 and was convinced that the exact location of the excavation could be found. Although some of the maps lacked important detail, Brown recalled vividly the "lay of the land" and with the help of never-aired National Geographic Society film footage the team was able to return to the exact findspot. This was vital because any new dating had to be directly associated with the original site. Traces of the original excavation were preserved, and the team was able to collect hundreds

of mammalian fossils. Confirmation that this was indeed the location of the initial finds came when a newly found shaft fragment of a femur perfectly fit the bottom end of a femur recovered in 1967.

Geological work at the locality proved particularly rewarding when a volcanic ash layer situated immediately below the level that yielded Omo I and II was found to have pumices containing mineral crystals which could be analyzed using argon dating methods. Argon-Argon dating technology produced an age estimate for the Kibish hominids of 195,000 years ago, making them the oldest evidence yet uncovered for *Homo sapiens*.

Homo sapiens, **Omo I** (see page 45). At 195,000 years, the skull and partial skeleton of Omo is one of the oldest known modern human fossils. Actual size. *Photograph courtesy of Michael Day.*

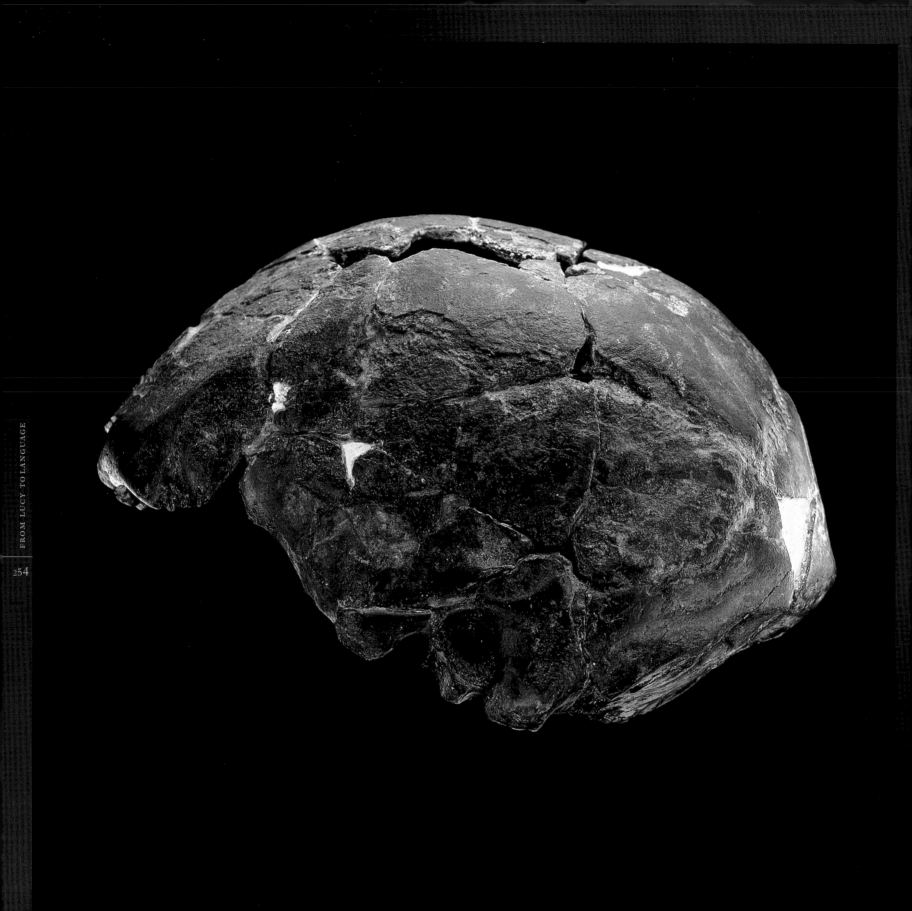

Homo sapiens, Omo II (see previous page).
The companion cranium from Omo-Kibish lacks
a face and is more heavily built with rugged
muscle markings and a receding forehead.
Actual size. *Photograph by David L. Brill;
courtesy of National Museum of Ethiopia.*

SPECIMEN	LOCALITY	AGE	DISCOVERERS	DATE	PUBLICATION
Adult female skull	Qafzeh cave, Israel	ca. 90,000-100,000 years	Bernard Vandermeersch	1969	Vandermeersch, B. 1969. Les nouveaux squellettes moustériens découverts à Qafzeh (Israël) et leur signification, C.R. Acad. Sci. Paris 268: 2562-2565

Homo sapiens
QAFZEH IX

The woman whose remains were found in Qafzeh cave, in Israel, in 1969 is one of the oldest known modern humans. Like Skhul V (see page 258), she may be a member of the population that gave rise to all anatomically modern *Homo sapiens* outside of Africa. This specimen is the most complete of the 21 skeletons of infants, children, adolescents, and adults that were buried in Qafzeh cave. Seven individuals were excavated in the 1930s, and further excavation outside the cave mouth between 1965 and 1980 uncovered the remains of at least 14 additional individuals represented by fragments and by eight partial skeletons, including Qafzeh IX.

The individual died at around age 20 and was part of a double burial. Her skeleton lay on its left side, and the skeleton of a very young child lay beside her flexed lower legs. After the burial's discovery, an Israeli Air Force helicopter gently lifted both skeletons, still inside a one-ton block of rock-hard breccia that had been wrapped in plaster for protection, and delivered them to a laboratory where the bones could be removed more carefully.

In the 1950s, paleoanthropologist Clark Howell recognized the significance of the Qafzeh specimens that were first excavated when he compared them to those from Skhul and described both populations as "proto Cro-Magnons," to emphasize their modern-looking anatomy. The Qafzeh fossils took on increased significance in the 1980s after two new dating techniques revealed the site's antiquity to be more than twice most previous estimates. The thermoluminescence technique dated Qafzeh to 92,000 years ago, and estimates obtained by the electron spin resonance technique place it between 100,000 and 120,000 years ago. The combination of these early dates and the skeletons' mostly modern morphology means that these specimens approach the ancestral form for our species, if genetic evidence is right that we emerged within the past 200,000 years (see page 43). Moreover, the Qafzeh cave is close to Africa, which lends support to the genetic and fossil evidence suggesting that modern humans first arose on that continent.

The reconstructed skull of Qafzeh IX shows some deformation due to pressure from surrounding sediment during burial, but much of the anatomy can be clearly observed and described. The cranial wall is thinner than a Neandertal's and comparable to the mean for modern Europeans. In comparison with contemporary and younger Neandertals from the Levant, such as those from Tabun, Kebara, and Amud, the cranium of Qafzeh IX has a high forehead, a high, parallel-sided braincase, and a reduced browridge. The facial skeleton and lower jaw are marked by the presence of a canine fossa, a flat midface, the partial development of a bony chin, and the absence of a gap behind the third molar—all features that characterize modern humans. As for Skhul V, the lower face of Qafzeh IX projects far forward to accommodate its large teeth. Both this specimen and the male individual Qafzeh VI have a cranial capacity of about 1,554 cc—higher than even the modern-day average for our species. The postcranial skeletons are essentially indistinguishable from our own.

***Homo sapiens*, Qafzeh IX.** The skeleton
young child lay beside the flexed legs of this
adult female, one of the earliest members of
our species, seen here in frontal and opposite
in lateral views. The high forehead, parallel-sided braincase, and reduced browridge of
Qafzeh can be contrasted with the Neandertal
from nearby Amud. Actual sizes. *Photographs
by David l. Brill; courtesy of Israel Antiquities
Authority, Rockefeller Museum.*

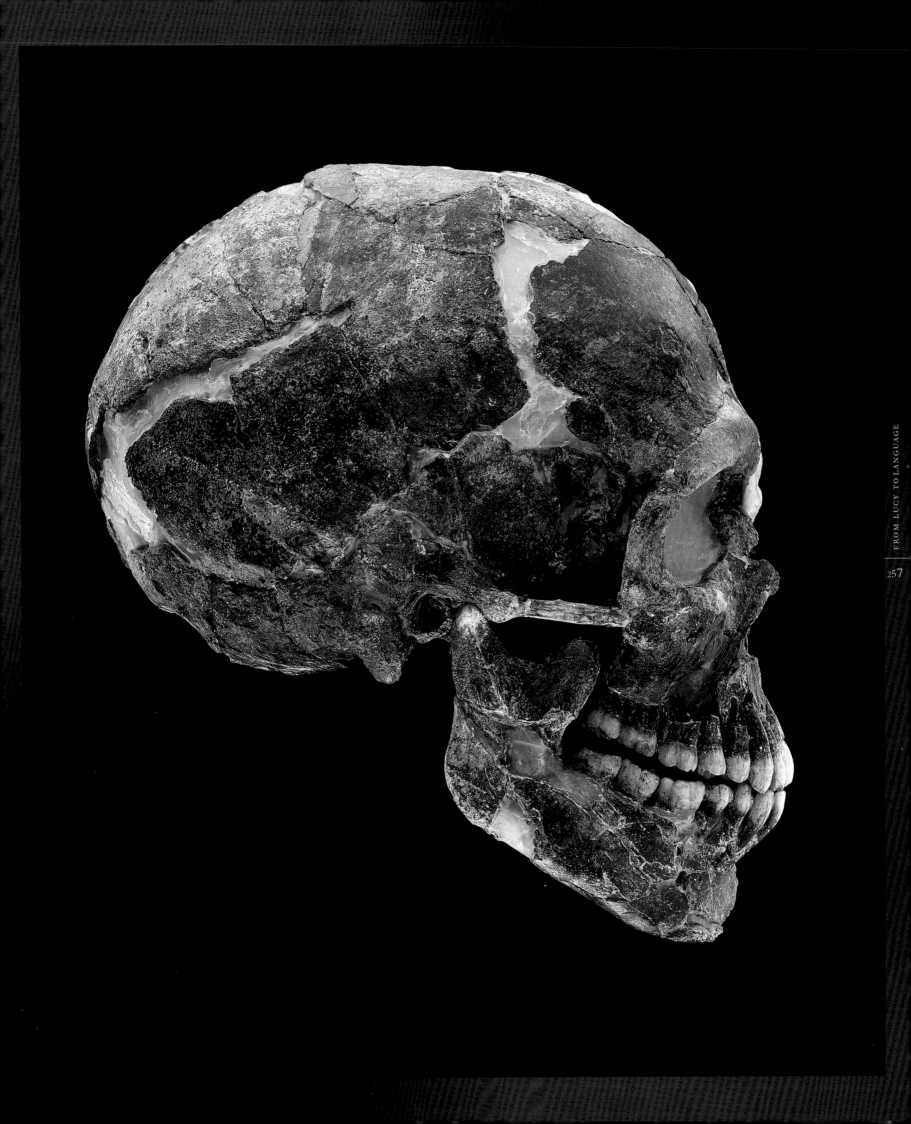

SPECIMEN	LOCALITY	AGE	DISCOVERERS	DATE	PUBLICATION
Adult male skull	Skhul cave, Mount Carmel, Israel	ca. 90,000 years	Theodore McCown and Hallum Movius Jr.	May 2, 1932	McCown, T.D., and A. Keith. 1939. The fossil remains from the Levalloiso-Mousterian. *The Stone Age of Mount Carmel*, vol. II. Oxford: The Clarendon Press

Homo sapiens
SKHUL V

The adult male specimen known as Skhul V and nine other adults and children excavated from the cave of Skhul, in Israel, may represent, along with the people from the nearby cave of Qafzeh (see page 255), the ancestral population of modern humans that later spread out to occupy the globe. The evidence from Skhul and Qafzeh shows that humans appeared in the Near East long before they set foot in Europe, and it will be crucial in solving the mystery of modern human origins.

Originally, Skhul was thought to date to only about 40,000 years ago, based on comparisons of the faunal remains and tools found at the site with those found at the adjacent site of Tabun. This relatively late date meant that the more ancient Neandertals who occupied Tabun might have had time to evolve into the more modern-looking people at Skhul and Qafzeh. In the late 1980s, the application of new dating techniques made it clear that the Skhul group was in fact at least twice as old as previously thought. These individuals were the Tabun people's contemporaries, and they preceded other Neandertals in the Near East and in much of Europe. It suddenly seemed far less plausible that the Neandertals had evolved into the modern populations.

The remains of eight males and two females were intentionally buried in the cave at a level that also harbored nearly 10,000 Mousterian stone tools. The Skhul V individual lay on his back, turned to the right, with the chin pressed against the chest and legs tightly flexed. The left arm stretched across the body, and the hand apparently once clasped the mandible of a wild boar, the only evidence from these burials of a grave offering.

Skhul V is the tallest male and has the most complete skull, but it is missing the middle part of the face. From the wear on the teeth (half of the first molar crowns are worn down to the dentine layer) and the sutures on the skull, the approximate age at death of this individual was 30 to 40 years.

The Skhul people combine modern and primitive features in the skull and skeleton, but the anatomy is overwhelmingly similar to that of modern humans and contrasts particularly with the anatomy of the "classic" European Neandertals. For example, Skhul V possesses a high vault in the front of the cranium, a rounded occipital (unlike the flattened nuchal area in Neandertals) at the rear, and a modern-like flex to the cranial base. The chin is less apparent than in other individuals from Skhul. Notable differences between this individual and an average modern human skull are the prominent browridge and the prognathic lower face, which projects like a muzzle from beneath the slender cheekbones. The teeth show signs of abscesses and gum disease, and there is evidence of rheumatoid arthritis at the temporomandibular joint where the lower jaw connects to the cranium.

Through a small opening on the top of the skull, a working space was created through which a chisel and drill could be inserted. The limestone breccia was laboriously removed from inside the skull in order to make an endocast of the cranial cavity. The cast reveals a brain that was basically modern in general shape and in the proportions of the brain lobes. Removing the matrix also permitted measurement of the cranial capacity, which was around 1,518 cc, slightly below the mean value for modern Europeans.

Besides the skull, the preserved remains of Skhul V include some vertebrae and ribs, most of the left scapula, the right clavicle and part of the left clavicle, both humeri, the right radius and a shaft fragment from the left, the right ulna and the left ulnar shaft, some bones from both hands, the right ilium and half of the ischium, most of the right femur and part of the left femur, parts of the right and left tibiae, most of the left fibula, and a few bones of the left foot. These limb bones tend to be long and slender rather than stout and curved as in Neandertals.

Theodore McCown and Sir Arthur Keith, who described the Skhul fossils, noted the differences from Neandertals in the Skhul sample. They viewed the Skhul anatomy as a forerunner to that seen in specimens such as Cro-Magnon 1 (see page 260), but McCown and Keith concluded that Skhul and Tabun sampled a single, variable population. Others have subsequently interpreted Skhul V's anatomy as showing signs of hybridization between modern and Neandertal populations. Hybridization is difficult to demonstrate in fossils, and even if it did happen rarely, it would not mean that Neandertals and modern humans were a single species. The Tabun individuals clearly differed from those at Skhul, who were undoubtedly on the cusp of becoming modern humans.

Homo sapiens, **Skhul V** (see page 10). Contemporaries of the people from Qafzeh, the inhabitants of Skhul cave likewise possessed a very modern human anatomy. Unlike humans today, this male skull lacks a chin and has a more projecting lower face. Actual size. *Photograph by David l. Brill; courtesy of Peabody Museum of Archaeology and Ethnology, Harvard University.*

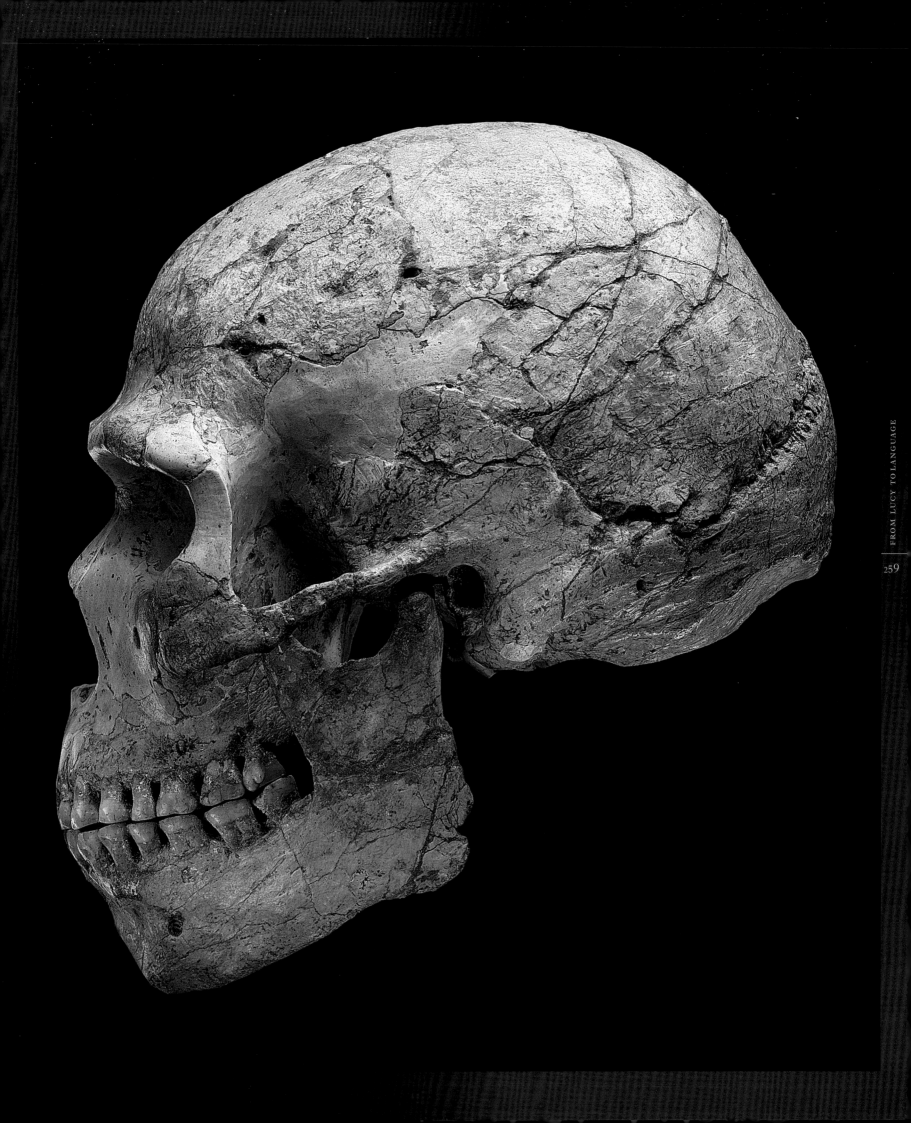

Homo sapiens
CRO-MAGNON I

To the public and to many anthropologists, the name of the Les Eyzies rockshelter has become synonymous with modern humans. The anatomy of the individuals found at Cro-Magnon, France, was essentially our own. They were, however, not the first modern humans to evolve, nor were they entirely representative of human populations across Europe at the time when they lived. Nonetheless, when we gaze at the fossils from Cro-Magnon, we clearly see a reflection of ourselves. These were the type of people who painted the walls of caves and carved delicate figures in ivory.

Workmen constructing a railway line and station for the town of Les Eyzies (Cro-Magnon means "big cliff," referring to the limestone massif that rises above the town), now often labeled the capital of prehistory, discovered five skeletons representing three adult males, an adult female, and one infant. The bodies apparently had been deliberately buried in a single grave, along with body adornments: pierced shells and animal teeth that had probably been worn as necklaces or pendants. Bones from reindeer, bison, woolly mammoth, and other mammals and stone blades and knives belonging to the Aurignacian industry (the earliest style of Upper Paleolithic tools) completed the trove, which geologist Louis Lartet and banker Henry Christy excavated at the rockshelter. These men sorted and studied the human bones that had been removed and inadvertently scrambled by the railway workers.

The earliest definitive modern human from western Europe, Cro-Magnon I, also dubbed the "Old Man," has a face pitted from a fungal infection and probably died in middle age. Except for the teeth and the condyles, or articular surfaces, of the lower jaw, his skull is complete. It shows such classic modern human features as a high, rounded cranium, a steep forehead, a very large brain capacity (over 1,600 cc), a short face with rectangular eye sockets, a tall and narrow nasal opening, a parabolic palate, and a prominent chin. The various Cro-Magnon specimens, however, do display variation in their skeletons, especially in the size and robustness of the browridge and the occipital bone of the skull.

Nonetheless, Cro-Magnon I was used to generalize about the physical features of all Cro-Magnons, or early Europeans other than the Neandertals and their ancestors. French paleoanthropologist Marcellin Boule, as an example, compared Cro-Magnon with the Neandertal from La Chapelle-aux-Saints (see page 238), who came across as a stumbling, brutish, primitive dead end. By comparison, Cro-Magnon, who walked upright, thought sentient thoughts, and created art, was unnecessarily valorized.

The Cro-Magnon skeletons reveal several ailments. A few have fused neck vertebrae, and the adult female survived with a fractured skull. Such injuries indicate not only that the individuals lived hard lives, but that their companions cared for them.

Present-day Europeans have departed in anatomy from the skull shapes found at Cro-Magnon. The initial modern human immigrants, including the people of Cro-Magnon, appear more similar to modern Africans and other subtropical populations than to those of temperate northern latitudes, based on analyses that used two dozen cranial measurements to characterize overall skull size and shape. This resemblance to subtropical populations suggests that Cro-Magnon I and his fellows migrated from some warmer clime, probably Africa or the Near East.

Homo sapiens, **Cro-Magnon I** (see page 179). The 30,000 year-old cranium of Cro-Magnon I, seen here in frontal and next page in lateral views, exhibits a number of definitive modern human traits, including the squat face with hollowed cheekbones and projecting nasal bones, sharply defined eye sockets, a high forehead, and curved parietal bones along the braincase, where the greatest cranial breadth occurs. Actual size. *Photograph by David l. Brill; courtesy of Musée de l'Homme.*

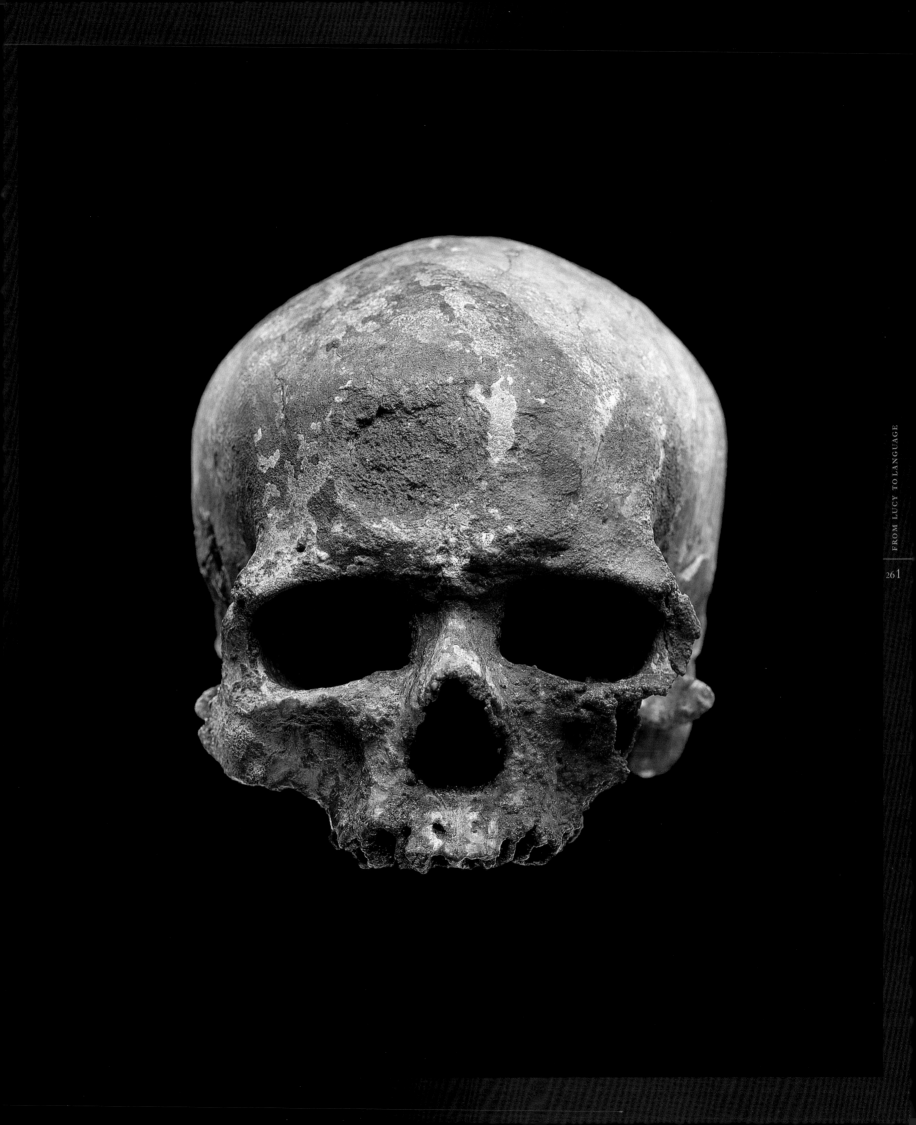

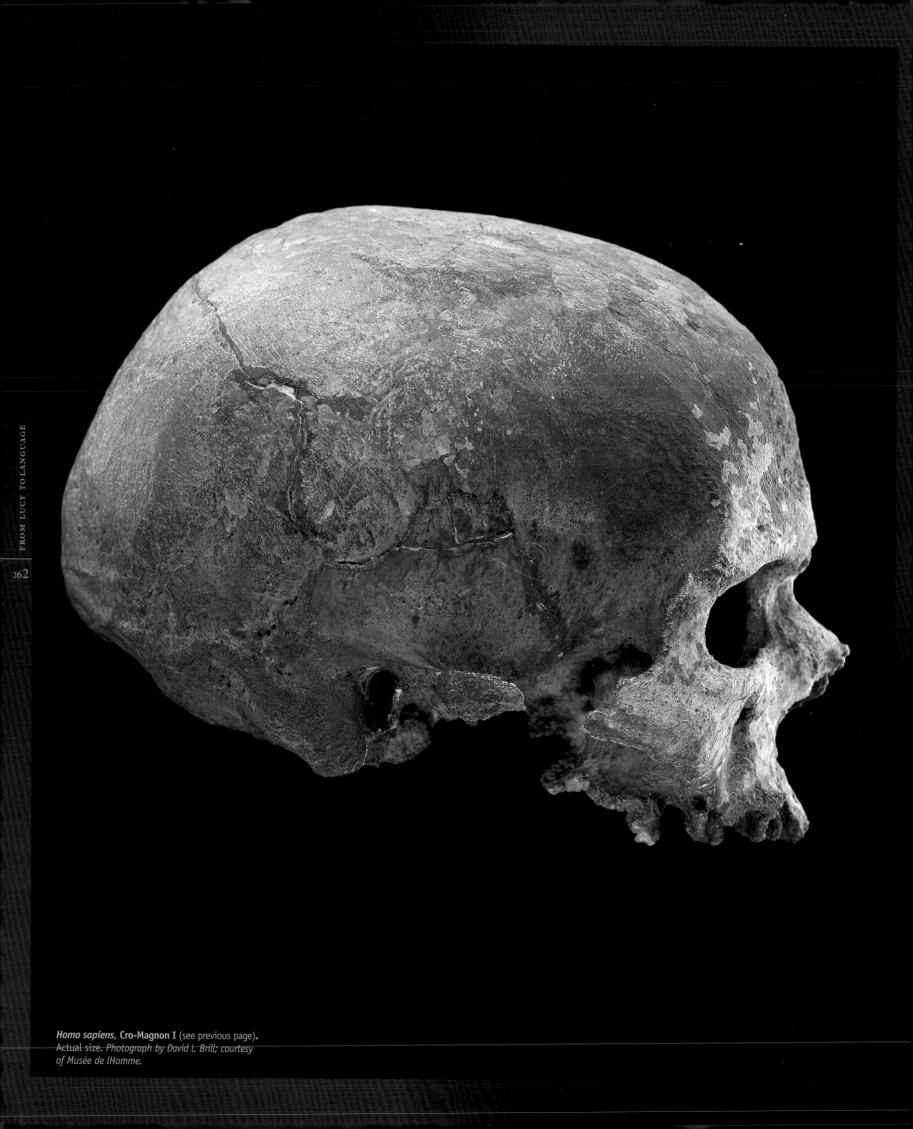

Homo sapiens, **Cro-Magnon I** (see previous page).
Actual size. Photograph by David L. Brill; courtesy
of Musée de lHomme.

SPECIMEN	LOCALITY	AGE	DISCOVERERS	DATE	PUBLICATION
Adult male skeleton	Kow Swamp, Victoria, Australia	ca. 10,000 years	Alan Thorne and Phillip Macumber	1967 and 1968	Thorne, A.G. and P.G. Macumber. 1972. Discoveries of Late Pleistocene Man at Kow Swamp, Australia. *Nature* 238: 316-319

Homo sapiens
KOW SWAMP 1

The primitive anatomy of the face and front half of the cranium of Kow Swamp 1 poses a puzzle for so recent a member of our species. The flat, sloping forehead, massive eyebrow ridges, broad cheekbones, and projecting lower face distinguish this specimen from modern Aborigines. Kow Swamp 1 was part of an isolated remnant population in Australia that may have retained some archaic *Homo sapiens* traits, or, as its discoverers contend, these distinctive features may link the Kow Swamp people in a direct line of descent from *Homo erectus* in Java, where similar-looking fossils date to nearly a million years ago.

The archaic features stood out when a carbonate-encrusted partial skeleton caught anthropologist Alan Thorne's eye at the National Museum of Victoria in 1967. He mounted an expedition the following year to Kow Swamp, a freshwater lake near the Murray River, where he found and unearthed the rest of the bones belonging to this robust male, Kow Swamp 1. Kow Swamp's windblown silts have since yielded the remains of at least 40 individuals who lived between 9,500 and 14,000 years ago. Adults, juveniles, and infants are represented in the burials, and some skeletons were adorned with animal bone, shell, ivory beads, and marsupial teeth. One body had been cremated, but the other skeletons lay in varying positions: fully extended or flexed partially or tightly. Kow Swamp 1 and another male individual, Kow Swamp 5, constitute the most complete skeletons from the site.

Thorne and colleague Milford Wolpoff, as part of their multiregional model of modern human origins, hypothesize that the Kow Swamp skeletons represent the recent end of an "out of Java" sequence of fossils that begins with the *Homo erectus* specimen Sangiran 17 (see page 205). Kow Swamp 1, Kow Swamp 5, and Sangiran 17 share a large but short projecting face and flat forehead, a thick cranial vault, and such details as a rounding of the lower side border of the eye socket and the absence of a marked division between the nasal floor and the lower face. These and other similarities suggest the presence of regional morphological continuity among Australasian hominids during the last million years.

Other parts of the skeleton seem to support a different story. In rear view, for instance, the Kow Swamp skulls have the high, straight sides of a typical modern human skull rather than the characteristic squat *erectus* shape seen in Sangiran 17. An analysis of thigh bones in the Kow Swamp collection found that the femur is completely modern in shape and has a thin-walled shaft. These femora lack both the thick cortical bone and the high "waist," or point of minimum breadth on the shaft that distinguish an *erectus* femur. Some studies of the Kow Swamp crania, particularly Kow Swamp 5, have concluded that the distinctive anatomy may stem from these individuals having undergone cranial deformation as part of a cultural ritual that involved prolonged pressing or binding of the head. Frontal bone compression has been practiced by modern populations in parts of Australia and the Pacific, and such cultural factors or perhaps unknown environmental factors could account for the Kow Swamp group's resemblance to Javanese *Homo erectus*.

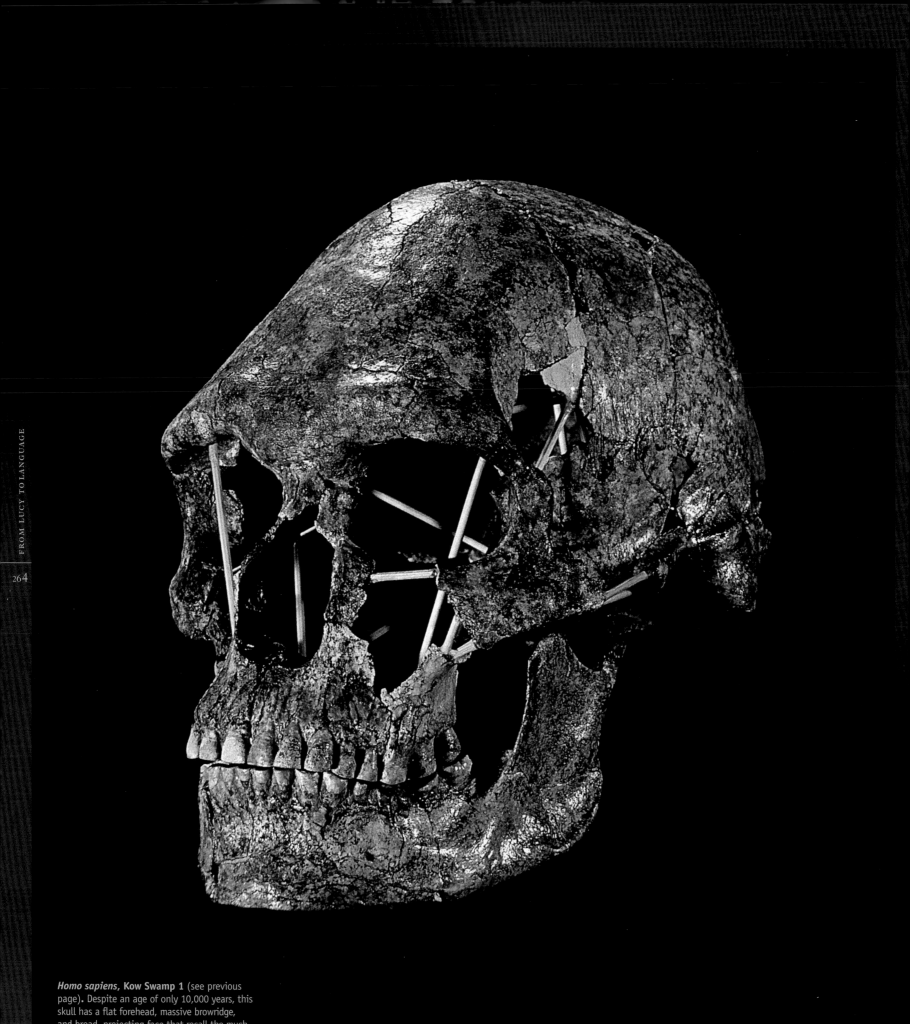

***Homo sapiens,* Kow Swamp 1** (see previous
page)**.** Despite an age of only 10,000 years, this
skull has a flat forehead, massive browridge,
and broad, projecting face that recall the much
more ancient Javan specimen Sangiran 17.
Actual size. *Courtesy of Alan Thorne, Australian
National University.*

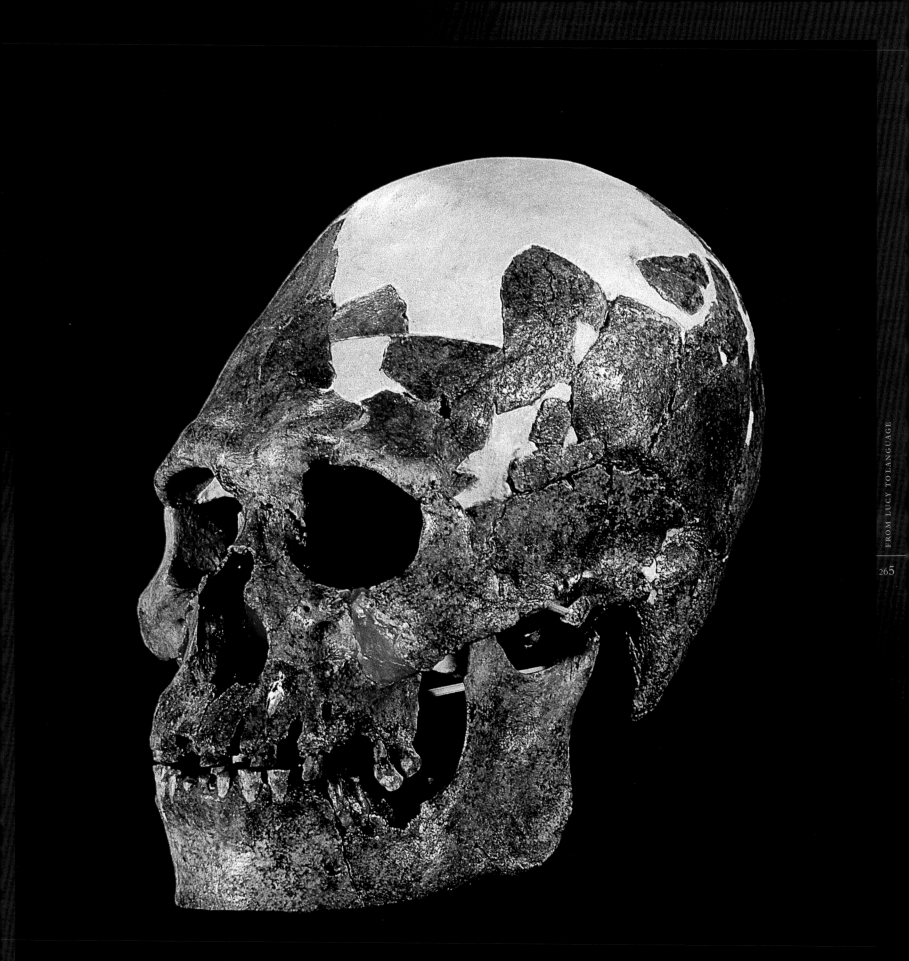

Homo sapiens, **Kow Swamp 5** (see page 263).
Another male specimen from Kow Swamp
displays a similar range of primitive features as
Kow Swamp 1. Are these due to certain cultural
or environmental factors or did this population
descend directly from *Homo erectus* in Indonesia?
Actual size. *Courtesy of Alan Thorne, Australian
National University.*

PALEOLITHIC TECHNOLOGY

Technology represents an important hallmark of humanness, and tracing the origin and evolution of technology is a critical part of paleoanthropology. Fewer than 20 animal species have been observed using tools in the wild, and among the primates, only humans and chimpanzees habitually use tools. Early humans employed sophisticated cognition in their selection and shaping of raw materials into stone tools, and in a broad sense stone technology proceeds from relatively rudimentary implements to more elaborate types that demanded increasingly complex manufacturing techniques.

The production and use of stone tools has not been associated with any of the earliest hominids, the australopithecines, although australopithecines may well have used other sorts of tools. For instance, bony horn cores of antelopes excavated from Swartkrans, South Africa, have strange striations running vertically and horizontally near each bone's rounded tip. These markings may have resulted from the use of these bones by some hominid as digging sticks to extract bulbs and tubers from rocky soil beneath scree slopes near the cave site, and the vast majority of hominid fossils from Swartkrans belong to *Australopithecus robustus*. But stone tools may be an innovation and adaptation that was confined to our genus, *Homo*. Early humans began to use stone tools as extensions of their bodies to modify or manipulate other objects or elements of their environment, and this became a significant part of our ecological adaptation, to the extent that paleolithic technology seems to have evolved in parallel with the expanding brain and enhanced social behavior of *Homo*. We summarize here the characteristics of the four major categories of paleolithic industry, their artifact types, and the processes by which these stone and bone tools were made.

Oldowan

The earliest record of human stone technology dates to about 2.4 million years ago and comes from Gona and the Omo Basin, both in Ethiopia. At 2.3 million years old, basalt and chert flakes and a *Homo* maxilla from the Makaamitalu region of Hadar, Ethiopia (see page 89), represent the oldest known association of

a hominid fossil with stone tools. Tools from each of these sites belong to the Oldowan industry, named and defined by Mary Leakey in the 1960s based on her extensive excavations at Olduvai Gorge, Tanzania. (Oldowan is the adjectival form of the word Olduvai.) Oldowan tools continue to appear in the archaeological record of East Africa until about 1.5 million years ago. They are most commonly attributed to *Homo habilis*, but Oldowan tools could have been made by other species, such as *Homo rudolfensis*, from Koobi Fora, Kenya.

These crude chopping and scraping implements and flakes represent the dawn of human culture and may constitute the first personal possessions. The most conspicuous artifacts among the Oldowan industry are simply flaked cores made from pebbles or chunks of rock, typically quartz or basalt. Mary Leakey described several types of these cores, from which a few flakes had been struck, and named them for their shape or presumed function. These types include hammerstones (often unmodified cobbles used to strike flakes from other rocks or to break open bones), unifacial and bifacial (flaked on one or both sides, respectively) choppers, heavy- and light-duty scrapers, and spheroids. A related industry known as the Developed Oldowan appeared about 1.5 million years ago and is noted for having relatively fewer choppers and bifaces and more scrapers, spheroids, and subspheroids. Oldowan and Developed Oldowan artifacts persist in the archaeological record of Africa and overlap extensively in time with the succeeding Acheulean industry.

Archaeologist Nicholas Toth has argued convincingly that Oldowan core forms were not the end products but rather the means to the end. Cores provided the raw material for producing sharp-edged flakes, and it was these flakes that became the primary tools for cutting open carcasses or slicing sinews in the quest for meat and bone marrow. The peculiar and sometimes ubiquitous spheroids may also not have been an intentional end product. Experiments with angular quartz chunks demonstrated that their repeated use as a hammerstone to flake other rocks converts the chunks into objects closely resembling spheroids after just a few hours, which suggests that hammerstones were reused and most likely kept and transported by early hominids.

If spheroids were an inadvertent result of rock bashing, other Oldowan tools bear specific marks of deliberate manufacture, such as the striking platform where a hammerstone hits a core, or the bulb of percussion on the inner surface of a flake taken from the core. But Oldowan technology involved minimum-effort techniques for flaking rock, without the apparent conventions and rules for manufacturing artifacts that characterize subsequent archaeological industries. The basic Oldowan flaking technique was a hard-hammer percussion technique in which a smooth, rounded stone was wielded as a hammer to remove a flake from a core. The flake scar left on the core then provided a relatively flat surface with an acute angle—the optimal striking platform for removing a second flake. Alternative means of making Oldowan artifacts were the anvil technique, which involved striking a core on a stationary stone anvil, and the bipolar technique, which used a hammerstone to flake a core held against an anvil.

Oldowan chopper, Gadeb, Ethiopia *(top)*.
Oldowan artifacts were manufactured between about 2.6 and 1.5 million years ago. One of the primary tools was cores like this one from which a few sharp flakes had been struck. Actual size. *Photograph by David L. Brill; courtesy of National Museum of Ethiopia.*
Horn core tool, Swartkrans, South Africa *(left)*.
Although bone tools do not become common until much later in the archeological record, this antelope horn core appears to have been used as a digging stick in rocky soil, since it has criss-crossing striations near its blunt tip that have been replicated in modern experiments. Actual size. *Photograph by David L. Brill; courtesy of Transvaal Museum.*
Bone point, Swartkrans, South Africa *(bottom)*.
Another bone tool from the same cave site may be as old as 2 million years. Actual size. *Photograph by David L. Brill; courtesy of Transvaal Museum.*

Oldowan end chopper, Olduvai Gorge, Tanzania *(top left).* The name Oldowan comes from Olduvai Gorge, where Mary Leakey defined and described the industry and classified tools, such as this end chopper from Olduvai Bed I, by their form or apparent function. Actual size. *Photograph by David L. Brill; courtesy of National Museum of Tanzania.*

Heavy-duty scraper, Olduvai Gorge, Tanzania *(top right).* The steep, flaked edge of this kind of stone artifact led to its classification as a scraper, but other research suggests that the flakes themselves, rather than these chunky rock cores, were the primary tools. Actual size. *Photograph by David L. Brill; courtesy of National Museum of Tanzania.*

Spheroid, Olduvai Gorge, Tanzania *(bottom).* Ubiquitous at some East African sites, spheroids are probably just the unintended by-product of repeatedly bashing an angular hunk of quartz as a hammerstone to remove flakes from other rocks. Actual size. *Photograph by David L. Brill; courtesy of National Museum of Tanzania.*

Acheulean hand ax, Olduvai Gorge, 1 **(see next page).** The signature Acheu is an ax flaked on both sides and sha_ that the butt fits snugly in the hand. form of this multi-use tool changed li million years. Actual size. *Photograph David L. Brill; courtesy of National Mus Tanzania.*

Acheulean

Named for the type archaeological site of Saint-Acheul, France, this industry specialized in large, heavy-duty, sharp-edged tools. Analysis of microscopic wear traces on the working edges of these tools suggests that they primarily served in butchery and woodworking tasks. Acheulean tools are usually flaked all over both sides of a rock (quartzite, lava, chert, and flint were commonly used), hence the generic term biface. At some sites these tools look crude and chunky, at others they appear elegant and elongate, but the Acheulean reveals a degree of standardization in production technique and artifact form lacking in the Oldowan industry. This technological uniformity lasted for more than a million years in a region stretching from Africa through much of Europe, the Near East, and India. This major difference from the Oldowan industry signals a shift from simple, sharp flakes to specialized, functional tools. The major specific tool types include pointed hand axes, picks, and flat-edged cleavers. Crafting these tools required a complex set of procedures at each step, from flaking a suitable core to the difficult process of flaking, trimming, and thinning the core to produce a pointed end for a hand ax or a truncated tip for a cleaver. Despite the elaborate manufacturing process, hand axes and cleavers make up a considerable percentage of the total archaeological assemblage at most Acheulean sites.

Acheulean tools first appeared around 1.4 million years ago. The oldest known occurrence is from Konso-Gardula, Ethiopia, where abundant Acheulean tools over 1.3 million years old were found in 1991 along with a mandible of *Homo ergaster*. Other sites with early Acheulean tools of similar age are Olduvai Gorge, Tanzania, and Gadeb, Ethiopia. Tools from the earliest archaeological sites in Europe more closely resemble the Oldowan than the Acheulean, suggesting that the first African migrants lacked a hand ax-based industry, but by 500,000 years ago the Acheulean industry proper had penetrated into Europe, where it continued until after 200,000 years ago. In France, Acheulean-like tools from a variant industry called the Tayacian are associated with remains of *Homo heidelbergensis* at the cave site of Arago (see page 212). Other Acheulean sites with fossils from this hominid species include Steinheim, Germany (see page 216), and Swanscombe, England. Acheulean tools apparently never reached East Asia, perhaps because more suitable raw materials for tools, such as bamboo, were plentiful, or because geographic barriers precluded biological and cultural diffusion between West and East Asia.

It took great skill and strength to master the making of Acheulean tools. The risky first step of striking a huge flake from a boulder with considerable force provided the raw form for a biface. This flake would then be worked with a hammerstone to remove smaller flakes alternately from each side, gradually whittling down the thickest areas and creating a more symmetrical shape. The toolmaker switched to a soft hammer of bone or antler or wood in order to remove thinner flakes and form a consistent edge to the biface.

Around a million years ago, more Acheulean assemblages became dominated by slim, symmetrical bifaces, especially the classic teardrop and lanceolate hand axes. Some incredibly oversized examples of Acheulean hand axes and picks come from the site of Isimila in Tanzania, tools so unwieldy that it taxes the imagination to picture them in practical use. At sites such as Olorgesailie, Kenya, where visitors on a raised wooden walkway can observe an archaeological floor littered with hand axes, cleavers, and other tools, the incredible abundance of artifacts conjures images of an Acheulean assembly line.

Acheulean cleaver, Bihorel oest, France *(left)*. Another standard bifacial tool is the cleaver, made from a single large stone flake that is thinned and truncated to form a straight, broad tip. Actual size. *Photograph by David L. Brill; courtesy of Denise de Sonneville-Bordes, Université de Bordeaux I.*
Acheulean lanceolate hand ax, Briqueterie, France *(right)*. Acheulean artifacts appear in Europe after 500,000 years ago and often adopt elongate, symmetrical shapes, such as this example crafted from flint. Actual size. *Photograph by David L. Brill; courtesy of Université de Bordeaux I.*

Mousterian

The sort of core preparation that typified Acheulean toolmaking continued in the succeeding Mousterian industry, although the resulting tools were much smaller and had a wider range of shapes and uses. Named for the site of Le Moustier, France, where the remains of an adult and infant Neandertal were found, this industry appeared around 200,000 years ago (and thus overlapped with the Acheulean for thousands of years) and persisted until about 40,000 years ago. It occurred in various regional forms throughout Europe and in the Near East and Africa in essentially the same area where Acheulean tools have been found. In Europe, these artifacts are most closely associated with *Homo neanderthalensis*, but elsewhere, such as at the Near Eastern sites of Kebara (see page 232), Tabun, Qafzeh (see page 255), and Skhul (see page 258), Mousterian tools were made by both Neandertals and early *Homo sapiens*.

Many Mousterian toolmakers utilized either of two main techniques to prepare a stone core for the removal of a single, large flake of predetermined shape and thickness, which could then be shaped into a versatile range of tools by the process of retouching, or by the removal of many tiny flakes to form a new functional edge or to resharpen a dulled edge. These flaking techniques are jointly known as Levallois, for a Paris suburb where the technique was first identified in excavated artifacts. Although sometimes regarded as wasteful because only one flake could be produced, Levallois techniques in fact aimed to get as much cutting edge as possible from a core. The first technique began with a discoidal rock that was flaked around its circumference to create a faceted platform for the removal of an oval flake, forming a Levallois tortoise core (so named because the rounded top and flat bottom of the finished core resemble a tortoise's carapace). The resulting flake would often be retouched along one or both sides to form a scraper, which was used to work hides and wood. The second Levallois core technique also focused on preparing the core for a single flake, but the flake was triangular and pointed and probably served as the business end of projectiles or spears.

Other tools found in the Mousterian industry include notched flakes, denticulates (flakes with a rough, serrated edge), and flake blades similar to what would come to characterize Upper Paleolithic industries. When they were available near a hominid site, fine-grained flints were often chosen as raw material for making these Mousterian artifacts. One of the more unusual Mousterian tools comes from a North African industry called the Aterian, which may date to between 75,000 and 100,000 years ago. Tough made with traditional Levallois techniques, Aterian points are notable for the squared-off tang protruding from the base that must have been used to haft this tool to the end of a wooden shaft as a spear point or arrowhead, an innovation that would be carried to new heights in the Upper Paleolithic.

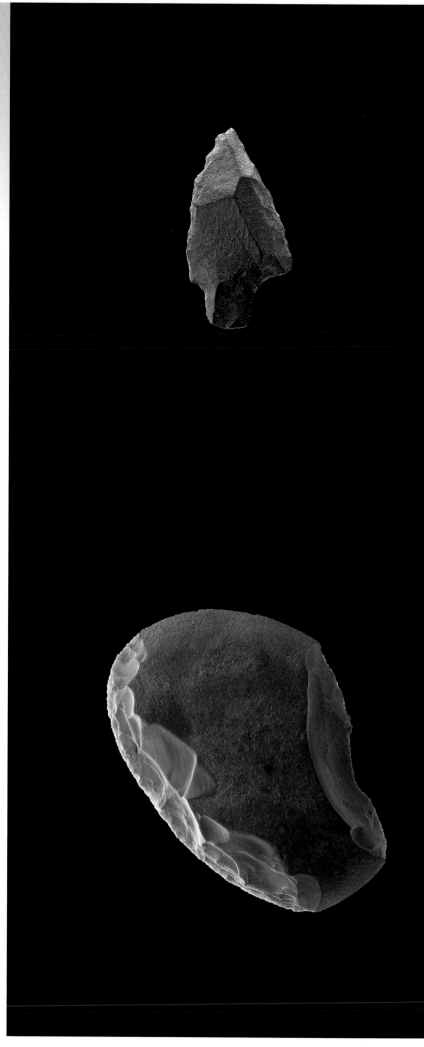

Aterian point, Colomb-Bechar, France *(top left)*. The tang at the base of this stone point identifies it as part of the variant Aterian industry. The point may have been hafted to the end of a stick to create a spear. Actual size. *Photograph by David L. Brill; courtesy of André Debenath, Université de Bordeaux I.*

Double-sided scraper with Quina retouch, Combe-Grenal, France *(bottom left)*. Up to eighty percent of tools in the Quina variant of the Mousterian industry consists of scrapers, often with convex edges that have been re-touched, or reshaped, by removing a series of very small flakes. Actual size. *Photograph by David L. Brill; courtesy of Universitee de Bordeaux I.*

Mousterian point, Houppeville, France *(top right)*. A large pointed flake struck from a carefully prepared stone core is a typical tool from the Mousterian industry, which lasted from 200,000 to 40,000 years ago. The points may have been fashioned into projectiles. Actual size. *Photograph by David L. Brill; courtesy of Denise de Sonneville-Bordes, Université de Bordeaux I.*

Levallois core and point, Le Tillet, France *(bottom right)*. This flint core reveals deliberate flaking around its edge to prepare a striking platform at the top so that this pointed flake would be removed without damage by the blow from another stone. Actual size. *Photograph by David L. Brill; courtesy of Denise de Sonneville-Bordes, Université de Bordeaux I.*

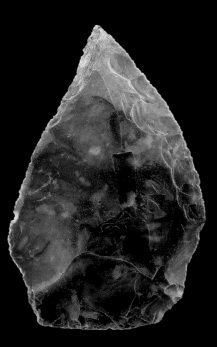

Upper Paleolithic

Diversity and specialization mark the archaeological record of the Upper Paleolithic, in stark contrast to the technology of the Mousterian and every other preceding industry. Upper Paleolithic tools are more easily categorized and recognizable as having specific functions than those of any other paleolithic industry. This industry was a wide technological departure from the past, corresponding with the first signs of accelerated cultural evolution (and physically these toolmakers looked much like we do today). The most characteristic Upper Paleolithic tools were made from thin blades struck from cores. Blades are defined as flakes that are at least twice as long as their width. This technique greatly increased the number of cutting edges that could be obtained from a single core. Often, blades were broken and reshaped into microliths—small, geometrically shaped flakes that were hafted. One common tool from this industry, the burin, possessed a chisel-like engraving edge that facilitated the working of bone. Specialized bone and antler tools appear frequently in Upper Paleolithic archaeological sites. Innovations in Upper Paleolithic technology and culture led to other aspects of human existence becoming more common or highly developed, such as the use of fire, burial of the dead, construction of clothing and shelter, and the creation of art.

The Upper Paleolithic spanned the period from 40,000 to 10,000 years ago in Europe. This industry probably started in western Asia and Africa at about the same time, although some African evidence suggests a much earlier origin there for the sorts of innovations that characterize the Upper Paleolithic. Barbed bone harpoons and points have been excavated from the Katanda site along the Democratic Republic of the Congo's Semliki River, and these artifacts, which strongly resemble classic Upper Paleolithic bone tools from western Europe, are estimated to be as old as 90,000 years, which would place them firmly within the preceding Middle Stone Age. Recent doubt has been cast on the veracity of the theromoluminesence and ESR dates for these artifacts, suggesting they might be much younger, although carbon dating of the bone tools themselves might resolve this issue.

The most reliable and compelling evidence for the occurrence of typical Upper Paleolithic bone tools in an ancient context derives from Blombos Cave, South Africa. Excavations here have recovered more than 30 elegant polished awls useful for piercing leather or as projectile points that were most likely hafted and used spears. Archaeologists surmise that these 78,000 year-old-old bone tools, nearly twice the age of the first occurrence in Europe, demanded remarkable craftsmanship to make, some perhaps taking several hours of work.

Unlike previous industries, the Upper Paleolithic quickly diversified, displaying distinctive regional styles. The earliest Upper Paleolithic industry was the Aurignacian (from 40,000 to 28,000 years ago), which was named after the Aurignac rockshelter in the Pyrenees foothills where such tools were first found in 1860. Aurignacian implements accompanied the modern human burials at Cro-Magnon and have been found at many sites across Europe. In the Near East, the Aurignacian spans the period from at least 32,000 to 17,000 years ago. Aurignacian assemblages can be characterized by the prevalence of end scrapers, burins, distinctive bone points, ivory beads and tooth necklaces, and some abstract human and animal figurines, including the recently discovered Dancing Venus of Gelgenberg, Austria, a flat figurine of a woman in green serpentine. France's earliest known cave paintings, from Chauvet and Cosquer, date to the late Aurignacian, 32,000 and 27,000 years ago, respectively.

Another Upper Paleolithic industry contemporary with the Aurignacian was the Châtelperronian, named for the French cave site of Châtelperron. This industry, which combined an evolved Mousterian technology with the use of blade and bone tools, can be confidently attributed to *Homo neanderthalensis* based on the association of both the Saint-Césaire skeleton (see page 246) and a juvenile Neandertal temporal bone from Arcy-sur-Cure, France, with Châtelperronian tools. This industry may have evolved in parallel with the Aurignacian or been adopted by Neandertals after the introduction of the Aurignacian to western Europe. The Châtelperronian industry apparently disappeared with the Neandertals around 34,000 years ago, while the Aurignacian and all later phases of Upper Paleolithic technology were created by *Homo sapiens*.

The Gravettian industry (from 28,000 to 22,000 years ago) added backed blades (backing refers to the broaden of one side of the blade by pressure-flaking that edge) and beveled-based bone points), which served as a new and more streamlined kind of spear tip, to the Upper Paleolithic tool kit. During this period, ivory beads became body adornments for burials, and the first of the Venus figurines, voluptuously depicted woman carved in ivory, probably appeared in eastern, central, and western Europe. The brief Solutrean period (from about 21,000 to 19,000 years ago) featured incredible flint craftsmanship, best exemplified by bifacially flaked, leaf-shaped knives that were heated over a flame to permit the necessary precision flaking without snapping the object in two. This period of spectacular stoneworking both appears and disappears abruptly in the archaeological record of France.

Perhaps the heyday of the Upper Paleolithic came with the Magdalenian industry (from 18,000 to 12,000 years ago), which is most notable for its wall art, the spectacular painted caves of Lascaux, Altamira, and other sites. Named for the French rockshelter La Madeleine, this period saw an increase in the use of microliths, which could be interchangeably hafted to make a range of composite tools, including arrows, as well as a prevalence of multibarbed bone harpoon heads, and spear throwers of wood, bone, or antler, which enhanced the speed and distance which a propelled spear could travel. All of these advances suggest much greater skill and sophistication for hunting many types of game.

Magdalenian biconical bone point, Abri Faustin France *(left)*. Bone points appear in the earliest Upper Paleolithic period, after 40,000 years ago, but this example comes from the Magdalenian period, which began 18,000 years ago. Actual size. *Photograph by David L. Brill; courtesy of Michael Lenoir, Université de Bordeaux I.*
Perigordian flint blade, Corbiac, France *(right)*. The most characteristic stone tool of the Upper Paleolithic, long blades provided a starting point for crafting microlithic flakes that were often combined into composite tools or used for a variety of specialized functions. Actual size. *Photograph by David L. Brill; courtesy of Denise de Sonneville-Bordes; Université de Bordeaux I.*

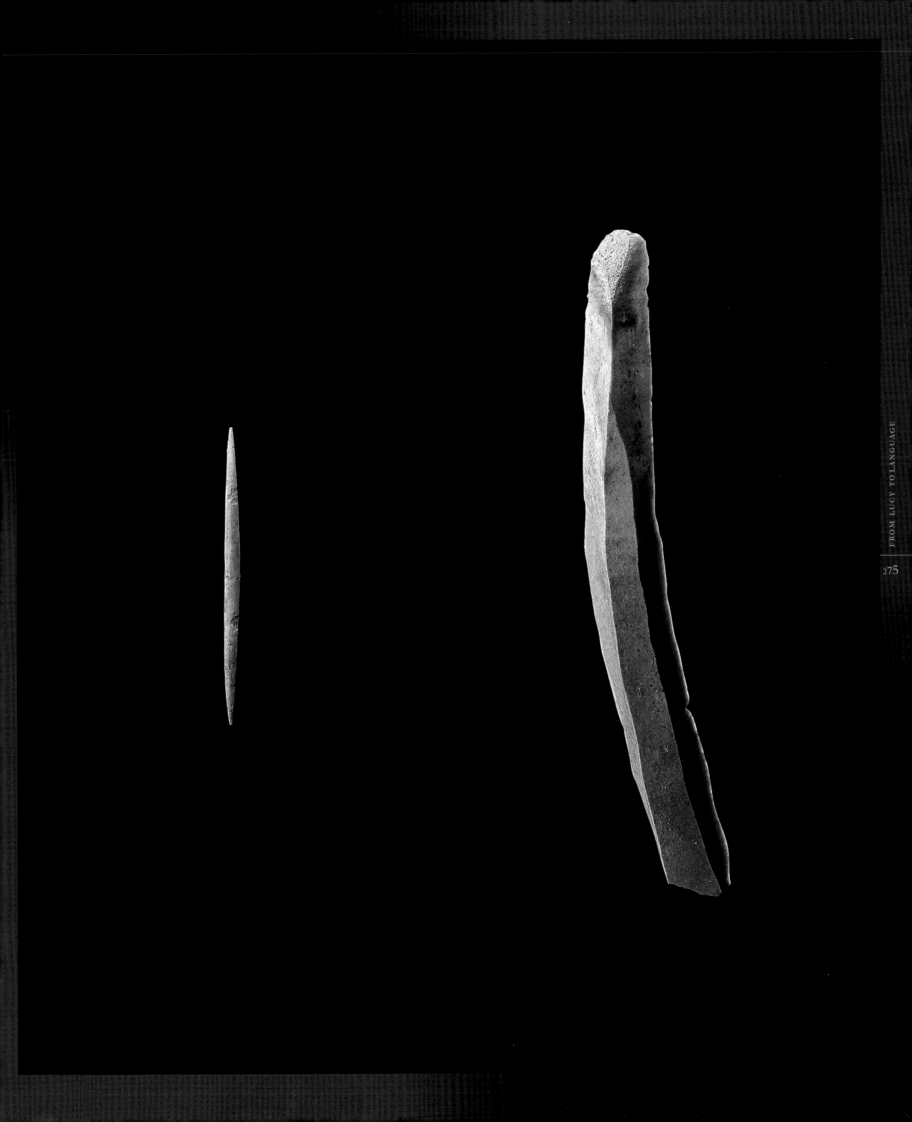

Prismatic blade core, Corbiac, France *(left)*.
One end was knocked off this core to make a
flat striking platform for removing a series of
narrow blades along its periphery. Actual size.
*Photograph by David L. Brill; courtesy of Denise
de Sonneville-Bordes, Université de Bordeaux.*
**Solutrean Willow leaf point, Fourneau du
Diable, France *(middle)*.** This beautiful,
bifacial knife dates to around 20,000 years
ago during a brief flowering of flint craftsmanship. It may have had greater ceremonial than
functional use. Actual size. *Photograph by David
L. Brill; courtesy of Université de Bordeaux I.*
**Magdalenian Double-row barbed harpoon,
Le Morin, France *(right)*.** Bone harpoons
became common toward the end of the Upper
Paleolithic and indicate that humans were
exploiting a range of new food resources. In addition to such fishing gear, arrows and specialized spear-throwers were prevalent. Actual size.
*Photograph by David L. Brill; courtesy of
R. Deffarge, Université de Bordeaux I.*

TYPE SPECIMENS FOR HOMINID SPECIES

SPECIES	SPECIMEN NO.	SPECIMEN	PAGE NO.
Sahelanthropus tchadensis	TM 266-01-060	Adult cranium	*117*
Orrorin tugenensis	BAR 1000'00	Adult mandible	*119*
Ardipithecus kadabba	ALA-VP-2/10	Adult mandible frag.; assoc. teeth	*120*
Ard. ramidus	ARA-VP-6/1	Assoc. set of adult teeth	*120*
Australopithecus anamensis	KNM-KP 29281	Adult mandible, temporal frag	*130*
A. afarensis	L.H.-4	Adult mandible	*141*
Kenyanthropus platyops	KNM-WT 40000	Adult cranium	*122-123*
A. africanus	Taung	Immature skull	*79 and 155*
A. aethiopicus	Omo 18	Adult mandible	*167*
A. garhi	BOU-VP-12/130	Partial adult cranium	*132*
Homo rudolfensis	KNM-ER 1470	Adult cranium	*192-193*
A. boisei	OH 5	Adult cranium	*5, 74, and 169*
H. habilis	OH 7	Adult mandible	*183*
A. robustus	TM 1517	Adult mandible	*137*
H. ergaster	KNM-ER 992	Adult mandible	*175*
H. georgicus	D 2600	Adult mandible	*193*
H. erectus	Trinil 2	Adult skullcap	*201*
A. crassidens	SK 6	Adolescent mandible	*161*
H. antecessor	TD6	Frag. adult cranium, teeth	*46*
H. heidelbergensis	Mauer 1	Adult mandible	*211*
H. neanderthalensis	Neandertal 1	Partial adult skeleton	*243*
H. floresiensis	LB 1	Partial adult skeleton	*249*
H. sapiens	none		

The type specimen of a species is a particular specimen to which the species name was first properly applied. The rules governing the proper procedure for naming a species are found in the International Code of Zoological Nomenclature. In the case of a single specimen from a site, such as the Chad hominid, there is no ambiguity about which specific individual was used to designate the new binomial (meaning two names, the genus and the species); this specimen is properly called the holotype.

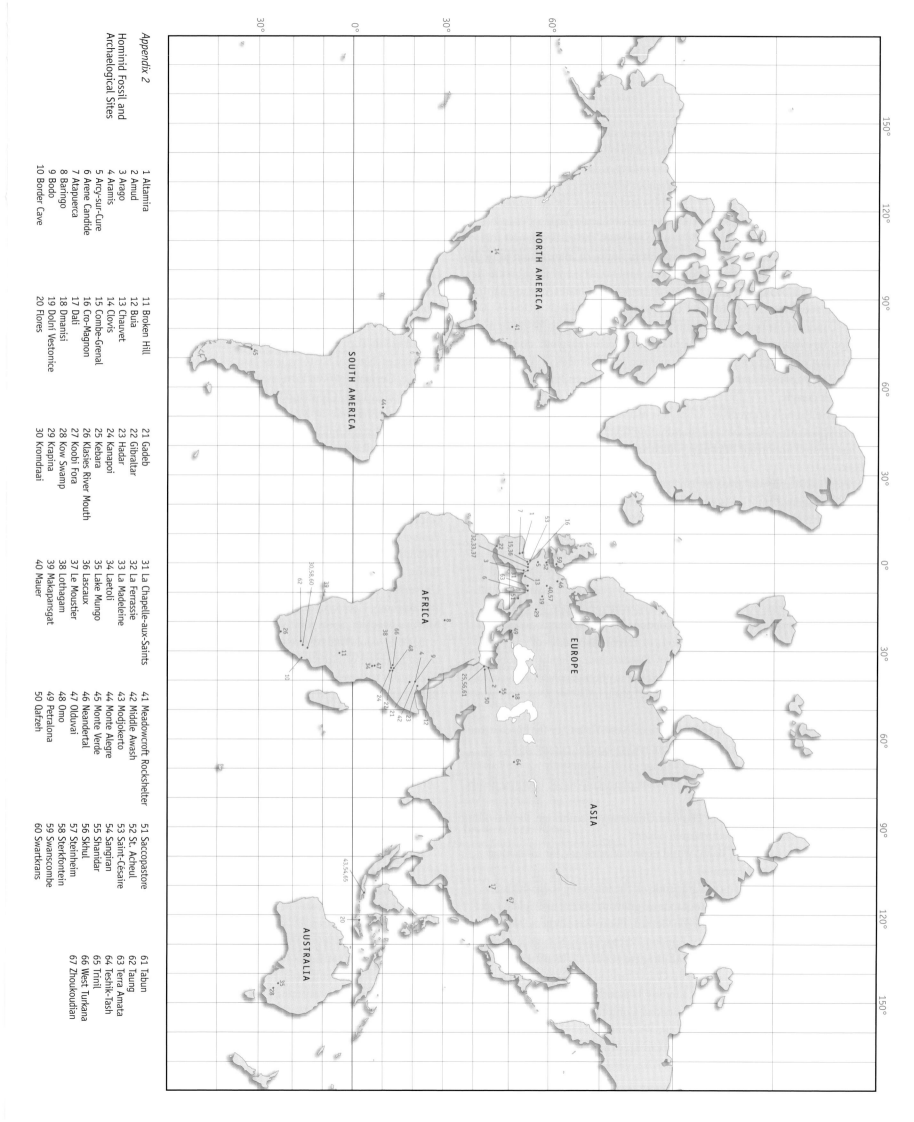

Appendix 2

Hominid Fossil and
Archaelogical Sites

1 Altamira
2 Amud
3 Arago
4 Aramis
5 Arcy-sur-Cure
6 Arene Candide
7 Atapuerca
8 Baringo
9 Bodo
10 Border Cave

11 Broken Hill
12 Buia
13 Chauvet
14 Clovis
15 Combe-Grenal
16 Cro-Magnon
17 Dali
18 Dmanisi
19 Dolní Vestonice
20 Flores

21 Gadeb
22 Gibraltar
23 Hadar
24 Kanapoi
25 Kebara
26 Klasies River Mouth
27 Koobi Fora
28 Kow Swamp
29 Krapina
30 Kromdraai

31 La Chapelle-aux-Saints
32 La Ferrassie
33 La Madeleine
34 Laetoli
35 Lake Mungo
36 Lascaux
37 Le Moustier
38 Lothagam
39 Makapansgat
40 Mauer

41 Meadowcroft Rockshelter
42 Middle Awash
43 Modjokerto
44 Monte Alegre
45 Monte Verde
46 Neandertal
47 Olduvai
48 Omo
49 Petralona
50 Qafzeh

51 Saccopastore
52 St. Acheul
53 Saint-Césaire
54 Sangiran
55 Shanidar
56 Skhul
57 Steinheim
58 Sterkfontein
59 Swanscombe
60 Swartkrans

61 Tabun
62 Taung
63 Terra Amata
64 Teshik-Tash
65 Trinil
66 West Turkana
67 Zhoukoudian

Olson, Steve. 2002. *Mapping Human History : Discovering the Past Through Our Genes.* Boston: Houghton Mifflin.

23. Defining Human Species

Dawkins, Richard. 1995. *River Out of Eden.* New York: Basic Books.

Eldredge, Niles. 1985. *Unfinished Synthesis.* New York: Oxford University Press.

*Kimbel, William H., and Lawrence B. Martin, editors. 1993. *Species, Species Concepts and Primate Evolution.* New York: Plenum Press.

*Mayr, Ernst. 1957. "Species concepts and definitions." In Ernst Mayr, editor, *The Species Problem.* Washington, D.C.: American Association for the Advancement of Science Publication.

*Meikle, W. Eric, and Sue Taylor Parker. 1994. *Naming our Ancestors.* Prospect Heights: Waveland Press.

Tattersall, Ian. 1986. "Species recognition in human paleontology." *Journal of Human Evolution* 15:165-75.

Wrangham, Richard W. 1987. "The Significance of African apes for reconstructing human social evolution." In Warren G. Kinzey, editor, *The Evolution of Human Behavior: Primate Models* 51-71. Albany: State University of New York Press.

24. Co-Existing Human Species

Bower, Bruce. 1995. "Pruning the family tree." *Science News* 148:154-155.

Henneberg, M., and J.F. Thackeray. 1995. "A single-lineage hypothesis of hominid evolution." *Evolutionary Theory* 11:31-38.

Mayr, Ernst. 1950. "Taxonomic categories in fossil hominids." *Cold Springs Harbor Symposia on Quantitative Biology* 15:109-118.

Sawyer, G.J., Esteban Sarmiento, and Ian Tattersall. 2005. *The Last Human.* New Haven: Yale University Press.

Tattersall, Ian. 2000. "Once we were not alone." *Scientific American* 282(1):56-62.

Walker, Alan C. 1976. "*Australopithecus, Homo erectus* and the single species hypothesis." *Nature* 261:572-574.

25. Human Diversity Today

*Marks, Jonathan. 1995. *Human Biodiversity: Genes, Race, and History.* New York: Aldine de Gruyter.

26. What is Race?

Coon, Carleton S. 1962.*The Origin of Races.* New York: Alfred A. Knopf.

Gould, Stephen Jay. 1981. *The Mismeasure of Man.* New York: W.W. Norton and Company.

Sarich, V. and E. Miele. 2004. *Race: The Reality of Human Differences.* Boulder, Colorado: Westview Press.

Wolpoff, M.H. and R. Caspari 1996. *Race and Human Evolution: A Fatal Attraction.* New York: Simon and Schuster.

27. The Size of Early Humans

*Jungers, William L., editor. 1985. *Size and Scaling in Primate Biology.* New York: Plenum Press.

McHenry, Henry M. 1992. "How big were early hominids?" *Evolutionary Anthropology* 1:15-20.

Ruff, C. 2002. "Variation in human body size and shape." *Annual Review in Anthropology* 32: 211-232.

28. Sexual Dimorphism

McHenry, Henry M. 1991. "Sexual dimorphism in *Australopithecus afarensis.*" *Journal of Human Evolution* 20:21-32.

Plavcan, J. Michael, and Carel P. van Schaik. 1993/94. "Canine dimorphism." *Evolutionary Anthropology* 2:208-214.

Plavcan, J.M., R.F. Kay, W.L. Jungers, and C.P. van Schaik, Editors. *Reconstructing Behavior in the Primate Fossil Record.* New York: Kluwer Academic/Plenum Publications.

29. Gestation

Kaplan, H., K. Hill, J. Lancaster, and A.M. Hurtado. 2000. "A theory of human life history evolution: Diet, intelligence, and longevity." *Evolutionary Anthropology* 9:156-185.

Ross, C. 1998. "Primate life histories." *Evolutionary Anthropology* 6:54-63.

30. Maturation

Beynon, Alan D., and Bernard A. Wood. 1987. "Patterns and rates of enamel growth in the molar teeth of early hominids." *Nature* 326:493-496.

Guatelli-Steinberg, D. 2001. "What can developmental defects of enamel reveal about physiological stress in nonhuman primates??" *Evolutionary Anthropology* 10:138-151.

31. Evolution of the Human Brain

Falk, D. 2004. *Braindance: New Discoveries About Human Origins and Brain Evolution.* Gainesville: University Press of Florida.

Holloway, R.L., D.C. Broadfield, M.S. Yuan, J.L. Schwartz,. and I. Tattersall. 2004. *The Human Fossil Record: Brain Endocasts: The Paleoneurological Evidence, Volume 3.* New York: Wiley-Liss.

Wills, Christopher. 1993. *The Runaway Brain: The Evolution of Human Uniqueness.* New York: Basic Books.

32. Reconstructing the Appearance of Early Humans

Johanson, Donald C. 1996. "Face-to-Face with Lucy's Family." *National Geographic Magazine* 189(3):96-117.

Rensberger, Boyce. 1981. "Facing the past." *Science* 81 2(8):40-51.

Sawyer, G.J., and B. Maley. 2005. "Neanderthal reconstructed." *The Anatomical Record* (Part B: New Anat.) 283B:23-31.

Waters, Tom. 1990. "Almost human." *Discover* 11(5):42-45.

Wells, Spencer. 2002. *The Journey of Man: A Genetic Odyssey.* Princeton: Princeton University Press.

33. Primate Societies and Early Human Social Behavior

de Waal, F.B.M. 2002. *Tree of Origin: What Primate Behavior Can Tell Us About Human Social Evolution.* Cambridge: Harvard University Press.

de Waal, F. B. M. 1997. *Bonobo: The Forgotten Ape.* Berkeley: University of California Press.

Penny, David. 2005. "Evolutionary biology: Relativity for molecular clocks." *Nature* 436:183-184.

Wrangham, Richard W. 1987. "The Significance of African apes for reconstructing human social evolution." In Warren G. Kinzey, editor, *The Evolution of Human Behavior: Primate Models* 51-71. Albany: State University of New York Press.

Wrangham, R. and D. Peterson. 1996. *Demonic Males: Apes and the Origins of Human Violence.* Boston: Houghton Mifflin.

34. Evidence for Bipedalism

Coppens, Yves, and Brigitte Senut, editors. 1991. *Origine(s) de la Bipédie chez les Hominidés.* Paris: Cahiers de Paléoanthropologie.

Leakey, Mary D., and John M. Harris, editors. 1987. *Laetoli: A Pliocene Site in Northern Tanzania.* Oxford: Clarendon Press.

Lovejoy, C. Owen. 1988. "Evolution of human walking." *Scientific American* 256(5):118-125.

Sellers, W.I., C.M. Cain, W. Wang, and R.H. Crompton. 2005. "Stride lengths, speed and energy costs in walking of *Australopithecus afarensis*: using evolutionary robotics to predict locomotion of early human ancestors." *Journal of the Royal Society Interface*, In Press.

Stern, Jr., J.T. 2000. "Climbing to the Top: A personal memoir of *Australopithecus afarensis.*" *Evolutionary Anthropology* 9:113-133.

Stern, Jack T. Jr., and Randall L. Susman. 1983. "The locomotor anatomy of *Australopithecus afarensis.*" *American Journal of Physical Anthropology* 60:279-317.

35. The Origins of Bipedalism

Hrdy, Sarah B., and William Bennett. 1981. "Lucy's husband: What did he stand for?" *Harvard Magazine* July-August:7-9, 46.

Kingdon, J. 2003. *Lowly Origin.* Princeton: Princeton University Press.

Lovejoy, C. Owen. 1981. "The origin of man." *Science* 211:341-350.

__________. 1993. "Modeling human origins: are we sexy because we're smart, or smart because we're sexy?" In D. Tab Rasmussen, editor. *The Origin and Evolution of Humans and Humanness.* Boston: Jones and Bartlett Publisher.

Shreeve, James. 1996. "Sunset on the savanna." *Discover* 17(7):116-125.

Stanford, C. 2003. *Upright: The Evolutionary Key to Becoming Human.* New York: Houghton Mifflin Co.

36. The Oldest Stone Tools

*Leakey, Mary D. 1971. *Olduvai Gorge: Excavations in Beds I and II, 1960-1963.* Cambridge: Cambridge University Press.

Panger, M.A., A.S. Brooks, B.G. Richmond, and B. Wood. 2002. "Older than the Oldowan? Rethinking the emergence of hominin tool use." *Evolutionary Anthropology* 11:235-245.

Plummer, T. 2004. "Flaked stones and old bones: biological and cultural evolution at the dawn of technology." *Yearbook of Physical Anthropology* 47:118-164.

Roche, H., A. Delanges, J.-P. Brugal, C. Feibel, M. Kibunjia, V. Ourre, and P.-J. Texier. 1999. "Early hominid stone tool production and technical skill 2.34 Myr ago in West Turkana, Kenya." *Nature* 399:57-60.

Roche, Hèléne, and Jean-Jacques Tiercelin. 1977. "Découverte d'une industrie lithique ancienne in situ dans la formation d'Hadar, Afar central, Éthiopie." *Comptes Rendus des séances de l'Academie des Sciences* (Paris) 284:1871-1874.

Semaw, S., M.J. Rogers, J. Quade, P. Renne, R. Butler, M. Domínguez-Rodrigo, D. Stout, W. Hart, T. Pickering, and S. Simpson. 2003. "2.6-million-year-old stone tools and associated bones from OGS-6 and OGS-7, Gona, Afar, Ethiopia." *Journal of Human Evolution* 45:169-177.

37. Hunters, Gatherers, or Scavengers?

Binford, Lewis R. 1981. *Bones: Ancient Men and Modern Myths.* New York: Academic Press.

__________. 1988. "Fact and fiction about the *Zinjanthropus* floor: data, arguments, and interpretations." *Current Anthropology* 29:123-149.

Brain, Charles K. 1981. *The Hunters or the Hunted?* Chicago: University of Chicago Press.

Domínguez-Rodrigo, M., and T.R. Pickering. 2003. "Early hominid hunting and scavenging: A Zooarcheological Review." *Evolutionary Anthropology* 12:275-282.

Issac, Glynn Ll. 1978. "Food sharing and human evolution: archaeological evidence from the Plio-Pleistocene of East Africa." *Journal of Anthropological Research* 34:311-325.

*Lee, Richard B., and Irven DeVore. 1968. *Man the Hunter.* Chicago: Aldine.

Theime, H. 1997. "Lower Palaeolithic hunting spears from Germany." *Nature* 385:807-810.

Tooby, John, and Irven DeVore. 1987. "The reconstruction of hominid behavioral evolution through strategic modeling." In Warren G. Kinzey, editor, *The Evolution of Human Behavior: Primate models* 183-237. Albany: State University of New York Press.

38. Diet

Scott, Robert S., P.S. Ungar, T.S. Bergstrom, C.A. Brown, F.E. Grine, M.F. Teaford, and A. Walker. 2005. "Dental microwear texture analysis shows within species diet variability in fossil hominins." *Nature* 436:693-695.

Stanford, C.B., and H.T. Bunn, editors. 2001. *Meat-Eating and Human Evolution.* Oxford: Oxford University Press.

Teaford, Mark F. 1994. "Dental microwear and dental function." *Evolutionary Anthropology* 3:17-30.

Ungar, P.S. and M. F. Teaford, editors. 2002. *Human Diet: Its Origin and Evolution.* London: Bergin and Garvey.

39. Cannibalism

Villa, Paola, and E. Mahieu. 1991. "Breakage patterns of human long bones." *Journal of Human Evolution* 21:27-48.

White, Tim D. 1986. "Cut marks on the Bodo cranium: A case of prehistoric defleshing." *American Journal of Physical Anthropology* 69:503-509.

__________. 1992. *Prehistoric Cannibalism at Mancos 5Mtumr-2346.* Princeton: Princeton University Press.

40. Fire

Brain, Charles K., and Andrew Sillen. 1988. "Evidence from the Swartkrans cave for the earliest use of fire." *Nature* 336:464-466.

Fischman, Joshua. 1996. "A fireplace in France." *Discover* 17(1):69.

Goren-Inbar, N. N. Alperson, M.E. Kislev, O. Simchoni, Y. Melamed, A.B-n., and E. Werker. 2004. "Evidence of hominin control of fire at Gesher Benot Ya`aqov, Israel." *Science* 304:725-727.

Gowlett, J.A.J., J.W.K. Harris, D. Walton, and B.A. Wood. 1981. "Early archaeological sites, hominid remains and traces of fire from Chesowanja, Kenya." *Nature* 294:125-129.

James, Steven R. 1989. "Hominid use of fire in the lower and middle Pleistocene." *Current Anthropology* 30(1):1-26.

Rigaud, J. P., J. F. Simek, and T. Ge. 1995. "Mousterian fires from Grotte XVI (Dordogne, France)." *Antiquity* 69:902-912.

41. Shelter

Bordes, François. 1972. *A Tale of Two Caves.* New York: Harper and Row.

Gladkih, Mikhail I., Ninelj L. Kornietz, and Olga Soffer. 1984. "Mammoth bone dwellings on the Russian plain." *Scientific American* 251(5):164-170.

Klein, Richard G. 1973. *Ice Age Hunters of the Ukraine.* Chicago: The University of Chicago Press.

42. Clothing

Spindler, Konrad. 1994. *The Man in the Ice.* New York: Harmony Books.

White, Randall. 1986. *Dark Caves, Bright Visions: Life in Ice Age Europe.* New York: W.W. Norton.

43. Burial

Chase, Philip G., and Harold C. Dibble. 1987. "Middle Paleolithic symbolism: A review of current evidence and interpretations." *Journal of Anthropological Archaeology* 6:263-296.

Gargett, R. H. 1999. "Middle Palaeolithic burial is not a dead issue: the view from Qafzeh, Saint-Césaire, Kebara, Amud, and Dederiyeh." *Journal of Human Evolution* 37:27-90.

Hovers, E. W.H. Kimbel, and Y. Rak. 2000. "The Amud 7 skeleton—still a burial. Response to Gargett. *Journal of Human Evolution* 39:253-260.

44. Art

*Bahn, Paul G., and Jean Vertut. 1988. *Images of the Ice Age.* New York: Facts on File.

Chauvet, Jean Marie, Eliette Brunel Deschamps, and Christian Hillaire. 1996. *Dawn of Art: The Chauvet Cave.* New York: Harry N. Abrams.

Clottes, Jean, and Jean Courtin. 1996. *The Cave Beneath the Sea.* New York: Harry N. Abrams.

Clottes, Jean, and David Lewis-Williams. 1998. *The Shamans of Prehistory: Trance and Magic in the Painted Caves.* New York: Harry N. Abrams.

Leroi-Gourhan, André. 1967. *Treasures of Prehistoric Art.* New York: Harry N. Abrams.

Lewis-Williams, J. David. 1981. *Believing and Seeing: Symbolic Meanings in Southern San Rock Paintings.* New York: Academic Press.

Lorblanchet, Michel. 1988. "From the cave art of the reindeer hunters to the rock art of the kangaroo hunters." *L'Anthropologie* 92:271-316.

Wong, K. 2005. "The morning of the modern mind." *Scientific American* 292(6):86-95.

45. The Origin of Language

Deacon, T. 1997. *The Symbolic species: The Co-evolution of Language and the Brain.* New York: W.W. Norton and Company.

Dunbar, Robin I.M. 1996. *Grooming, Gossip, and the Evolution of Language.* Cambridge: Harvard University Press.

INDEX

Homo neanderthalensis, Amud 1

Shea, J.J. 2003. "Neandertals, competition, and the origin of modern humans in the Levant." *Evolutionary Anthropology* 12:173-187.

Suzuki, Hisashi, and F. Takai, editors. 1970. *The Amud Man and His Cave Site.* Tokyo: The University of Tokyo.

Homo neanderthalensis, Amud 7

Hovers, E., W.H. Kimbel, and Y. Rak. 1999. "The Amud 7 skeleton—still a burial. Response to Gargett." *Journal of Human Evolution* 39:253-260.

Rak, Yoel, William H. Kimbel, and Erella Hovers. 1994. "A Neandertal infant from Amud Cave, Israel." *Journal of Human Evolution* 26:313-324.

Homo neanderthalensis, La Chapelle-aux-Saints

Boule, Marcellin. 1911-1913. "L'homme fossile de La Chapelle-aux-Saints." *Annales de Paléontologie* 6-8.

Straus, William L., Jr., and A.J.E. Cave. 1957. "Pathology and the posture of Neanderthal man." *Quarterly Review of Biology* 32(4):348-363.

Homo neanderthalensis, Ferrassie 1

Heim, Jean-Louis. 1976. "Les hommes fossiles de la Ferrassie I." *Archives de l'Institut de Paléontologie Humaine* 35:1-331.

Puech, Pierre-François. 1981. "Tooth wear in La Ferrassie man." *Current Anthropology* 22(4):424-430.

Homo neanderthalensis, Neandertal 1

Churchill, S.E. 1998. "Cold adaptation, heterochrony, and Neandertals." *Evolutionary Anthropology* 7:46-71.

King, William. 1864. "The reputed fossil man of the Neanderthal." *Quarterly Journal of Science* 1:88-97.

Krings, M., A. Stone, R.-W. Schmitz, H. Krainitzki, M. Stoneking, and S. Pääbo. 1997. "Neandertal DNA sequences and the origin of modern humans." *Cell* 90:19-30.

Schaafhausen, Hermann. 1858. "Zur Kenntnis der ältesten Rassenschädel." *Archiv für Anatomie, Physiologie und wissenschaftliche Medicin* 453-478.

Schmitz, R.W., D. Serre, G. Bonani, S. Feine, F. Hillgruber, H Krainitzki, S. Pääbo, and F.H. Smith. 2002. "The Neandertal type site revisited: Interdisciplinary investigation of skeletal remains from the Neander Valley, Germany." *Nature* 99:13342-13347.

Stringer, C., and R. McKie. 1996. *African Exodus: The Origins of Modern Humanity.* London: Cape.

Wolpoff, M.H., B. Mannheim, A. Mann, J. Hawks, R. Caspari, K.R. Rosenberg, D.W. Frayer, G.W. Gill, and G.A. Clark. 2004. "Why *not* the Neandertals?" *World Archaeology* 36(4):527-546.

Homo neanderthalensis, Gibraltar 1

Keith, Arthur 1911. "The early history of the Gibraltar cranium." *Nature* 87:313-314.

Stringer, C.B, R.N.E. Barton, and J.C. Finlayson, editors. 2000. *Neanderthals on the Edge.* Oxford: Oxbow Books.

Homo neanderthalensis, St.-Césaire

Apsimon, A.M. 1980. "The last neanderthal in France?" *Nature* 287:271-272.

Duarte, C., Maurício, J., Pettitt, P.B., Souto, P., Trinkaus, E., s van der Plicht, H., and J. Zilhão. 1999. "The early Upper Paleolithic human skeleton from the Abrigo do Lagar Velho (Portugal) and modern human emergence in Iberia." *Proceedings of the National Academy of Sciences,* 96:7604-7609.

Mardis, Scott E. 1995. "The last Neanderthals." *Archaeology* 48(6):12-13.

Mercier, N., H. Valladas, J-L. Joron, J-L. Reyss, F. Leveque, and B. Vandermeersch. 1991. "Thermoluminescence dating of the late Neanderthal remains from Saint-Césaire." *Nature* 351:737-739.

Stringer, Christopher B., and Rainer Grün. 1991. "Time for the last Neanderthals." *Nature* 351:701-702.

Tattersall, Ian, and J.L. Schwartz. 1999. "Hominids and hybrids: The place of Neanderthals in human evolution." *Proceedings of the National Academy of Sciences* 96:7117-7119.

Homo floresiensis, LB1

Barham, L., editor. 2005. "Some initial informal reactions to publication of the discovery of *Homo floresiensis* and replies from Brown & Morwood." *Before Farming* 4:1-7.

Falk, D., C. Hildebolt, K. Smith, M. J. Morwood, T. Sutikna, P. Brown, Jatmiko, E. Wayhu Saptomo, B. Brunsden, and F. Prior. 2005. "The Brain of LB1, *Homo floresiensis.*" *Science* 308:242-245.

Lieberman, Daniel E. 2005. "Further fossil finds from Flores." *Nature* 437:957-958.

Morwood, M.J., P. Brown, Jatmiko, T. Sutikna, E. WahyuSaptomo, K.E. Westaway, RokusAweDue, R.G. Roberts, T. Maeda, S. Wasisto, and T, Djubiantono. 2005. "Further evidence for small-bodied hominins from the Late Pleistocene of Flores, Indonesia." *Nature* 437:1012-1017.

Morwood, M. J., R. P. Soejono, R. G. Roberts, T. Sutikna, C. S. M. Turney, K. E. Westaway, W. J. Rink, J.-x. Zhao, G. D. van den Bergh, Rokus Awe Due, D. R. Hobbs, M. W. Moore, M. I. Bird, and L. K. Fifield. 2004. Archaeology and age of a new hominin from Flores in eastern Indonesia. *Nature* 431:1087-91.

Wong, K. 2005. "The littlest human." *Scientific American* 292(2):56-66.

Homo sapiens, Dali

Stringer, Christopher B. 1992. "Reconstructing recent human evolution." *Philosophical Transactions of the Royal Society of London* B 337:217-224.

Wu, Xinzhi, and Frank E. Poirier. 1995. *Human Evolution in China: A Metric Description of the Fossils and a Review of the Sites.* New York: Oxford University Press.

Homo sapiens, Omo I

Leakey, Richard E.F., Karl W. Butzer, and Michael H. Day. 1969. "Early *Homo sapiens* remains from the Omo River Region of South-west Ethiopia." *Nature* 222:1132-1138.

Tattersall, Ian, Francis H. Brown, and John G. Fleage. 2005. "Stratigraphic placement and age of modern humans from Kibish, Ethiopia." *Nature* 433:733-736.

Homo sapiens, Qafzeh IX

Vandermeersch, Bernard. 1981. *Les Hommes Fossiles de Qafzeh.* Paris: Editions du Centre National de la Recherche Scientifique.

Homo sapiens, Skhul V

McCown, Theodore D., and Sir Arthur Keith. 1939. *The Stone Age of Mount Carmel, Volume II.* Oxford: The Clarendon Press.

Homo sapiens, Cro-Magnon 1

Lartet, Eduard, and Henri Christy. 1868. In T.R. Jones, editor, *Reliquiae aquitanicae.* London: Williams and Norgate.

Stringer, C.B., J.-J. Hublin, and B. Vandermeersch. 1984. In F.H. Smith and F. Spencer, editors, *The Origin of Modern Humans: A World Survey of the Fossil Evidence.* New York: Alan R. Liss.

Homo sapiens, Kow Swamp 1

Kennedy, Gail E. 1984. "Are the Kow Swamp hominids 'Archaic'?" *American Journal of Physical Anthropology* 65:163-168.

Thorne, Alan G. and P.G. Macumber. 1972. "Discoveries of Late Pleistocene man at Kow Swamp, Australia." *Nature* 238:316-319.

Paleolithic Technology

Asfaw, Berhane, Y. Beyene, G. Suwa, R.C. Walter, T.D. White, G. WoldeGabriel, and T. Yemane. 1992. "The earliest Acheulean from Konso-Gardula." *Nature* 360:732-734.

Gutin, JoAnn. 1995. "Do Kenya tools root birth of modern thought in Africa?" *Science* 270:1118-1119.

Henshilwood, C.S., F.E. d'Errico, C.W. Marean, R.G. Milo, R. Yates. 2001. An early bone tool industry from the Middle Stone Age at Blombos Cave, South Africa: implications for the origins of modern human behaviour, symbolism and language. *Journal of Human Evolution* 41:631-678.

Hublin, Jean-Jacques, F. Spoor, M. Braun, F. Zonneveld, and S. Condemi. 1996. "A late Neanderthal associated with Upper Palaeolithic artefacts." *Nature* 381:224-226.

Odell, George H. 2004. *Lithic Analysis.* Norwell, Massachusetts: Plenum Publishers.

★Schick, Kathy D., and Nicholas Toth. 1993. *Making Silent Stones Speak: Human Evolution and the Dawn of Technology.* New York: Simon and Schuster.

Wymer, John. 1982. The Paleolithic Age. London: Croom Helm.

★RECOMMENDED READING

Aiello, Leslie, and M. Christopher Dean. 1990. *An Introduction to Human Evolutionary Anatomy.* London: Academic Press.

Alexander, R.D. 1979. *Darwinism and Human Affairs.* Seattle: University of Washington Press.

—————. 1987. *The Biology of Moral Systems.* New York: Aldine de Gruyter.

Alexander, R. McNeill. 2005. *Human Bones: A Scientific and Pictorial Investigation.* New York: Pi Press.

Ciochon, Russell, and John G. Fleagle, editors. 1987. *Primate Evolution and Human Origins.* New York: Aldine de Gruyter.

—————. 1993. *The Human Evolution Source Book.* Englewood Cliffs: Prentice Hall.

Day, Michael H. 1986. *Guide to Fossil Man* (Fourth Edition). Chicago: University of Chicago Press.

Darwin, Charles. 1859. *On the Origin of Species by Means of Natural Selection, or the Preservation of Favoured Races in the Struggle for Life.* London: John Murray.

—————. 1871. *The Descent of Man, and Selection in Relation to Sex,* 2 Volumes. London: John Murray.

Dawkins, Richard. 2004. *The Ancestor's Tale: A Pilgrimage to the Dawn of Evolution.* Boston: Houghton Mifflin.

Delson, Eric, editor. 1985. *Ancestors: The Hard Evidence.* New York: Alan R. Liss.

Diamond, Jared. 1998. *Guns, Germs, and Steel: The Fates of Human Societies.* New York: W. W. Norton & Company.

Hart, Donna and Robert W. Sussman. 2005. *Man the Hunted: Primates, Predators and Human Evolution.* Boulder: Westview Press.

Hartwig, W., editor. 2002. *The Primate Fossil Record.* Cambridge: Cambridge University Press.

Johanson, Donald, Lenora Johanson, and Blake Edgar. 1994. *Ancestors: In Search of Human Origins.* New York: Villard Books.

Jones, Steve, Robert Martin, and David Pilbeam, editors. 1992. *The Cambridge Encyclopedia of Human Evolution.* New York: Cambridge University Press.

Mai, L.L., M.Y. Owl, and M.P. Kersting. 2005. *The Cambridge Dictionary of Human Biology and Evolution.* New York: Cambridge University Press.

Mayr, Ernst. 2002. *What Evolution Is.* New York: Basic Books.

Meikle, W. Eric, F. Clark Howell, and Nina G. Jablonski, editors. 1996. *Contemporary Issues in Human Evolution.* San Francisco: California Academy of Sciences.

Mellars, P. 1996. *The Neanderthal Legacy: An Archaeological Perspective from Western Europe.* Princeton: Princeton University Press.

Morell, Virginia. 1995. *Ancestral Passions : The Leakey Family and the Quest for Humankind's Beginnings.* New York: Simon & Schuster.

Reader, John. 1981. *Missing Links: The Hunt for Earliest Man.* London: Collins.

Rennie, J. editor. 2003 "New Look at Human Evolution." *Scientific American:* vol. 13.

Ridley, Matt. 1999. *Genome: The Autobiography of a Species in 23 Chapters.* New York: HarperCollins.

Rightmire, G. Philip. 1990. *The Evolution of* Homo erectus*: Comparative anatomical studies of an extinct human species.* Cambridge: Cambridge University Press.

Sherratt, Andrew, editor. 1980. *The Cambridge Encyclopedia of Archaeology.* New York: Crown Publishers.

Shreeve, James. 1995. *The Neandertal Enigma: Solving the Mystery of Modern Human Origins.* New York: William Morrow.

Sloan, Chris. 2004. *The Human Story: Our Evolution from Prehistoric Ancestors to Today.* Washington, D.C.: National Geographic.

Stahl, A.B., editor. 2005. *African Archaeology.* Oxford: Blackwell Publishing.

Stringer, Christopher, and Peter Andrews. 2005. *The Complete World of Human Evolution.* London: Thames & Hudson.

Stringer, Christopher, and Clive Gamble. 1993. *In Search of the Neanderthals.* London: Thames and Hudson.

Stringer, Christopher, and Robin McKie. 1996. *African Exodus: The Origins of Modern Humanity.* London: Cape.

Tattersall, Ian. 1995. *The Last Neandertal: The Rise, Success, and Mysterious Extinction of Our Closest Human Relatives.* New York: Macmillan.

—————. 1995. *The Fossil Trail: How We Know What We Think We Know about Human Evolution.* New York: Oxford University Press.

Tattersall, Ian, Eric Delson, John Van Couvering, and Alison Brooks, editors. 1988. *Encyclopedia of Human Evolution and Prehistory.* New York: Garland Publishing.

Trinkaus, Erik, and Pat Shipman. 1992. *The Neandertals: Changing the Image of Mankind.* New York: Alfred A. Knopf.

Walker, A. and P. Shipman. 2005. *The Ape in the Tree: An Intellectual and Natural History of* Proconsul. Cambridge: Harvard University Press.

Wright, Robert. 1994. *The Moral Animal: The New Science of Evolutionary Psychology.* New York: Pantheon Books.

WEBSITES

American Museum of Natural History
www.amnh.org
Atapuerca
www.ucm.es/info/paleo/ata/english/
Darwiniana and Evolution
http://members.aol.com/darwinpage/hominid.htm
Institute of Human Origins
www.becominghuman.org
Mapping Humanity's Genetic Journey Through the Ages
ww5.nationalgeographic.com/genographic
Museum of Paleontology, University of California, Berkeley
www.ucmp.berkeley.edu
National Center for Science Education
www.ncseweb.org
National Geographic Society
www.nationalgeographic.com
The Smithsonian Institution
www.mnh.si.eduanthro/humanorigins
Talk.Origins
www.talkorigins.org

Australopithecus africanus, STS 14

Abitbol, M. Maurice. 1995. "Reconstruction of the STS 14 (*Australopithecus africanus*) pelvis." *American Journal of Physical Anthropology* 96:143-158.

Robinson, John T. 1972. *Early Hominid Posture and Locomotion*. Chicago: University of Chicago Press.

Australopithecus africanus, Sts 71 and Sts 36

Rak, Yoel. 1983. *The Australopithecine Face*. New York: Academic Press.

Australopithecus africanus, Taung Child

Berger, Lee R., and Ronald J. Clarke. 1995. "Eagle involvement in accumulation of the Taung child fauna." *Journal of Human Evolution* 29:275-299.

Broom, Robert. 1925. "Some notes on the Taungs skull." *Nature* 115:569-571.

Dart, Raymond A. 1967. *Adventures with the Missing Link*. Philadelphia: The Institutes Press.

Tobias, Phillip V. 1992. "New researches at Sterkfontein and Taung with a note on Piltdown and its relevance to the history of paleo-anthropology." *Transactions of the Royal Society of South Africa* 48:1-14.

Australopithecus africanus, TM 1517

Rak, Yoel. 1983. *The Australopithecine Face*. New York: Academic Press.

Robinson, John T. 1956. *The Dentition of the Australopithecinae. Transvaal Museum Memoir* 9:1-179.

Australopithecus africanus, StW 252

Clarke, Ron J. 1996. "The genus *Paranthropus*: What's in a name?" In Meikle, W. Eric, F. Clark Howell, and Nina G. Jablonski, editors, *Contemporary Issues in Human Evolution*. San Francisco: California Academy of Sciences.

Australopithecus crassidens, SK6, SK 48, and SK 79

Rak, Yoel. 1983. *The Australopithecine Face*. New York: Academic Press.

Robinson, John T. 1956. *The Dentition of the Australopithecinae. Transvaal Museum Memoir* 9:1-179.

Australopithecus aethiopicus, KNM-WT 17000

Arambourg, Camille, and Yves Coppens. 1968. "Sur la découverte dans le Pléistocène inférieur de la vallé de l'Omo (Éthiopie) d'une mandibule d'Australopithécien. *Comptes Rendus des séances de l'Academie des Sciences*, Paris 265:589-590.

__________. 1968. "Découverte d'un Australopithécien nouveau dans les gisements de l'Omo (Ethiopie)." *South African Journal of Science* 64:58-59.

Kimbel, William H., Tim D. White, and Donald C. Johanson. 1988. "Implications of KNM-WT 17000 for the evolution of "Robust" *Australopithecus*." In Fred E. Grine, editor, *Evolutionary History of the "Robust" Australopithecines* 259-268. New York: Aldine de Gruyter.

Australopithecus boisei, OH 5

Suwa, G., B. Asfaw, Y. Beyene, T.D. White, S. Katoh, S. Nagaoka, H. Nakaya, K. Uzawa, P. Renne, and G. Wolde Gabriel. 1997. "The first skull of *Australopithecus boisei*." *Nature* 389:489-492.

*Tobias, Phillip V. 1968. *The Cranium and Maxillary Dentition of* Australopithecus (Zinjanthropus) boisei. *Olduvai Gorge*, Volume 2. Cambridge: Cambridge University Press.

Australopithecus boisei, KNM-ER 406 and KNM-ER 732

Walker, Alan C., and Richard E.F. Leakey. 1978. "The hominids of East Turkana." *Scientific American* 239(2):54-66.

Homo

Bramble, D.M., and D.E. Lieberman. 2004. "Endurance running and the evolution of *Homo*." *Nature* 432:345-352.

Dunsworth, H., and A. Walker. 2002. "Early Genus *Homo*." In W. Hartwig, editor, *The Primate Fossil Record* 83-96. Cambridge: Cambridge University Press.

Kramer, Andrew, Steven M. Donnelly, James H. Kidder, Stephen D. Ousley, and Stephen M. Olah. 1995. "Craniometric variation in large-bodied hominoids: testing the single-species hypothesis for *Homo habilis*." *Journal of Human Evolution* 29:443-462.

Miller, Joseph A. 1991. "Does brain size variability provide evidence for multiple species in *Homo habilis*?" *American Journal of Physical Anthropology* 84:385-398.

Stringer, Chris B. 1986. "The credibility of *Homo habilis*." In Bernard A. Wood, Lawrence Martin and Peter Andrews, editors, *Major Topics in Primate and Human Evolution* 266-294. Cambridge: Cambridge University Press.

Tattersall, Ian, and Schwartz, J.L. 2001. *The Human Fossil Record, Volume 1, Terminology and Craniodental Morphology of Genus* Homo *(Europe)*. New York: Wiley Liss.

__________. 2005. *The Human Fossil Record, Volume 2, Craniodental Morphology of Genus* Homo *(Africa and Asia)*. New York: Wiley Liss.

*Tobias, Phillip V. 1991. *Olduvai Gorge: Volume 4. The skulls, endocasts and teeth of* Homo habilis. Cambridge: Cambridge University Press.

*Wood, Bernard A. 1991. *Koobi Fora Research Project. Volume 4: Hominid Cranial Remains*. Oxford: Clarendon Press.

__________. 1992. "Origin and evolution of the genus *Homo*." *Nature* 355:783-790.

__________. 1993. "Early *Homo*. How many species?" In William H. Kimbel and Lawrence B. Martin, editors. *Species, Species Concepts, and Primate Evolution* 485-522. New York: Plenum Press.

Wood. B.A., and Collard, M. 1999. "The changing face of genus *Homo*." *Evolutionary Anthropology* 8(6):195-207.

Homo habilis, OH 7 and OH 24

Blumenschine, R.J., C.R. Peters, F. T. Masao, R.J. Clarke, A.L. Deino, R.L. Hay, C.C. Swisher, I.G. Stanistreet, G.M. Ashley, L.J. McHenry, N.E. Sikes, N.J. van der Merwe, J.C. Tactikos, A.E. Cushing, D.M. Deocampo, J.K. Jnau, and J.I. Ebert. 2003. "Late Pliocene *Homo* and hominid land use from western Olduvai Gorge, Tanzania." *Nature* 299:1217-1221.

Leakey, Louis S.B., P.V. Tobias, and J.R. Napier. 1964. "A new species of genus *Homo* from Olduvai Gorge." *Nature* 202(4927):7-9.

Leakey, Mary D., Ronald J. Clarke, and Louis S.B. Leakey. 1971. "New hominid skull from Bed I, Olduvai Gorge, Tanzania." *Nature* 232:308-312.

Rightmire, G. Philip. 1993. "Variation Among early *Homo* crania from Olduvai Gorge and the Koobi Fora region." *American Journal of Physical Anthropology* 90(1):1-33.

Tobias, Phillip V. 1991. *Olduvai Gorge: The Skulls, Endocasts and Teeth of* Homo habilis, *Volume 4*. Cambridge: Cambridge University Press.

Homo habilis, KNM-ER 1813

Leakey, R.E.F. 1974. "Further evidence of Lower Pleistocene hominids from East Rudolf, North Kenya, 1973." *Nature* 248: 653-656.

Leakey, Richard, and Roger Lewin. 1992. *Origins Reconsidered: In Search of What Makes Us Human*. New York: Doubleday.

Wood, Bernard A. 1991. *Koobi Fora Research Project Volume 4: Hominid Cranial Remains*. Oxford: Oxford University Press.

Homo habilis, OH 62

Hartwig-Scherer, S. 1993. "Body weight prediction in early fossil hominids: Towards a taxon-"independent" approach." *American Journal of Physical Anthropology* 92:17-36.

Johanson, Donald C., and James Shreeve. 1989. *Lucy's Child: The Discovery of a Human Ancestor*. New York: William Morrow.

Homo rudolfensis, KNM-ER 1470

Leakey, Richard E.F. 1973. "Evidence for an advanced Plio-Pleistocene hominid from East Rudolf, Kenya." *Nature* 242:447-450.

Wood, Bernard A. 1991. *Koobi Fora Research Project, Volume 4: Hominid Cranial Remains*. Oxford: Oxford University Press.

Homo georgicus, D 2600

Fishman, J. 2005. "Family ties". *National Geographic Magazine* 202(4):16-27.

Gabunia, L. S.C. Antón, D. Lordkipanidze, A. Vekua, A. Justus, and C.C. Swisher III. 2001. "Dmanisi and dispersal." *Evolutionary Anthropology* 10:158-170.

Vekua, A., D. Lordkipanidze, G. P. Rightmire, J. Agusti, R. Ferring, G. Maisuradze, A. Mouskhelishvili, M. Nioradze, M. Ponce de Leon, Martha Tappen, M. Tvalchrelidze, C. Zollikofer. 2002. "A new skull of early *Homo* from Dmanisi, Georgia." *Science* 297:85-89.

Gore, R. 2002. "New find". *National Geographic Magazine* 202(2).

Wong, K. 2003. "Stranger in a new land." *Scientific American* 289(5):74-83.

Homo ergaster, KNM-ER 3733

Leakey, Richard E.F. 1976. "New hominid fossils from the Koobi Fora Formation in Northern Kenya." *Nature* 261:574-576.

Homo ergaster, KNM-WT 15000

Brown, Frank, John Harris, Richard Leakey, and Alan Walker. 1985. "Early *Homo erectus* skeleton from west Lake Turkana, Kenya." *Nature* 316: 788-792.

*Walker, Alan, and Richard Leakey, editors. 1993. *The Nariokotome* Homo erectus *Skeleton*. Cambridge: Harvard University Press.

Walker, A.C., and P. Shipman. 1996. *The Wisdom of the Bones*. New York: Alfred A. Knopf.

Homo ergaster, SK 847

Clarke, Ronald J., and F. Clark Howell. 1972. "Affinities of the Swartkrans 847 hominid cranium." *American Journal of Physical Anthropology* 37(3):319-335.

Clarke, R.J., F. Clark Howell, and C.K. Brain. 1970. "More evidence of an advanced hominid at Swartkrans." *Nature* 225:1219-1222.

Homo erectus, Trinil 2

Dubois, Eugène 1926. "On the principal characters of the cranium and the brain, the mandible and the teeth of *Pithecanthropus erectus*." *Proceeding of the Academy of Science, Amsterdam* 27:265-278.

Shipman, P. 2001. *The Man Who Found the Missing Link: Eugene Dubois and His Lifelong Quest to Prove Darwin Right*. New York: Simon and Schuster.

Homo erectus, Zhoukoudian (Peking Man)

Rukang, Wu, and Lin Shenglong. 1983. "Peking Man." *Scientific American* 248(6):86-94.

Sawyer, Gary J., and Ian Tattersall. 1995. "A New Reconstruction of the Skull of *Homo erectus* from Zhoukoudian, China." *Proceedings of The International Conference of Human Paleontology* 277-286.

Tattersall, Ian, and Gary J. Sawyer. 1996. "The skull of *Sinanthropus* from Zhoukoudian, China: A new reconstuction." *Journal of Human Evolution* 31:311-314.

Weidenreich, Franz. 1943. "The Skull of *Sinanthropus pekinensis*: a comparative study on a primitive hominid skull." *Palaeontologia Sinica*, New Series D 10:1-485.

Homo erectus, Sangiran 17

Jacob, Teuku. 1975. "Morphology and paleoecology of early man in Java." In Russell H. Tuttle, editor. *Paleoanthropology, Morphology and Paleoecology*. 311-326. The Hague: Mouton Publishers.

Sartono, S. 1972. "Discovery of another hominid skull at Sangiran, Central Java." *Current Anthropology* 13(1):124-125.

__________. 1975. "Implications arising from *Pithecanthropus* VIII." In Russell H. Tuttle, editor. *Paleoanthropology, Morphology and Paleoecology*. 327-360. The Hague: Mouton Publishers.

Homo heidelbergensis, Bodo

Clark, J.D., J. de Heinzelin, K.D. Schick, W.K. Hart, T.D. White, G. Woldegabriel, R.C. Walter, G. Suwa, B. Asfaw, E. Vrba , and Y. H-Selassie. 1994. "African *Homo erectus* : Old radiometric ages and young Oldowan assemblages in the Middle Awash Valley, Ethiopia." *Science* 264:1907-1909.

Conroy, Glenn C., C.J. Jolly, D. Cramer, and J.E. Kalb. 1978. "Newly discovered fossil hominid skull from the Afar depression, Ethiopia." *Nature* 276:67-70.

Homo heidelbergensis, Mauer 1

Kraatz, Reinhart. 1985. In Eric Delson, editor, *Ancestors: The Hard Evidence*. New York: Alan R. Liss.

Rightmire, G.P. 1998. "Human evolution in the Middle Pleistocene: The role of *Homo heidelbergensis*." *Evolutionary Anthropology* 8:218-227.

Schoetensack, Otto. 1908. *Der Unterkiefer des* Homo heidelbergensis *aus den Sanden von Mauer bei Heidelberg*. Leipzig: Wilhelm Engelmann.

Homo heidelbergensis, Arago XXI

Lumley, Henry de, editor. 1979. *L'Homme de Tautavel. Dossiers de l'Archéologie* 36.

Homo heidelbergensis, Petralona 1

Kokkoros, P., and A. Kanellis. 1960. "Découverte d'un crane d'homme paléolithique dans la peninsule Chalcidique." *Anthropologie* 64:132-147.

Poulianos, A.N. 1971. "Petralona: A Middle Pleistocene cave in Greece." *Archaeology* 24:6-11.

Stringer, Christopher B. 1974. "A multivariate study of the Petralona skull." *Journal of Human Evolution* 3:397-404.

Homo heidelbergensis, Steinheim

Adam, Karl Dietrich. 1985. "The chronological and systematic position of the Steinheim skull." In Eric Delson, editor, *Ancestors: The Hard Evidence*. 272-276. New York: Alan R. Liss.

Homo heidelbergensis, Atapuerca 5

Arsuaga, Juan-Luis, I. Martínez, A. Gracia, J.-M. Carretero, and E. Carbonell. 1993. "Three new human skulls from the Sima de los Huesos Middle Pleistocene site in Sierra de Atapuerca, Spain." *Nature* 362:534-537.

Bermúdez de Castro, J.M., M. Martinón-Torres, E. Carbonell, S. Sarmiento, A. Rosas, J. vand der Made, and M. Lozano. 2004. "The Atapuerca sites and their contribution to the knowledge of human evolution in Europe." *Evolutionary Anthropology* 13:25-41.

Stringer, Christopher B 1993. "Secrets of the pit of the bones." *Nature* 362:501-502.

Homo heidelbergensis, Broken Hill 1

Keith, Arthur. 1926. *The Antiquity of Man*, Second Edition. London: Williams and Norgate

Klein, Richard G. 1973. "Geological antiquity of Rhodesian Man." *Nature* 244:311-312.

Woodward, Arthur S. 1921. "A new cave man from Rhodesia, South Africa." *Nature* 108(2716):371-372.

Homo neanderthalensis, Krapina C

Smith, Fred H. 1976. *The Neandertal remains from Krapina: A descriptive and comparative study*. University of Tennessee Department of Anthropology Report of Investigations 15:1-359.

Homo neanderthalensis, Saccopastore 1

Manzi, Giorgio, and Pietro Passarello. 1991. "Anténéandertaliens et Néandertaliens du Latium (Italie Centrale)" *L'Anthropologie* 95(2/3):501-522.

Homo neanderthalensis, Teshik-Tash

Movius, Hallam L. Jr. 1953. "The Mousterian cave of Teshik-Tash, southeastern Uzbekistan, Central Asia." *American School of Preshistoric Research Bulletin* 17:11-71.

Homo neanderthalensis, Kebara 2

Bar-Yosef, Ofer, B. Vandermeersch, B. Arensburg, A. Belfer-Cohen, P. Goldberg, H. Laville, L. Meignen, Y. Rak, J.D. Speth, E. Tchernov, A-M. Tillier, and S. Weiner. 1992. "The Excavations in Kebara Cave, Mt. Carmel." *Current Anthropology* 33(5):497-550.

Enard, W., M. Przeworski, S. E. Fisher, C. S. L. Lai, V. Wiebe, T. Kitano, A. P. Monaco, and Svante Pääbo 2002. "Molecular evolution of FOXP2, a gene involved in speech and language." *Nature* 418:869-872.
Noble, William, and Iain Davidson. 1991. "The evolutionary emergence of modern human behaviour: Language and its archaeology." *Man* 26(2):223-254.
Pinker, Steven. 1994. *The Language Instinct: How the Mind Creates Language*. New York: William Morrow.

46. The Problem of Consciousness

Dennett, Daniel C. 1991. *Consciousness Explained*. Boston: Little, Brown.
Flinn, M.V., D.C. Geary, and C.V. Ward. 2004. "Ecological dominance, social competition, and coalitionary arms races: Why humans evolved extraordinary intelligence." *Evolution and Human Behavior* 26:10-46.
Geary, D.C. 2004. *The Origin Of The Mind: Evolution Of Brain, Cognition, And General Intelligence*. Washington, D.C.: American Psychological Association.
Horgan, John. 1994. "Can science explain consciousness?" *Scientific American* 268(7):88-94.
Humphrey, Nicholas. 1992. *A History of the Mind*. New York: Simon and Schuster.
Klein, Richard and Blake Edgar. 2002. *The Dawn of Human Culture*. New York: John Wiley and Sons.
Koch, C. 2004. *The Quest for Consciousness: A Neurobiological Approach*. Greenwood Village, Colorado: Roberts and Co. Publishers.
Mithen, Steven. 1966. *The Prehistory of the Mind: The Cognitive Origins of Art and Science*. New York: Thames and Hudson.
Pinker, S. 1997. *How the Mind Works*. New York: W.W. Norton and Company.
Tomasello, Michael 1999. *The Cultural Origins of Human Cognition*. Cambridge: Harvard University Press.

47. Will Humans Become Extinct?

Eldredge, Niles. 1998. *Life In the Balance*. Princeton: Princeton University Press.
McKibben, Bill. 1989. *The End of Nature*. New York: Random House.
Nitecki, Matthew H., editor. 1984. *Extinctions*. Chicago: The University of Chicago Press.
Tattersall, Ian, and Jeffrey L. Schwartz. 2001. *Extinct Humans*. Denver: Westview Press.
Ward, Peter. 1994. *The End of Evolution: On Mass Extinctions and the Preservation of Biodiveristy*. New York: Bantam Books.
Wilson, Edward O. 1992. *The Diversity of Life*. Cambridge: Harvard University Press.

48. Place of Humans in Nature

Huxley, Thomas H. 1863. *Evidence as to Man's Place in Nature*. London: Williams and Norgate.
Margulis, Lynn, and Dorion Sagan. 1995. *What Is Life?*. Berkeley: University of California Press.

PART 2:
Encountering the Evidence

***Sahelanthropus tchadensis*, TM 266-01-60-1**

Brunet, Michel, Alain Beauvilain, Yves Coppens, Emile Heintz, Aladji H.E. Moutaye, and David Pilbeam. 1995. "The first australopithecine 2,500 kilometres west of the Rift Valley (Chad). *Nature* 378:273-275.
__________ 1996. "*Australopithecus bahrelghazali*, une nouvelle espèce d'Hominidé ancien de la région de Koro Toro (Tchad)." *Comptes Rendus des séances de l'Academie des Sciences*. Paris 322:907-913.
Wolpoff, M.H., B. Senut, M. Pickford, and J. Hawks. 2002. "Palaeoanthropology:*Sahelanthropus* or '*Sahelpithecus*'?" *Nature* 419:581-582.
Zollikofer, C. P. E., M. S. Ponce de León, D.E. Lieberman, F. Guy, D. Pilbeam, A. Likius, Hassane T. Mackaye, P. Vignaud & M. Brunet. 2005. "Virtual cranial reconstruction of *Sahelanthropus tchadensis*." *Nature* 434:755-759.

***Orrorin tugenensis*, BAR 1000'00**

Galik, K. B. Senut, M. Pickford, and J. Gommery. 2004. "External and Internal Morphology of the BAR 1002 00 *Orrorin tugenensis* Femur." *Science* 305:1450-1453.
Wong, K. 2003. "An ancestor to call our own". *Scientific American* 288(1):54-63.

***Ardipithecus ramidus*, ARA-VP-6/129**

Day, Michael H. 1995. "Remarkable delay." *Nature* 376:111.
Haile-Selassie, Y., G. Suwa, and T.D. White. 2004. "Late Miocene Teeth from Middle Awash, Ethiopia, and Early Hominid Dental Evolution." *Science* 303:1503-1505.
Kalb, Jon E., Clifford J. Jolly, Elizabeth B. Oswald, and Paul F. Whitehead. 1984. "Early hominid habitation in Ethiopia." *American Scientist* March-April 72:168-178.
Kappelman, John, and John G. Fleagle. 1995. "Age of early hominids." *Nature* 376:558-559.
Semaw, S., S.W. Simpson, J. Quade, P.R. Renne, F. Butler, W.C. McIntosh, N. Levin, M. Dominguez-Rodgrio, and M.J. Rogers. 2005. "Early Pliocene hominids from Gona, Ethiopia." *Nature* 433:301-305.
White, Tim D., Gen Suwa, and Berhane Asfaw. 1995. "*Australopithecus ramidus*, a new species of early hominid from Aramis, Ethiopia." *Nature* 375:88.
WoldeGabriel, Giday, Tim D. White, Gen Suwa, Paul Renne, Jean de Heinzelin, William K. Hart, and Grant Heiken. 1994. "Ecological and temporal placement of early Pliocene hominids at Aramis, Ethiopia." *Nature* 371:330-333.

***Kenyanthropus platyops*, KNM-WT 40000**

Lieberman, D.E. 2001. "Another face in our family tree." *Science* 410: 419-420.
White, T.D., 2003. "Early hominids—diversity or distortion?" *Science* 299: 1194-1997.

Australopithecines

Broom, Robert, and John T. Robinson 1950. *Further Evidence of the Structure of the Sterkfontein Ape-Man* Plesianthropus. *Transvaal Museum Memoir* 4:11-84.
Broom, Robert, and G.W.H. Schepers. 1946. *The South African Fossil Ape-Men. The Australopithecinae. Transvaal Museum Memoir* 2:1-271.
Broom, Robert, and John T. Robinson. 1952. *Swartkrans Ape-Man:* Paranthropus crassidens. *Transvaal Museum Memoir* 6:1-123.
Clark, W.E. Le Gros. 1967. *Man-Apes or Ape-Men? The Story of Discoveries in Africa*. New York: Holt, Rinehart and Winston.
*Grine, Fred E. 1993. "Australopithecine taxonomy and phylogeny: historical background and recent interpretation." In Russell L. Ciochon and John G. Fleagle, editors, *The Human Evolution Source Book* 145-175. Englewood Cliffs, New Jersey: Prentice Hall.
Grine, Fred E., editor. 1988. *Evolutionary History of the "Robust" Australopithecines*. New York: Aldine de Gruyter.
Howell, F. Clark. 1978. "Hominidae." In Vincent J. Maglio and H.B.S. Cooke, editors, *Evolution of African Mammals* 154-428. Cambridge: Harvard University Press.
Johanson, Donald C., and Tim D. White. 1979. "A systematic assessment of early African hominids." *Science* 202:321-330.
McHenry, H.M. and K.E. Coffing. 2000. "*Australopithecus* to *Homo*: Transformations of body and mind." *Annual Review of Anthropology* 2000 29:125-166.
Rak, Yoel. 1983. *The Australopithecine Face*. New York: Academic Press.
*Reed, Charles A. 1983. "A short history of the discovery and early study of the Australopithecines: the first find to the death of Robert Broom (1924-1951)." In Kathleen J. Reichs, editor, *Hominid Origins: Inquiries Past and Present* 1-77. Washington, D.C.: University Press of America.
Robinson, John T. 1956. *The Dentition of the Australopithecinae. Transvaal Musuem Memoir* NO. 9:1-179.
Tattersall, Ian, and Schwartz, J.L. 2005. *The Human Fossil Record, Volume 4, Draniodental Morphology of Early Hominids and Overview*. New York: Wiley-Liss.
Tobias, Phillip V. 1968. *The Cranium and Maxillary Dentition of* Australopithecus (Zinjanthropus) boisei. *Olduvai Gorge. Vol. 2*. Cambridge: Cambridge University Press.
__________. 1988. "*Australopithecus afarensis*" and *A. africanus*: critique and an alternative hypothesis." *Palaeontologia Africana* 23:1-17.
Wallace, John A. 1972. *The Dentition of the south African Early Hominids: A study of form and Function*. Ph.D. Dissertation, The University of the Witwatersrand.
__________1974. "Dietary adaptations of *Australopithecus* and early *Homo*. In Russell H. Tuttle, editor, *Paleoanthropology, Morphology and Paleoecology* 204-223. The Hague: Mouton Publishers.
White, Tim D., Donald C. Johanson, and William H. Kimbel. 1981. "*Australopithecus africanus*: its phyletic position reconsidered." *South African Journal of Science* 77:445-470.
Wood, B. and B.G. Richmond. 2000. "Human Evolution: Taxonomy and paleobiology." *Journal of Anatomy* 196:19-60.

***Australopithecus anamensis*, KNM-KP 29281**

Coffing, Katherine, Craig Feibel, Meave Leakey, and Alan Walker. 1994. "Four-million-year-old hominids from east Lake Turkana, Kenya." *American Journal of Physical Anthropology* 93:55-65.
Kimbel, W.H., C.A. Lockwood, C.V. Ward, M.G. Leakey, Y. Rak, and D.C. Johanson. In Press. "Was *Australopithecus anamensis* ancestral to *A. afarensis*. A case of anagenesis in the hominin fossil record." *Journal of Human Evolution*.
Leakey, Meave. 1995. "The farthest horizon." *National Geographic Magazine* 188:38-51.
Ward, C.V, M.G. Leakey, and A. Walker. 2001. "Morphology of *Australopithecus anamensis* from Kanapoi and Allia Bay, Kenya." *Journal of Human Evolution* 41:255-368.

***Australopithecus garhi*, BOU-VP-12/130**

Asfaw, B., T. White, O. Lovejoy, B. Latimer, S. Simpson and G. Suwa. 1999. "Cladistics and early hominid phylogeny: Response". *Science* 285:1209.
McCollum, M. 1999. "Cladistics and early hominid phylogeny: Response". *Science* 285:1209.
Strait, D.S. and F.E. Grine. 1999. "Cladistics and early hominid phylogeny". *Science* 285:1209.

***Australopithecus afarensis*, A.L. 288-1, Lucy**

Alemseged, Z, J.G. Wynn, W.H. Kimbel, D. Reed, D. Geraads and R. Bobe. 2005. "A new hominin from the Basal Member of the Hadar Formation, Dikika, Ethiopia, and its geological context." *Journal of Human Evolution*. 49:499-514.
Häusler, M. and P. Schmid. 1995. "Comparison of the pelves of Sts 14 and AL 288-1: implications for birth and sexual dimorphism in australopithecines." *Journal of Human Evolution* 29:363-383.
Johanson, D.C. 2004. "Lucy, Thirty years later: An expanded view of *Australopithecus afarensis*." *Journal of Anthropological Research* 60(4):465-486.
Johanson, Donald C., and Maitland A. Edey. 1981. *Lucy: The Beginnings of Humankind*. New York: Simon and Schuster.
Johanson, Donald C., C. Owen Lovejoy, William H. Kimbel, Tim D. White, Steven C. Ward, Michael E. Bush, Bruce M. Latimer, and Yves Coppens. 1982. "Morphology of the Pliocene partial hominid skeleton (A.L. 288-1) from the Hadar Formation, Ethiopia." *American Journal of Physical Anthropology* 57:403-451.
Lockwood, C. A., W.H. Kimbel and D.C. Johanson. 2000. "Temporal trends and metric variation in the mandible and dentition of *Australopithecus afarensis*." *Journal of Human Evolution* 39:23-55.
Tague, R.G. and C.O. Lovejoy. 1998. "AL 288-1 Lucy or Lucifer: gender confusion in the Pliocene." *Journal of Human Evolution* 35:75-94

***Australopithecus afarensis*, A.L. 333-105, The First Family**

Johanson, Donald C. 1976. "Ethiopia yields first "family" of early man." *National Geographic Magazine* 150:791-811.
Kimbel, William H., Donald C. Johanson, and Yves Coppens. 1982. "Pliocene hominid cranial remains from the Hadar Formation, Ethiopia." *American Journal of Physical Anthropology* 57:453-499.
White, Tim D., and Donald C. Johanson. 1989. "The hominid composition of Afar locality 333: some preliminary observations. In Giacomo Giacobini, editor, *Hominidae* 97-101. Milan: Jaca Book.

***Australopithecus afarensis*, A.L. 444-2**

Johanson, Donald C. 1996. "Face-to-face with Lucy's family." *National Geographic Magazine* 189(3):96-117.
Kimbel, W.H., Y. Rak, and D.C. Johanson. 2004. *The Skull of* Australopithecus afarensis. New York: Oxford University Press.
Kimbel, William H., Tim D. White, and Donald C. Johanson. 1984. "Cranial morphology of *Australopithecus afarensis*: a comparative study based on composite reconstruction of the adult skull." *American Journal of Physical Anthropology* 64:337-388.

***Australopithecus afarensis*, A.L. 129-1a+1b**

Johanson, Donald C., C. Owen Lovejoy, and Kingsbury G. Heiple. 1976. "Functional implications of the Afar knee joint. *American Journal of Physical Anthropology* 45:188.
Lovejoy, C.O., R.S. Meindl, J.C. Ohman, K.G. Heiple, and T.D. White. 2002. "The Maka femur and its bearing on the antiquity of human walking: applying contemporary concepts of morphogenesis to the human fossil record." *American Journal of Physical Anthropology* 119:97-133.
Tardieu, Christine. 1991. "Étude comparative des déplacements du centre de gravité du corps pendant la marche par une nouvelle méthode d'analyse tridemensionnelle. Mise à l'épreuve d'une hypothèse évolutive. In Yves Coppens, and Brigitte Senut, editors. 49-58 *Origine(s) de la Bipédie chez les Hominidés*. Paris: Cahiers de paléoanthropologie.

***Australopithecus afarensis*, LH 4 / fossil footprints**

Feibel, Craig S., Neville Agnew, Bruce Latimer, Martha Demas, Fiona Marshall, Simon A.C. Waane, and Peter Schmid. 1995. "The Laetoli hominid footprints—a preliminary report on the conservation and scientific restudy." *Evolutionary Anthropology* 4:149-154.
Hay, Richard L., and Mary D. Leakey. 1982. "The fossil footprints of Laetoli." *Scientific American* 246(2):50-57.
Johanson, Donald C., Tim D. White, and Yves Coppens. 1978. "A new species of the genus *Australopithecus* (Primates: Hominidae) from the Pliocene of Eastern Africa." *Kirtlandia* NO. 28:1-14.
*Leakey, Mary D., and John M. Harris, editors. 1987. *Laetoli: A Pliocene Site in Northern Tanzania*. Oxford: Clarendon Press.
White, Tim D. 1977. "New fossil hominids from Laetoli, Tanzania." *American Journal of Physical Anthropology* 46:197-230.

***Australopithecus africanus*, StW 573**

Clarke, Ron. 1998. "First ever discovery of a well-preserved skull and associated skeleton *Australopithecus*." *South African Journal of Science* 94:460-463.

***Australopithecus africanus*, Sts 5**

Rak, Yoel. 1983. *The Australopithecine Face*. New York: Academic Press.